The IMA Volumes in Mathematics and its Applications

Volume 122

Series Editor
Willard Miller, Jr.

Springer
New York
Berlin
Heidelberg
Barcelona
Hong Kong
London
Milan
Paris
Singapore
Tokyo

Institute for Mathematics and its Applications IMA

The **Institute for Mathematics and its Applications** was established by a grant from the National Science Foundation to the University of Minnesota in 1982. The IMA seeks to encourage the development and study of fresh mathematical concepts and questions of concern to the other sciences by bringing together mathematicians and scientists from diverse fields in an atmosphere that will stimulate discussion and collaboration.

The IMA Volumes are intended to involve the broader scientific community in this process.

Willard Miller, Jr., Professor and Director

* * * * * * * * * *

IMA ANNUAL PROGRAMS

1982–1983	Statistical and Continuum Approaches to Phase Transition
1983–1984	Mathematical Models for the Economics of Decentralized Resource Allocation
1984–1985	Continuum Physics and Partial Differential Equations
1985–1986	Stochastic Differential Equations and Their Applications
1986–1987	Scientific Computation
1987–1988	Applied Combinatorics
1988–1989	Nonlinear Waves
1989–1990	Dynamical Systems and Their Applications
1990–1991	Phase Transitions and Free Boundaries
1991–1992	Applied Linear Algebra
1992–1993	Control Theory and its Applications
1993–1994	Emerging Applications of Probability
1994–1995	Waves and Scattering
1995–1996	Mathematical Methods in Material Science
1996–1997	Mathematics of High Performance Computing
1997–1998	Emerging Applications of Dynamical Systems
1998–1999	Mathematics in Biology
1999–2000	Reactive Flows and Transport Phenomena
2000–2001	Mathematics in Multimedia
2001–2002	Mathematics in the Geosciences
2002–2003	Optimization

Continued at the back

Christopher K.R.T. Jones Alexander I. Khibnik
Editors

Multiple-Time-Scale Dynamical Systems

With 76 Illustrations

Springer

Christopher K.R.T. Jones
Division of Applied Mathematics and Lefschetz Center for Dynamical Systems
Brown University
Providence RI 02912
USA
ckrtj@cfm.brown.edu

Alexander I. Khibnik
United Technologies Research Center
East Hartford CT 06108
USA

Series Editor:
Willard Miller, Jr.
Institute for Mathematics and its Applications
University of Minnesota
Minneapolis, MN 55455, USA

Mathematics Subject Classification (2000): 34-02, 34A26, 34A34, 34C215, 34C20, 34C23, 34C26, 34C28, 34C29, 34C37, 34D15, 34D35, 34E10, 34E15, 35-01, 35B05, 35B25, 35B32, 35B35, 35B40, 92C20

Library of Congress Cataloging-in-Publication Data
Jones, C.K.R.T. (Christopher K.R.T.)
Multiple-time-scale dynamical systems / Christopher K.R.T. Jones, Alexander Khibnik.
p. cm. — (The IMA volumes in mathematics and its applications ; 122)
Includes bibliographical references.
ISBN 0-387-95126-1 (alk. paper)
1. Perturbation (Mathematics) I. Khibnik, A.I. (Aleksandr Iosifovich) II. Title.
III. IMA volumes in mathematics and its applications ; v. 122.
QA871 .J58 2000
515′.35—dc21 00-059469

Printed on acid-free paper.

Production managed by A. Orrantia; manufacturing supervised by Jerome Basma.
Camera-ready copy prepared by the IMA.
Printed and bound by Sheridan Books, Inc., Ann Arbor, MI.
Printed in the United States of America.

9 8 7 6 5 4 3 2 1

ISBN 0-387-95126-1 SPIN 10778223

Springer-Verlag New York Berlin Heidelberg
A member of BertelsmannSpringer Science+Business Media GmbH

FOREWORD

This IMA Volume in Mathematics and its Applications

MULTIPLE-TIME-SCALE DYNAMICAL SYSTEMS

is based on the proceedings of a workshop with the same title. The workshop was an integral part of the 1997–98 IMA program on "EMERGING APPLICATIONS OF DYNAMICAL SYSTEMS."

I would like to thank the organizing committee: Freddy Dumortier (Limburgs Universitair Centrum), Christopher Jones, Chair (Brown University), Alexander Khibnik (United Technologies Research Center), David Terman (Ohio State University), and Steve Wiggins (Caltech) for their excellent work as organizers of the meeting. Special thanks to Christopher Jones and Alexander Khibnik for their role as editors of the proceedings.

I also take this opportunity to thank the National Science Foundation (NSF), and the Army Research Office (ARO), whose financial support made the workshop possible.

Willard Miller, Jr., Professor and Director
Institute for Mathematics and its Applications
University of Minnesota
400 Lind Hall, 207 Church St. SE
Minneapolis, MN 55455-0436
612-624-6066, FAX 612-626-7370
miller@ima.umn.edu
World Wide Web: http://www.ima.umn.edu

PREFACE

Physical problems are often posed with many time or space scales. Reactors containing different chemical reactions occurring simultaneously may be modelled by equations with many time scales, each corresponding to a different reaction. In electro-optics, the electrical and optical components will exhibit rates that are vastly different; in neurobiology, different biochemical mechanisms and electrical effects result in a variety of time scales occurring in the models. The mathematical area that focuses on the study of these multiple-time scale systems is singular perturbation theory.

The physical significance is fortuitous as the multiplicity of scales allows us to reduce the order of the system. This reduction of systems can bring a seemingly complicated system to a surprisingly manageable form. The reduction of multiple-time scale systems offers more than other types of reduction. Reductions onto different sub-systems can be made, namely those living on fast and slow time-scales respectively, and each of these may be analyzed by virtue of their lower dimensional character. There is an added extra in this case, however, which is that intrinsically high-dimensional behavior may result from piecing the different reductions together. Suppose, for example, that the full system is four-dimensional and there are two fast and two slow variables. This would lead to fast and slow subsystems that are each two-dimensional. Since such low dimensional systems can often be analyzed, the reduction by singular perturbation techniques may afford a fairly complete analysis of the full system. However, the behavior resulting can be much richer than that found in two dimensions. Indeed, behavior that is characteristic of the full four dimensional phase space may appear. The extraordinary feature of this analysis is that it will have been analyzed by appealing mainly to two dimensional systems. The four-dimensional structure will come in through matching that is carried out in piecing together the different two-dimensional subsystems. There are many reduction methods in dynamical systems; for instance, Hamiltonian systems with symmetries or integrals can afford a reduction of the phase space by up to one half of the total dimension. In these cases, however, the behavior seen in the full system will then be that of the lower dimensional system merely embedded in the full phase space. Seen in this light, the singular perturbation strategy of multiple time scale system reduction is potentially a much richer theory.

A traditional singular perturbation phenomenon is the relaxation-oscillation that was first observed in the van der Pol equations. This involves an oscillation with two phases: one slow and one fast. It exists in a two-dimensional phase space (the reductions to fast and slow systems are both to one dimension). A fascinating phenomenon was discovered by Marc and Francine Diener (see the paper by Dumortier et al. for a reference) in which an unstable slow part could interfere with the fast jump. This became known as a "canard" solution because its representation in phase space looked like a cartoon duck. The Dieners used non-standard analysis and the phenomenon was subsequently analyzed by standard methods (see, again, the paper by Dumortier et al.) The generalization of relaxation oscillations and the related canard phenomena to higher dimensions awaited the development of new analysis. This has been carried out, to a great extent, by Dumortier and Roussarie. Their paper in this volume gives an overview of this work. Recently, Szmolyan and collaborators have developed a geometric approach that gives an alternative view. In particular, Szmolyan's approach is quite effective and in many ways simpler than its predecessors.

Trajectories of a dynamical system with specific dynamic properties, such as periodic or homoclinic orbits, are often the focus of attention in a given analysis. Because of the reduction strategy discussed above, singularly perturbed systems offer a context in which such complicated recurrent orbits can be isolated by appealing to fast/slow subsystems and their matching. There is a complex interaction between the fast and slow reduced subsystems and a general strategy has been formulated in work over the past decade. The main result has become known as the Exchange Lemma and a primer for its use is given in the paper by Kaper and myself. Moreover, a completely new direction for the theory is developed in the paper by Soto-Trevino. In that paper, she formulates general conditions for the construction of periodic orbits in fast/slow systems.

Mathematical biology has been an area that has supplied considerable motivation over the years for the development of singular perturbation techniques. In particular, the theory of bursting and synchronous oscillations have their roots in singular perturbation theory. The geometry lying behind the van der Pol type relaxation oscillations is the basis of both these theories. Terman describes some of the developments of synchronous oscillations and their application to neurobiology in his paper.

A phenomenon that has challenged the theory of multiple-scale systems is the presence of exponentially small effects resulting from singular perturbations. This tends to occur in systems with reduced flows on manifolds that are normally elliptic, such as occur in the theory of averaging, rather than normally hyperbolic manifolds which occur in the cases mentioned above. There have been tremendous advances in this area over the past decade and a number of papers in this volume address this issue. Delshams and Gutierrez focus on the exponential splitting of invariant manifolds that occurs in the problem of Arnol'd diffusion. They develop a new theory in terms of Melnikov potentials for measuring this splitting distance. The paper by Su surveys the phenomenon of delayed bifurcation that can occur in the presence of differing time scales. Ward discusses the use of exponential asymptotics in assessing slow motion that occurs in a singularly perturbed partial differential equation. This paper is unique in that it goes beyond the context of finite-dimensional dynamical systems and illustrates how these tools can be effective in spatially distributed systems.

Many people made this conference possible and I want to thank them all warmly. In particular, I want to acknowledge the support of Willard Miller and Bob Gulliver of the IMA whose encouragement at such events is always a critical ingredient for success. The organizing committee of the special year in dynamical systems deserve thanks for suggesting the topic and their help in the planning process. In particular, John Guckenheimer played a key role in keeping the focus on this topic as one of the main events of the year. I want to thank the organizing committee of Freddy Dumortier, Alex Khibnik, Dave Terman, and Steve Wiggins for all their help and support as the workshop was planned. Many people contributed to the success of the meeting whose work is not represented here. In particular, I want to thank the following for their lectures: Dave McLaughlin, Nigel Goldenfeld, Rich Haberman, Konstantin Mischaikow, Mike Davis, Linda Petzold, Xiao-Biao Lin, Mohammed Ziane, Anatoly Neishtadt, Rob Gardner, Chongchun Zeng, Yulij Ilyashenko, George Haller, Pavol Brunovsky, and Petar Kokotovic. Patricia V. Brick put this volume together at IMA and I thank her for all the work in ensuring the volume came together so well.

Christopher K.R.T. Jones
Providence, RI

CONTENTS

HOMOCLINIC ORBITS TO INVARIANT TORI IN HAMILTONIAN SYSTEMS*

AMADEU DELSHAMS† AND PERE GUTIÉRREZ‡

Abstract. We consider a perturbation of an integrable Hamiltonian system which possesses invariant tori with coincident whiskers (like some rotators and a pendulum). Our goal is to measure the splitting distance between the perturbed whiskers, putting emphasis on the detection of their intersections, which give rise to homoclinic orbits to the perturbed tori. A geometric method is presented which takes into account the Lagrangian properties of the whiskers. In this way, the splitting distance is the gradient of a splitting potential. In the regular case (also known as a priori-unstable: The Lyapunov exponents of the whiskered tori remain fixed), the splitting potential is well-approximated by a Melnikov potential. This method is designed as a first step in the study of the singular case (also known as a priori-stable: The Lyapunov exponents of the whiskered tori approach to zero when the perturbation tends to zero).

Key words. Hamiltonian systems, KAM and Nekhoroshev theory, whiskered tori, splitting of separatrices, Arnold diffusion.

AMS(MOS) subject classifications. 58F05, 34C37, 58F36, 34C30.

1. Introduction and results.

1.1. Nearly-integrable Hamiltonians. The Hamiltonian dynamical systems that are close to integrable ones appear in a natural way as models of a wide class of real systems, and thus constitute a very active field of research in classical mechanics. The behavior of these systems is far from being completely understood, and one of the most relevant questions is the stability or instability of their trajectories. This problem remains unsolved for systems with more than 2 degrees of freedom.

The unperturbed rôle is played by a (completely) integrable Hamiltonian with n degrees of freedom. The Liouville–Arnold theorem (see for instance [3]) establishes, under certain hypotheses, the existence on some region of the phase space of canonical *action–angle variables* $(\varphi, I) = (\varphi_1, \ldots, \varphi_n, I_1, \ldots, I_n) \in \mathbb{T}^n \times G \subset \mathbb{T}^n \times \mathbb{R}^n$, in which the Hamiltonian only depends on the action variables: $h(I)$. The associated Hamiltonian equations for a trajectory $(\varphi(t), I(t))$ are

$$\dot{\varphi} = \omega(I), \qquad \dot{I} = 0,$$

where $\omega = \partial_I h$. Hence the dynamics is very simple: Every n-dimensional torus $I = \text{const}$ is invariant, with linear flow $\varphi(t) = \varphi(0) + \omega(I)t$, and thus

*This work was supported in part by the EC grant ERBCHRXCT940460.

†Departament de Matemàtica Aplicada I, Universitat Politècnica de Catalunya, Diagonal 647, 08028 Barcelona, SPAIN (amadeu@ma1.upc.es). Also supported in part by the Spanish grant DGICYT PB94–0215 and the Catalan grant CIRIT 1996SGR–00105.

‡Departament de Matemàtica Aplicada II, Universitat Politècnica de Catalunya, Pau Gargallo 5, 08028 Barcelona, SPAIN (gutierrez@ma2.upc.es).

all trajectories are stable. The motion on a torus is called quasiperiodic, with associated *frequencies* given by the vector $\omega(I) = (\omega_1(I), \ldots, \omega_n(I))$.

Every n-dimensional invariant torus can be nonresonant or resonant, according to whether its frequencies are rationally independent or not. A nonresonant torus is densely filled by any of its trajectories. On the other hand, a resonant torus is foliated into a family of lower dimensional tori.

A *nearly-integrable* Hamiltonian can be written in the form

$$H(\varphi, I) = h(I) + \varepsilon f(\varphi, I), \tag{1.1}$$

where ε is a small perturbation parameter. Then the Hamiltonian equations are

$$\dot{\varphi} = \omega(I) + \varepsilon \partial_I f(\varphi, I), \qquad \dot{I} = -\varepsilon \partial_\varphi f(\varphi, I).$$

For $\varepsilon \neq 0$, the dynamics can be very complicated: There can exist, in principle, chaotic trajectories or even unstable (in the sense that they can wander very far from their initial conditions). Note that the variation of the actions $|I(t) - I(0)|$ could grow slowly but unboundedly.

Although there may exist unstable trajectories, two very relevant results on stability have been established for nearly-integrable Hamiltonian systems. These are the *KAM theorem* and the *Nekhoroshev theorem*. We are going to give a brief description of both of them, putting emphasis on the fact that they lead to two different notions of stability.

1.2. KAM theorem. We can say that KAM (Kolmogorov–Arnold–Moser) theory is concerned with the preservation of quasiperiodic motions under small perturbations: The theorem states, under a suitable nondegeneracy condition, that most of the n-dimensional invariant tori $I = \text{const}$ of the integrable Hamiltonian h survive in (1.1), with some deformation, for $|\varepsilon|$ small enough. In fact, KAM theorem guarantees the preservation of the tori having "sufficiently" nonresonant frequencies, for which the influence of the *small divisors* $\langle k, \omega(I) \rangle$ can be overcome. This is expressed by means of a *Diophantine condition* on the frequency vector $\omega(I)$: For some constants τ and γ,

$$|\langle k, \omega(I) \rangle| \geq \gamma \, |k|^{-\tau} \quad \forall k \in \mathbb{Z}^n \setminus \{0\},$$

where $|k| = \sum_{j=1}^{n} |k_j|$. The vectors satisfying this condition for given $\tau > n-1$ and $\gamma > 0$ fill a Cantorian set of relative measure $1 - \mathrm{O}(\gamma)$ in $\mathbb{R}^n$. In this way, one obtains *perpetual stability* in the perturbed Hamiltonian, but only for initial conditions on the perturbed or *KAM tori* (the surviving ones), which form a Cantorian set not containing any open subset although its measure is large.

The KAM theorem was "discovered" in the 50's by Kolmogorov [30], and constituted one of the most relevant advances about stability in nearly-integrable Hamiltonian systems. In his first theorem, Kolmogorov established the preservation of only one fixed Diophantine torus (see also [4]).

Some years later, Arnold [1] proved, in the analytic case, the existence of a large family of invariant tori, giving an estimate for the measure of their complementary set (see also [48]). Also, Moser [42] proved an analogous theorem, in the differentiable case, for area-preserving maps satisfying a twist condition. As general references on KAM theory, see [9, 11].

The nondegeneracy condition required in KAM theorem is imposed on the frequency map $\omega = \partial_I h$ to ensure that a large set of actions I have Diophantine frequencies $\omega(I)$. There are two usual types of conditions:

$$\det\left(\partial_I \omega(I)\right) \neq 0, \qquad \det\begin{pmatrix} \partial_I \omega(I) & \omega(I) \\ \omega(I)^\top & 0 \end{pmatrix} \neq 0,$$

(for all $I \in G$), that can be called, respectively, *Kolmogorov nondegeneracy* and *isoenergetic nondegeneracy.* The second one is also called Arnold nondegeneracy. Under any of these conditions, for a given $\varepsilon > 0$ the invariant tori whose frequencies satisfy the Diophantine condition with $\tau > n - 1$ and $\gamma = \mathrm{O}\left(\sqrt{\varepsilon}\right)$ are preserved, and the measure of the complement is small: $\mathrm{O}\left(\sqrt{\varepsilon}\right)$ (see [48]).

Although the two versions (Kolmogorov and isoenergetic) of KAM theorem are known since the works by Kolmogorov, Arnold and Moser, there are only a few published proofs of the isoenergetic version. Complete proofs are given in [23, 10], indirectly from the Kolmogorov version. A direct proof of the isoenergetic KAM theorem, without using the Kolmogorov version, can be found in [19].

It has to be pointed out that the isoenergetic condition is more significant from the point of view of the stability, since it ensures the existence of a large family of KAM tori on each energy level $H = \text{const}$. For $n = 2$, one deduces stability even for trajectories which do not lie on these tori, because these 2-dimensional invariant tori separate the 3-dimensional energy levels. This reasoning does not hold for $n > 2$, and one cannot guarantee stability. So in this case there could exist unstable trajectories, although it seems clear that the KAM tori should constitute stronger barriers to instability.

1.3. Nekhoroshev theorem and effective stability. The other relevant result to be commented here is Nekhoroshev theorem, which leads to the concept of *effective stability.* Nekhoroshev theorem [44], first stated in 1977, establishes for all the trajectories of the system (1.1) that the variation of the action variables remains small for a very long time, exponentially large with respect to the parameter $\varepsilon > 0$. For every initial condition $(\varphi(0), I(0))$ one has an estimate of the type

$$|I(t) - I(0)| \leq r_0\, \varepsilon^b \qquad \text{for } \ |t| \leq T_0 \, \exp\left\{(\varepsilon_0/\varepsilon)^a\right\}.$$

The constants $a, b > 0$ are called *stability exponents.*

For the validity of Nekhoroshev theorem one imposes a *steepness condition* (see [44]) on the integrable Hamiltonian h. This is a very general condition, avoiding the quick escape of trajectories along certain directions related to resonances of the frequency vector $\omega(I) = \partial_I h(I)$. The simplest case is that of *quasiconvex* functions: One says h to be quasiconvex if, for any $I \in G$ and $v \in \mathbb{R}^n$,

$$\langle v, \omega(I)\rangle = 0 \quad \Longrightarrow \quad \langle v, \partial_I^2 h(I)\, v\rangle \neq 0.$$

For a perturbation of a quasiconvex Hamiltonian, the estimate of Nekhoroshev theorem holds with the stability exponents

$$a = b = \frac{1}{2n}.$$

The proof of the theorem with these exponents has been given in [37, 49]. However, Chirikov [14] had predicted several years before that the exponent $a = 1/2n$ should be optimal, a fact that was overlooked until it was explained in Lochak's survey [34] (see also [35]), where the exponent $a = 1/(2n+1)$ was already obtained.

Comparing Nekhoroshev theorem to KAM theorem, we see that perpetual stability has been replaced by finite time stability, but on the other hand the estimates are valid for all trajectories in the phase space. Since (despite the theoretical importance of KAM theorem) it is not possible to know whether a given trajectory lies on a KAM torus, the effective stability estimates are of obvious interest from the point of view of the applications. Some recent results are intending to fill the gap between KAM and Nekhoroshev theorems. These results concern the "stickiness" of KAM tori [46], as well as the remarkable results about "superexponential stability" [40], or the existence of "quasi-invariant tori" [19].

1.4. Arnold diffusion: Whiskered tori, splitting, and transition chains. Nothing is said in KAM theorem about the stability of the trajectories close to unperturbed resonant invariant tori. Nekhoroshev theorem does not exclude the existence of unstable trajectories, but predicts for them an exponentially long stability time. This extremely slow phenomenon of instability is called *Arnold diffusion,* and a first description of it was given in 1964 by Arnold [2] by means of his famous example. As noticed in Section 1.2, diffusion can only take place for more than 2 degrees of freedom.

The *Arnold's example* proposed in [2] is a nonautonomous Hamiltonian, periodic in the time variable t: In canonical variables $(x, y, \varphi_1, I_1) \in \mathbb{T} \times \mathbb{R} \times \mathbb{T} \times \mathbb{R}$,

$$H(x, y, \varphi_1, I_1, t) = \frac{1}{2}\left(y^2 + I_1^2\right) + \varepsilon(\cos x - 1)\left(1 + \mu(\sin\varphi_1 + \cos t)\right),$$

with ε and μ as two independent parameters. Taking t as a new angular variable, the Hamiltonian becomes the following 3-degrees-of-freedom Hamiltonian: In canonical variables $(x, y, \varphi, I) \in \mathbb{T} \times \mathbb{R} \times \mathbb{T}^2 \times \mathbb{R}^2$,

$$H(x,y,\varphi,I) = \frac{1}{2}\left(y^2 + I_1^2\right) + I_2 + \varepsilon(\cos x - 1)\big(1 + \mu(\sin\varphi_1 + \cos\varphi_2)\big). \tag{1.2}$$

We shall ignore the variable I_2 which does not take part in the Hamiltonian equations, considering a 5-dimensional phase space.

The frequency vector $\omega(y, I_1) = (y, I_1, 1)$ has, on $y = 0$, a *single resonance* (double for rational I_1), and for $\varepsilon = 0$ the 3-dimensional tori $y, I_1 = \text{const}$ are foliated into 2-dimensional invariant tori. On the resonance, KAM theorem cannot be applied.

A first perturbative step is to consider $\varepsilon > 0$ but $\mu = 0$, still having an integrable Hamiltonian (1.2) which is an *uncoupled* system formed by one pendulum (variables (x, y)) and 2 rotors (variables $(\varphi_1, \varphi_2, I_1)$). The 3-dimensional tori on the resonance $y = 0$ have been destroyed, but a 1-parameter family of 2-dimensional tori $\mathcal{T}_{I_1}$ still remains along the resonance: $x = y = 0$, $I_1 = \text{const}$, $(\varphi_1, \varphi_2) \in \mathbb{T}^2$, with associated frequencies $(I_1, 1)$. These tori come from the hyperbolic equilibrium point $x = y = 0$ of the pendulum, and hence they are *hyperbolic invariant tori,* also called *whiskered tori,* with associated 3-dimensional *stable* and *unstable whiskers* (invariant manifolds) $\mathcal{W}^{\pm}_{I_1}(\varepsilon, 0)$, inherited from the *separatrices* of the pendulum. Since these separatrices constitute a homoclinic connection given by the equation $y^2/2 + \varepsilon(\cos x - 1) = 0$, it turns out that for $\varepsilon > 0$, $\mu = 0$, the whiskers coincide: $\mathcal{W}^{+}_{I_1}(\varepsilon, 0) = \mathcal{W}^{-}_{I_1}(\varepsilon, 0)$. Notice that the characteristic exponents of the origin of the pendulum (and therefore the Lyapunov exponents of the whiskered tori $\mathcal{T}_{I_1}$) are $\pm\sqrt{\varepsilon}$, and therefore the whiskered tori are *weakly hyperbolic* for $|\varepsilon|$ small. Notice also that there are elliptic invariant tori on $x = \pi$, $y = 0$, which come from the elliptic point of the pendulum.

For $\varepsilon > 0$ and $\mu \neq 0$, the Hamiltonian (1.2) is no longer integrable. The *same* tori $\mathcal{T}_{I_1}$ are still whiskered tori, independently of μ, but their stable and unstable whiskers $\mathcal{W}^{\pm}_{I_1}(\varepsilon, \mu)$ do depend on μ and are not expected to coincide for $\mu \neq 0$. Indeed, Arnold [2] applied the Poincaré's perturbative method [47, §19] for detecting *splitting* of separatrices and showed that, for μ small enough: $0 < |\mu| < \mu_0(\varepsilon)$, the whiskers do not coincide and intersect transversely. The intersections are obtained as zeros of a distance function between the whiskers of the Hamiltonian (1.2). The maximum of this distance can be estimated as

$$\Delta \sim \mu \exp\left(-1/\sqrt{\varepsilon}\right) + \mathrm{O}\left(\mu^2\right), \tag{1.3}$$

where $\mu \exp\left(-1/\sqrt{\varepsilon}\right)$ is a bound of some integrals provided by Poincaré's method, which are usually known as *Poincaré–Melnikov integrals* (or, more briefly, *Melnikov integrals*). An important feature is that the Melnikov integral is exponentially small in ε, as a result of the weak hyperbolicity of the unperturbed whiskers (the power $1/2$ of ε in the exponential is due to the fact that the Hamiltonian (1.2) has only a finite number of harmonics in φ), and is the dominant term in (1.3) only if μ is chosen exponentially small with respect to ε:

$$\mu = \mathrm{o}\left(\exp\left(-1/\sqrt{\varepsilon}\right)\right).$$

The intersections above between whiskers are called *homoclinic* because they involve only one torus. These homoclinic intersections can be extended to *heteroclinic* intersections between the unstable whisker of a given torus $\mathcal{T}_{I_1}$ and stable whiskers of other tori $\mathcal{T}_{I_1'}$ in a neighborhood, as long as they are chosen exponentially close with respect to ε (see (1.3)):

$$|I_1' - I_1| \preceq \mu \exp\left(-1/\sqrt{\varepsilon}\right).$$

It is then possible to construct a *transition chain* along the resonance $y = 0$. This notion was introduced in [2], and means a finite sequence of tori $\mathcal{T}_{I_1^{(0)}}, \ldots, \mathcal{T}_{I_1^{(N)}}$ having *nonresonant* frequencies, with $\left|I_1^{(N)} - I_1^{(0)}\right| = \mathrm{O}(1)$, and such that the unstable whisker of each torus intersects transversely the stable whisker of the next one. Besides, it was asserted in [2] that the tori in the chain are *transition tori*: For every torus, arbitrarily small neighborhoods of points in its stable and unstable whiskers are connected by trajectories of the Hamiltonian. From these facts, Arnold concluded that diffusion can take place along neighborhoods of all the successive tori of a transition chain.

A feature that makes Arnold's example very particular is the fact that the perturbation and its derivatives vanish on the whiskered tori $\mathcal{T}_{I_1}$. This implies that all these tori survive in the perturbation for $\mu \neq 0$, remaining even unchanged. This is very convenient for the purpose of finding transition chains of whiskered tori, since it makes possible to choose two of them as close as necessary.

For a general perturbation, the survival of all the tori associated to a single resonance will not be true, and an adapted version of KAM theorem (see Section 1.7 for references) has to be applied. This *hyperbolic KAM theorem* only ensures the preservation of those whiskered tori with Diophantine frequencies, constituting a Cantorian set near the resonance. Thus, the set of surviving whiskered tori has a large relative measure but also "many gaps", which are due to the presence of double resonances. These gaps and the fact that the splitting is expected to be exponentially small can make the unstable whisker of a torus with frequency on this given set not to intersect the stable whisker of the "next" surviving torus, and prevents the construction of a transition chain in this way.

A generalization of Arnold's example to the case of an arbitrary number of degrees of freedom, was introduced by Lochak [33, 34] in order to point out the main difficulties concerning the detection of the diffusion. In canonical variables $(x, y, \varphi, I) \in \mathbb{T} \times \mathbb{R} \times \mathbb{T}^n \times \mathbb{R}^n$, *Lochak's example* is:

$$H(x, y, \varphi, I) = \frac{1}{2}\left(y^2 + \langle I, I\rangle\right) + \varepsilon(\cos x - 1) + \varepsilon\mu F(x, \varphi), \tag{1.4}$$

where F is an analytic function on a complex neighborhood of $\mathbb{T}^{n+1}$. As a new feature (see [34, V§2]), the fact that the function F is allowed to have arbitrarily high harmonics in the angles φ brings out the problem of the small divisors inside the Melnikov integrals, as an essential ingredient in order to decide which the relevant harmonics are in the study of these integrals. It turns out that the Melnikov integrals can be estimated only for Diophantine whiskered tori, and they are again exponentially small in ε. Their asymptotic behavior is no longer $\exp(-1/\sqrt{\varepsilon})$, but depends on the arithmetic properties of the frequencies of the invariant tori, as was noticed in [34, page 119], where the connection between the size of the splitting and the speed of diffusion was first discussed. (See [17, 18, 52, 53] for some explicit examples.)

Recently, it has become a point of great interest the question of the validity of the predictions of the splitting given by the Melnikov integrals for a power-like relation $\mu = \varepsilon^p$, with $p > 0$. Unfortunately, this is not an easy problem because of the exponentially small character of these integrals with respect to ε.

We remark that Arnold's and Lochak's examples are important not only as specific examples designed to detect Arnold diffusion but, mainly, because they are special cases of the general model near a single resonance of a nearly integrable Hamiltonian system. We suspend here this report of results related to Arnold diffusion to deduce such model. In passing, the actual kind of relation $\mu = \varepsilon^p$ needed for the real applications will become transparent.

1.5. One step of normal form near a single resonance. Near a single resonance, under some generic hypotheses, we are going to carry out *one* step of (resonant) normal form procedure and try to obtain a new expression of the Hamiltonian that generalizes the examples considered in Section 1.4. To reach a *hyperbolic normal form*, we will have to impose some nondegeneracy conditions on the perturbation f, since no hyperbolicity can be obtained from the unperturbed Hamiltonian h. Essentially, we follow Eliasson [25] and Niederman [45] (see also [51]).

To simplify the notation, the number of degrees of freedom in (1.1) will be, from now on, $n+1$ instead of n. Thus, the angle–action variables are $(\varphi, I) \in \mathbb{T}^{n+1} \times \mathbb{R}^{n+1}$.

Having selected an action $I^* = 0$, the unperturbed Hamiltonian h in (1.1) can be written (disregarding the additive constant) as:

$$h(I) = \langle \lambda^*, I \rangle + \frac{1}{2} \langle QI, I \rangle + \mathrm{O}_3(I).$$

Assume, for the selected action, that the associated frequency vector $\lambda^* = \partial_I h(0) \in \mathbb{R}^{n+1}$ has a *single resonance* ($\langle k^*, \lambda^* \rangle = 0$ for a certain $k^* \in \mathbb{Z}^{n+1} \setminus \{0\}$ and $\langle k, \lambda^* \rangle \neq 0$ for any $k \in \mathbb{Z}^{n+1}$ not co-linear to k^*). By a classical algebraic result, we can assume λ^* of the form

$$\lambda^* = (0, \omega^*),$$

where $\omega^* \in \mathbb{R}^n$ is nonresonant. In fact, we shall assume a Diophantine condition on ω^*.

Rewrite $(\varphi, I) \in \mathbb{T}^{n+1} \times \mathbb{R}^{n+1}$ as $(x, y, \varphi, I) \in \mathbb{T} \times \mathbb{R} \times \mathbb{T}^n \times \mathbb{R}^n$, and the matrix $Q = \partial_I^2 h(0)$ as

$$\begin{pmatrix} \beta^2 & q^\top \\ q & Q \end{pmatrix},$$

where we have put $\beta^2 > 0$ in order to fix ideas, $q \in \mathbb{R}^n$, and the new matrix Q is $n \times n$. We will assume $\beta = 1$; this can be achieved replacing y, I by y/β, I/β (changing in this way the time scale by a factor β), and rewriting ω^*/β, q/β^2, Q/β^2 as ω^*, q, Q respectively, and redefining also the function f. Then, we can write our Hamiltonian in the form

$$\begin{aligned} H(x, y, \varphi, I) &= h(y, I) + \varepsilon f(x, y, \varphi, I), \\ h(y, I) \quad &= \langle \omega^*, I \rangle + \frac{y^2}{2} + \langle q, I \rangle\, y + \frac{1}{2} \langle QI, I \rangle + \mathrm{O}_3(y, I). \end{aligned}$$

We now perform *one* step of resonant normal form procedure: Following the Lie method, we seek for functions $S(x, \varphi)$ and $R(x, y, \varphi, I) = \mathrm{O}(y, I)$ such that

$$\{S, h\} + V + R = f, \tag{1.5}$$

where $V(x)$ is the periodic function obtained by averaging with respect to the angles φ:

$$V(x) = \overline{f(x, 0, \cdot, 0)} = \frac{1}{(2\pi)^n} \int_{\mathbb{T}^n} f(x, 0, \varphi, 0) \mathrm{d}\varphi, \qquad x \in \mathbb{T}.$$

The construction of S and R is easily carried out: One first solves the equation

$$\langle \omega^*, \partial_\varphi S \rangle + V = f(\cdot, 0, \cdot, 0)$$

with the help of standard small divisors estimates, and then one takes R simply by fitting Equation (1.5). The time-1 symplectic flow Φ of the generating Hamiltonian εS leads to

$$H \circ \Phi = H + \{H, \varepsilon S\} + \mathrm{O}\left(\varepsilon^2\right) = h + \varepsilon(V + R) + \mathrm{O}\left(\varepsilon^2\right) = H_0 + H_1,$$

with

$$\begin{aligned} H_0(x, y, I; \varepsilon) &= \langle \omega^*, I \rangle + \frac{y^2}{2} + \varepsilon V(x) + \langle q, I \rangle\, y + \frac{1}{2} \langle QI, I \rangle, \\ H_1(x, y, \varphi, I; \varepsilon) &= \varepsilon R(x, y, \varphi, I) + \mathrm{O}_3(y, I) + \mathrm{O}\left(\varepsilon^2\right). \end{aligned}$$

This expression generalizes that of Lochak's example (1.4), with H_0 (the truncated normal form) playing the rôle of the integrable Hamiltonian, and H_1 being the perturbation of some size $\varepsilon\mu$ where μ is going to be determined in terms of ε.

It can be assumed, except for degenerate cases, that the function $V(x)$ has a unique and nondegenerate maximum x_0; we denote $\alpha^2 = -V''(x_0) > 0$. Then, for $\varepsilon > 0$, the 1-degree-of-freedom Hamiltonian

$$P(x, y; \varepsilon) = \frac{y^2}{2} + \varepsilon V(x),$$

has a hyperbolic point in $(x_0, 0)$, with (homoclinic) separatrices. The case $\varepsilon < 0$ is analogous, provided one considers a minimum instead of a maximum. Then, the Hamiltonian H_0 has whiskered tori with coincident whiskers associated to this hyperbolic point. Note that H_0 constitutes a Hamiltonian situated between the unperturbed Hamiltonian h and the perturbed one H, which possesses hyperbolic invariant tori but their whiskers still coincide. Note also that, in general, H_0 is not an uncoupled Hamiltonian because of the *coupling term* $\langle q, I\rangle\, y$. This seems an additional complication, but it cannot be avoided if one starts with an arbitrary unperturbed Hamiltonian h.

The Lyapunov exponents of the hyperbolic point of P are $\pm\sqrt{\varepsilon}\alpha$, which tend to zero for $\varepsilon \to 0^+$. To have fixed Lyapunov exponents, we can perform the following (non-canonical) linear change: We replace y, I by $\sqrt{\varepsilon}y$, $\sqrt{\varepsilon}I$. The new system is still Hamiltonian if we divide the Hamiltonian by ε (making in this way a change of time scale by a factor $\sqrt{\varepsilon}$). We obtain for $H = H_0 + H_1$ the expressions:

$$H_0 = \langle \omega, I\rangle + \frac{y^2}{2} + V(x) + \langle q, I\rangle\, y + \frac{1}{2}\langle QI, I\rangle, \tag{1.6}$$

$$H_1 = R\left(x, \sqrt{\varepsilon}y, \varphi, \sqrt{\varepsilon}I\right) + \frac{1}{\varepsilon}\, \mathrm{O}_3\left(\sqrt{\varepsilon}y, \sqrt{\varepsilon}I\right) + \mathrm{O}(\varepsilon) = \mathrm{O}(\mu), \tag{1.7}$$

where

$$\omega = \frac{\omega^*}{\sqrt{\varepsilon}}, \qquad \mu = \sqrt{\varepsilon}.$$

We close this section by remarking that an essential point in the approach described in this work is to have an integrable H_0. A similar procedure could have been carried out for double resonances (or higher multiplicity) instead of single ones, but then H_0 would not be, in general, integrable.

1.6. Regular and singular hyperbolic Hamiltonians. For $\varepsilon \to 0^+$, the study of the Hamiltonian (1.6–1.7) is a singular perturbation problem, due to the *fast frequencies* $\omega = \omega^*/\sqrt{\varepsilon}$ in the unperturbed Hamiltonian

H_0. We are thus confronted with a *singular* system, often referred to as *weakly hyperbolic*, and also called *a-priori stable* according to the terminology introduced in [13]. In fact, this case can be referred to as *totally singular*, because *all* the frequencies are fast.

The singular problem can be avoided if one considers independent parameters, namely a *fixed* $\varepsilon > 0$ (that is, a *fixed* ω in (1.6)) and $\mu \to 0$. In such a case, the system (1.6–1.7) has the property that the hyperbolicity and the homoclinic orbits are present in the unperturbed Hamiltonian ($\mu = 0$), and are simply perturbed for $|\mu|$ small. In this case, we are confronted with a *regular* or *strongly hyperbolic* system, or also *a-priori unstable.*

This strategy of keeping $\varepsilon > 0$ fixed and letting $\mu \to 0$ (having in this way a regular system) was introduced by Arnold in order to avoid dealing with a singular perturbation problem. As already remarked in Section 1.4, the exponentially small splitting of separatrices predicted by the Melnikov integrals could then be justified only for μ exponentially small in ε.

This regular situation ($\varepsilon > 0$ fixed and μ exponentially small in ε) has been considered in some papers through the construction of appropriate Melnikov integrals [39, 56], but their applicability to the existence of Arnold diffusion [29, 58, 59] has overlooked the existence of the gaps between the whiskered tori, as well as the transition properties needed for the invariant tori. (This transition property has been recently studied in [16, 24, 38, 45].) Also inside this regular situation there is a recent approach [5–7], reminiscent of the parallel shooting method in numerical analysis, which is based on looking directly for the connecting paths of diffusion, using variational methods.

Considering it only from the point of view of bifurcation theory, the relation μ exponentially small in ε is anything but a generic relation between these two parameters. Moreover, from the analysis of the previous section, a power-like relation $\mu = \varepsilon^p$, with $p > 0$, is the natural one from the point of view of perturbation theory (the smaller p, the better). In this singular case, the problem of detecting the splitting of separatrices from the Melnikov integrals is much more intricate because of its exponentially small character with respect to ε.

Concerning the singular problem for 2 degrees of freedom ($n = 1$ in Section 1.5), or equivalently area preserving maps, the first results [31, 43] appeared in 1984, and now it seems fairly well understood (see, for instance, [22] for a survey of results).

For more than 2 degrees of freedom, the existence of intersection between the whiskers can be detected by geometrical and topological methods [8, 25], but the effective measure of such intersections involve, in general, small divisors into the expressions for the Melnikov integrals, and arithmetic problems take place in giving asymptotics or lower bounds for such integrals.

As a way to avoiding such small divisors, one can consider different time scales in the frequencies, like for instance $\omega = (\omega_1^*, \omega_2^*/\sqrt{\varepsilon})$ in (1.6).

These different time scales take place in some problems of celestial mechanics, and were first considered in [12, 13], where the splitting was estimated by direct Lindstedt expansions, and was wrongly stated as some power of ε, due to some mistaken computations in the third order of such expansions. This error was recently corrected in [27] for for the case of *one* fast frequency $\omega_2^*/\sqrt{\varepsilon}$ (one 'singular' frequency), and an *even* perturbation $F(x,\varphi)$ with only a finite number of harmonics in the angles x, φ. In this case, the splitting of the separatrices turns out to be $O(\exp(-1/\sqrt{\varepsilon}))$, as in the Arnold's example, and no problems related with small divisors appear in the exponent of this expression.

The difficult "totally singular" case has been lately receiving special attention. The first formal discussion about the relation between the size of the splitting and the speed of diffusion can be found in [34]. Upper bounds appear in [26], and the first explicit computations showing the arithmetic problems for a concrete frequency were carried out in [53] for fast quasiperiodically forced Hamiltonians ("isochronous systems").

The paradigmatic example of a pendulum under fast quasiperiodic forcing (the Hamiltonian (1.6–1.7) for $q=0$, $Q=0$, $V(x)=\cos x-1$, and $H_1=\varepsilon^p m(\varphi)\cos x$, $p>3$) was studied in [17] for the case of a nonresonant two-dimensional external frequency $\omega^*=\left(1,(\sqrt{5}+1)/2\right)$, and an asymptotic behavior for the splitting of the type $\Delta\sim\varepsilon^{p/2-1}\exp\left(-c(\log\varepsilon)/\varepsilon^{1/4}\right)$ was proved, where $c(u)$ is a positive periodic function of u that depends on the arithmetic properties of ω^* and the analyticity properties of $m(\varphi)$.

Lower bounds for the anisochronous Lochak's example (1.4) with the *same* exponents as in the Nekhoroshev theorem have been recently stated in [52] for $\mu=\varepsilon^p$ and $p>0$ big enough, assuming an *even* perturbation $F(x,\varphi)$ (ensuring in this way the existence of a homoclinic orbit), but with a proof containing essential errors.

For the obtainment of asymptotic or lower bounds for the splitting of separatrices, the perturbations H_1 and F are assumed in [17, 52] to possess arbitrarily high harmonics in the angles φ, to ensure that the estimated dominant Fourier mode of the Melnikov integral is not zero. When the perturbation has only a finite number of harmonics (for instance, in the Arnold's example [2]), the Poincaré–Melnikov theory applied to the original system cannot give the correct asymptotics, since the phenomenum of small divisors does not show up in the Melnikov function. In this case, before applying the results of [17, 52], at least one step of averaging is required to produce a new unperturbed Hamiltonian and a new perturbation with arbitrarily high harmonics (see [53] for an explicit example).

Finally, let us point out that an *essential* assumption in [26, 27, 52] is the reversibility of the perturbation. This hypothesis is required to ensure the existence of a perturbed homoclinic orbit, to begin the iterative approximation of the whiskers through the study of variational-like equations. This assumption is not required in [17], where instead an invariant splitting function is defined in some flow-box coordinates provided by a normal form, convergent in a full neighborhood of the whiskered torus, which is constructed with a precise statement of the loss of complex domain in the variable φ.

We stop here this account of results, and refer to Lochak [36] for a better organized compendium of results concerning Arnold diffusion.

1.7. Description of the results. According to Section 1.5, we concentrate our efforts on the Hamiltonian H given in (1.6, 1.7), including the coupling term $\langle q, I\rangle\, y$.

The main goal of this note is to introduce a *geometrical* method to measure the splitting of the whiskers of a hyperbolic invariant torus of the anisochronous Hamiltonian (1.6, 1.7), relating it with the Melnikov function. Our main interest is the singular case $\mu = \varepsilon^p$ for some $p > 0$, but for the presentation of the method in this note, we will be mostly concerned with the regular case $\varepsilon > 0$ fixed, $\mu \to 0$. The proofs of the announced results are not included, and will appear elsewhere [20].

In our first results, in Section 3, we establish the preservation of the whiskered torus having frequency vector ω. This is ensured by means of a version of the KAM theorem, adapted to the preservation of whiskered tori under perturbations. We call it *hyperbolic KAM theorem.*

We point out that most papers about preservation of whiskered tori, like [28, 32, 55, 60], deal mainly with a local point of view. A significantly new approach was introduced by Eliasson [25], by expressing the hyperbolic KAM theorem in the original variables, global in $x \in \mathbb{T}$, instead of some local variables around the whiskered torus. Our version of this result (Theorem 3.2) can be considered a refinement of the one given in [25]. Thus, we provide a symplectic transformation Φ such that the new Hamiltonian $\widetilde{H} = H \circ \Phi$ (the *normal form*) has a simpler expression and has clearly an invariant torus with frequencies ω.

Moreover, to deal in the next future with the singular case $\mu = \varepsilon^p$, we obtain a quantitative normal form in Theorem 3.1, with a precise statement of the loss of complex domain in the variable φ, which can be applied later on an extension theorem like in [17, 22, 52].

This global point of view allowed Eliasson to introduce in a very natural way a (vector) *splitting function* measuring the distance between the perturbed whiskers. In fact, the paper [25] is concerned with the existence of (at least $n+1$) homoclinic intersections between the whiskers, in both regular and singular systems, although the splitting is not computed. This is deduced from the use of *exact symplectic* transformations to normal form in the hyperbolic KAM theorem.

Proceeding as in [25], but going farther with the use of exact symplectic transformations we see that the splitting function $\mathcal{M}(\varphi)$ is the gradient of a (scalar) function $\mathcal{L}(\varphi)$, which is called *splitting potential* (Theorem 4.1). It is an important remark that, to get this result, the splitting function is expressed in the normal form variables provided by the hyperbolic KAM theorem, where its properties appear transparently, instead of using the original variables. Besides, we point out that the existence of the splitting potential is also a reflection of the Lagrangian properties of the whiskers.

In Section 5, we will follow Treschev [56] for the Poincaré–Melnikov method, to obtain first order approximations for both the functions $\mathcal{L}(\varphi)$ and $\mathcal{M}(\varphi)$ in Theorem 5.1. The first order approximations $L(\varphi)$ and its gradient $M(\varphi)$ are called, respectively, the *Melnikov potential* (like in [21], although it deserves the name of Poincaré function [47, §19]), and the *Melnikov function.* In particular, when the Melnikov potential is a Morse function (a generic situation), there exist 2^n transverse homoclinic orbits for $|\mu|$ small enough (in the *regular* case).

The Melnikov potential $L(\varphi)$ and its gradient $M(\varphi)$ are given by absolutely convergent integrals, and are designed in order to be useful also for a coupled Hamiltonian H_0 (with a nonzero coupling term). To illustrate this, a computable coupled example is shown in Section 6.

2. Setup and the unperturbed Hamiltonian. In view of the analysis made in Section 1.5, we will consider a (real analytic) perturbation of a hyperbolic integrable Hamiltonian, with $n+1 \geq 3$ degrees of freedom, that in canonical variables $z = (x, y, \varphi, I) \in D \subset \mathbb{T} \times \mathbb{R} \times \mathbb{T}^n \times \mathbb{R}^n$, is of the form

$$H(x, y, \varphi, I; \mu) = H_0(x, y, I) + \mu H_1(x, y, \varphi, I), \tag{2.1}$$

$$H_0(x, y, I) = \langle \omega, I \rangle + \frac{y^2}{2} + V(x) + \langle q, I \rangle\, y + \frac{1}{2} \langle QI, I \rangle\,, \tag{2.2}$$

where μ is a perturbation parameter. The given ingredients of H_0 are the vectors $\omega, q \in \mathbb{R}^n$, a symmetric $(n \times n)$-matrix Q, and a function $V(x)$ of $x \in \mathbb{T}$ that is required to have a unique and nondegenerate global maximum. To fix ideas, we require

$$\begin{aligned} &V(0) = 0, \qquad V'(0) = 0, \qquad V''(0) < 0, \\ &V(x) < 0 \qquad \forall x \neq 0 \pmod{2\pi}, \end{aligned} \tag{2.3}$$

and we denote

$$\alpha = \sqrt{-V''(0)}. \tag{2.4}$$

We also assume the following nondegeneracy condition:

$$\det \begin{pmatrix} 1 & q^\top \\ q & Q \end{pmatrix} \neq 0. \tag{2.5}$$

Let us write

$$P(x, y) = \frac{y^2}{2} + V(x).$$

On every plane $I = \text{const}$, the Hamiltonian H_0 reduces to the 1-degree-of-freedom Hamiltonian

$$h(x, y; I) = P(x, y + \langle q, I \rangle) = \frac{(y + \langle q, I \rangle)^2}{2} + V(x).$$

This is a *generalized pendulum* (the standard pendulum being given by $V(x) = \cos x - 1$). It has $(x, y) \doteq (0, -\langle q, I \rangle)$ as a hyperbolic equilibrium point, with (homoclinic) separatrices given by $y + \langle q, I \rangle = \pm\sqrt{-2V(x)}$.

If we take $q = 0$ in (2.2), we have an *uncoupled* unperturbed Hamiltonian H_0, which is somewhat simpler since it is formed by one pendulum and n rotors. However, our approach is designed for the more general *coupled* case, for whose study we have given the motivation in Section 1.5.

We see that the Hamiltonian H_0 has an n-parameter family of n-dimensional hyperbolic invariant tori given by the equations $I = \text{const}$, $y = -\langle q, I \rangle$, $x = 0$. The associated stable and unstable whiskers of each torus coincide, and hence this (unique) whisker is made of homoclinic orbits, biasymptotic to the torus. Our aim is to give conditions for the splitting of the whiskers for $\mu \neq 0$.

We will focus our attention on a concrete hyperbolic torus, that we assume located at the origin: $I = 0$, $x = y = 0$, with flow $\dot{\varphi} = \omega$. The frequency vector ω will be assumed to satisfy a Diophantine condition: For some $\tau \geq n - 1$ and $\gamma > 0$,

$$|\langle k, \omega \rangle| \geq \gamma \, |k|^{-\tau} \qquad \forall k \in \mathbb{Z}^n \setminus \{0\}. \tag{2.6}$$

When the frequency vector ω is fixed, we say that the system (2.1, 2.2) is *regular*. Eventually, we can allow ω to depend on an additional parameter: $\omega = \omega(\varepsilon)$ (for instance, $\omega = \omega^*/\sqrt{\varepsilon}$ according to (1.6), and then $\gamma = \gamma^*/\sqrt{\varepsilon}$). Then, if we let $\varepsilon \to 0^+$ and consider $\mu = \varepsilon^p$, it becomes *singular*, but if we keep ε fixed and let $\mu \to 0$, the system is still regular.

We denote $\mathcal{T}_0$ the n-dimensional hyperbolic invariant torus, of frequency vector ω, of our unperturbed Hamiltonian (2.2). Since this torus is given by the equations $I = 0$, $x = y = 0$, we can obviously parameterize it as

$$\mathcal{T}_0 : \qquad z_0^*(\varphi) = (0, 0, \varphi, 0), \qquad \varphi \in \mathbb{T}^n.$$

Note that the trajectories on $\mathcal{T}_0$ are $t \mapsto z_0^*(\varphi + \omega t)$ for any given φ.

As mentioned in Section 2, the stable and unstable whiskers of the torus $\mathcal{T}_0$ coincide; this homoclinic whisker is given by the equations $I = 0$, $P(x, y) = 0$. We denote $\mathcal{W}_0$ the positive part $(y > 0)$ of the whisker. To give a suitable parameterization for $\mathcal{W}_0$, we consider the 1-degree-of-freedom Hamiltonian $P(x, y)$, and denote $(x_0(t), y_0(t))$ the associated homoclinic trajectory, biasymptotic (for $t \to \pm\infty$) to the hyperbolic point $(0, 0)$, with $x_0(0) = \pi$, $y_0(0) > 0$. We can then give the whisker $\mathcal{W}_0$ the parameterization

$$\mathcal{W}_0 : \; z_0(t, \varphi) = (x_0(t), y_0(t), \varphi + \delta_0(t) + \omega t, 0), \qquad \varphi \in \mathbb{T}^n, \quad t \in \mathbb{R},$$

where

$$\delta_0(t) = \left(\int_0^t y_0(s)\mathrm{d}s\right) q = (x_0(t) - \pi)q.$$

Note that $z_0(\cdot, \varphi)$ is a trajectory on $\mathcal{W}_0$ for any given φ. Besides, this trajectory is biasymptotic to the invariant torus $\mathcal{T}_0$, and $\delta_0(t)$ is just the *phase shift* undergone by any trajectory when traveling along $\mathcal{W}_0$. Since $\lim_{t\to\pm\infty} \delta_0(t) = \pm\pi q$, we have

$$\lim_{t\to\pm\infty} [z_0(t,\varphi) - z_0^*(\varphi \pm \pi q + \omega t)] = 0, \tag{2.7}$$

with exponentially decreasing bounds.

3. On the hyperbolic KAM theory: The perturbed hyperbolic torus and its local whiskers. We recall in this section the KAM theorem for hyperbolic invariant tori, which states that the torus $\mathcal{T}_0$ of H_0 survives, with its local stable and unstable whiskers, in the perturbation H for μ small enough. This result, which we call the hyperbolic KAM theorem, follows from a KAM-like iterative scheme providing, after a symplectic transformation, a *convergent local normal form* in a neighborhood of the hyperbolic torus. A first version was proved in [28] for a regular hyperbolic Hamiltonian. Subsequent papers, like [13, 51, 55], provide proofs that hold also for a singular hyperbolic Hamiltonian.

We consider here the approach of [25, 45], which are the closest to our purpose of carrying out a global control of the whiskers in order to study their splitting. Moreover, the use of exact symplectic transformations in the construction of the normal form will be very suitable in order to put the splitting function as the gradient of a splitting potential.

To formulate the hyperbolic KAM theorem, it is convenient to consider local canonical variables in a neighborhood of the torus, in such a way that the unperturbed local whiskers become coordinate planes. The new variables $w = (u, v, \varphi, I)$, referred from now on as *hyperbolic variables,* are defined by an exact symplectic transformation $z = \Psi(w)$, $\Psi = \Psi_1 \circ \Psi_2$, with

$$\Psi_1(x, y, \varphi, I) = (x, y - \langle q, I\rangle, \varphi + qx, I), \tag{3.1}$$

$$\Psi_2(u, v, \varphi, I) = (\psi(u, v), \varphi, I), \tag{3.2}$$

where ψ is a local (exact) symplectic transformation taking the pendulum P to a function of uv:

$$P(\psi(u,v)) = g(uv) = -\alpha uv + g_{\geq 2}(uv),$$

with $g_{\geq 2} = \mathrm{O}_2(uv)$ (recall that α is the Lyapunov exponent, defined in (2.4)). The existence of this transformation and its analyticity in a neighborhood of the origin is a well-known Moser's result [41] on the convergence of the Birkhoff normal form in a neighborhood of a hyperbolic equilibrium point (see also [13, §A3]).

We denote G_0, G, the Hamiltonians H_0, H, expressed in the hyperbolic variables:

$$G_0(u,v,I) = H_0 \circ \Psi = \langle \omega, I \rangle - \alpha uv + g_{\geq 2}(uv) + \frac{1}{2}\left\langle \widehat{Q}I, I \right\rangle, \tag{3.3}$$

$$G(u,v,\varphi,I;\mu) = H \circ \Psi = H_0 \circ \Psi + \mu\, H_1 \circ \Psi = G_0 + \mu G_1, \tag{3.4}$$

where

$$\widehat{Q} = Q - qq^\top.$$

Working in these hyperbolic variables, the hyperbolic KAM theorem will provide us a transformation $\widehat{\Phi}$ to normal form, and we will denote $\widetilde{G} = G \circ \widehat{\Phi}$ the new Hamiltonian.

In order to formulate quantitative statements, we introduce some notations. As a neighborhood of the whiskered torus, we define for $s, r, \rho > 0$ the complex domain

$$B_{s,r,\rho} = \Big\{ (u,v,\varphi,I) : |u|\,,|v| \leq s, |I| \leq r, |\mathrm{Im}\,\varphi| \leq \rho \Big\}.$$

For a function f analytic on some bounded domain D and continuous on its boundary, we consider the norm

$$|f|_D = \sup_D |f(u,v,\varphi,I)|\,.$$

The notation $f_1 \equiv f_2$ means that $f_1 - f_2 = \mathrm{const}$ (so the associated Hamiltonian equations are the same). In general, our functions depend also on μ as an additional parameter; the word "constant" will not exclude dependence on μ.

A statement of the hyperbolic KAM theorem to be applied in this work is given below. In fact this is a standard hyperbolic KAM theorem, with some minor changes due to that one is seeking for an exact symplectic transformation to normal form. The exactness is important in order to detect intersections between the whiskers, with the help of a generating function, as we show in Section 4. The proof is a refinement of Eliasson's proof (see [20]).

THEOREM 3.1 (hyperbolic KAM theorem). *Let $G = G_0 + \mu G_1$ as in (3.3, 3.4), analytic on $B_{s,r,\rho}$. Assume the frequency vector ω satisfies the Diophantine condition (2.6) for some $\tau > n-1$ and $\gamma > 0$. Assume also the nondegeneracy conditions*

$$\alpha \neq 0, \qquad \det \widehat{Q} \neq 0.$$

Let $0 < \eta < s$, $0 < \zeta < r$ and $0 < \delta < \rho$ given, and define

$$\mu_0 = \frac{\delta^{4\tau+4}\gamma^4}{C_1}. \tag{3.5}$$

Then, for $|\mu| \le \mu_0$ there exist an exact symplectic transformation $\widehat{\Phi}$: $B_{s-\eta,r-\zeta,\rho-\delta} \longrightarrow B_{s,r,\rho}$ (depending on μ), and constants $a = a(\mu)$, $b = b(\mu)$ such that $\widetilde{G} = G \circ \widehat{\Phi}$ takes the form

$$\widetilde{G} \equiv \langle \omega, I - a(\mu) \rangle - b(\mu)\alpha uv + \widehat{R}(u, v, \varphi, I; \mu),$$
$$\widehat{R} = \mathrm{O}_2\left(uv, I - a(\mu)\right).$$

The transformation $\widehat{\Phi}$ and the constants a, b are analytic on their domain with respect to (u, v, φ, I) and μ. For any $|\mu| \le \mu_0$ the following estimates hold:

$$\left|\widehat{\Phi} - \mathrm{id}\right|_{B_{s-\eta,r-\zeta,\rho-\delta}} \le \frac{C_2}{\gamma^2\delta^{2\tau+2}} |\mu|,$$
$$|a|, \ |b-1| \le \frac{C_3}{\gamma\delta^{\tau+1}} |\mu|.$$

The constants C_j do not depend on μ, δ, ω.

Notice that the threshold given in (3.5) is proportional to γ^4, as in [45], differing from the one given in [51], proportional to γ^2. This seems due to the different methods of proof that follow the Kolmogorov's and Arnold's approaches to KAM theorem.

An additional remark is that the validity of this version of KAM theorem in the singular case can be established by regarding how the involved parameters (mainly the threshold μ_0) depend on ω. In this way one can rewrite ω as $\omega^*/\sqrt{\varepsilon}$ without affecting, for $\varepsilon \to 0^+$, the smallness condition and the estimates (or rather making them better).

The next statement is a more refined version of Eliasson's result [25, p. 65]. From the hyperbolic variables w, this result comes back to the original variables z through the transformation Ψ introduced in (3.1, 3.2). The transformation to normal form is then $\Phi = \Psi \circ \widehat{\Phi} \circ \Psi^{-1}$ and the new Hamiltonian $\widetilde{H} = \widetilde{G} \circ \Psi^{-1} = H \circ \Phi$. This strategy of expressing the KAM theorem in the original variables will be very useful to our purposes. Concerning the domains, notice that, for any $s, r, \rho > 0$,

$$\Psi_1\left(B_{s,r,\rho}\right) = \Big\{(x, y, \varphi, I) : |x| \le s, |y + \langle q, I \rangle| \le s,$$
$$|I| \le r, |\mathrm{Im}(\varphi - qx)| \le \rho\Big\}.$$

THEOREM 3.2. *Let $H = H_0 + \mu H_1$ as described in (2.1–2.3), analytic on $\Psi_1\left(B_{s,r,\rho}\right)$. Assume the frequency vector ω satisfies the Diophantine* *condition (2.6) for some $\tau > n-1$ and $\gamma > 0$. Assume also the nondegeneracy condition (2.5). Let $0 < \delta < \rho$ given, and define*

$$\mu_0 = \frac{\delta^{4\tau+4}\gamma^4}{C_1}.$$

Then, for $|\mu| \le \mu_0$ and some $0 < \kappa < 1/2$ there exist an exact symplectic transformation $\Phi : \Psi_1\left(B_{\kappa s, r/2, \rho-\delta}\right) \longrightarrow \Psi_1\left(B_{s,r,\rho}\right)$ (depending on μ), and constants $a = a(\mu)$, $b = b(\mu)$ such that $\widetilde{H} = H \circ \Phi$ takes the form

$$\begin{aligned} &\widetilde{H} \equiv \langle \omega, I - a(\mu)\rangle + b(\mu)P(x, y + \langle q, I\rangle) + R(x,y,\varphi,I;\mu), \\ &R = \mathrm{O}_2\Big(P(x, y+\langle q, I\rangle), I - a(\mu)\Big). \end{aligned}$$

The transformation Φ and the constants a, b are analytic on their domain with respect to (x,y,φ,I) and μ. For any $|\mu| \le \mu_0$ the following estimates hold:

$$\begin{aligned} |\Phi - \mathrm{id}|_{\Psi_1\left(B_{\kappa s, r/2, \rho-\delta}\right)} &\le \frac{C_2}{\gamma^2\delta^{2\tau+2}}\,|\mu|, \\ |a|,\ |b-1| &\le \frac{C_3}{\gamma\delta^{\tau+1}}\,|\mu|. \end{aligned}$$

The constants C_j do not depend on μ, δ, ω.

The most important point about Theorem 3.2 is, in our opinion, the fact that the local normal form $\widetilde{H}$ is put in terms of $P(x, y+\langle q, I\rangle)$. Using this expression, it is constructed in [25] a "global" Hamiltonian (in x) having exactly the same hyperbolic torus and its (local) whiskers as $\widetilde{H}$, as well as the dynamics on them. These whiskers are global (and coincident) in this new Hamiltonian. We are going to use intensively this feature in Section 4.

From now on, in order to keep a more readable notation, we shall usually omit the μ-dependence of our functions.

It is clear that the Hamiltonian $\widetilde{H}$ has a hyperbolic invariant torus of frequencies ω. This torus and its associated local whiskers can be parameterized as follows:

$$\begin{aligned} \widetilde{\mathcal{T}} = \widetilde{\mathcal{T}}(\mu): &\quad \tilde{z}^*(\varphi) = (0, -\langle q, a\rangle, \varphi, a), \quad \varphi \in \mathbb{T}^n, \\ \widetilde{\mathcal{W}}^{\pm}_{\mathrm{loc}} = \widetilde{\mathcal{W}}^{\pm}_{\mathrm{loc}}(\mu): &\quad \tilde{z}(t,\varphi) = \Big(x_0(bt), y_0(bt) - \langle q, a\rangle, \varphi + \delta_0(bt) + \omega t, a\Big), \\ &\qquad\qquad \varphi \in \mathbb{T}^n, \quad \pm t \ge t_0, \end{aligned}$$

with a suitable $t_0 = t_0(s)$. We then have, for the original Hamiltonian H, a hyperbolic torus and its associated local stable and unstable whiskers (for $|\mu| \le \mu_0$):

$$\begin{aligned} \mathcal{T} = \mathcal{T}(\mu): &\quad z^*(\varphi) = \Phi\left(\tilde{z}^*(\varphi)\right), \qquad \varphi \in \mathbb{T}^n, \\ \mathcal{W}^{\pm}_{\mathrm{loc}} = \mathcal{W}^{\pm}_{\mathrm{loc}}(\mu): &\quad z^{\pm}(t,\varphi) = \Phi\left(\tilde{z}(t,\varphi)\right), \quad \varphi \in \mathbb{T}^n, \quad \pm t \ge t_0. \end{aligned}$$

These parameterizations of the whiskers $\mathcal{W}^{\pm}_{\text{loc}}$ can be extended to further values of t in a natural way, as trajectories associated to our Hamiltonian H. We denote $\mathcal{W}^{\pm} = \mathcal{W}^{\pm}(\mu)$ the extended or *global whiskers,* and our aim is to measure the distance between them.

4. The splitting potential and the splitting function. As mentioned in the previous section, the special formulation of Theorem 3.2 allows us to carry out a more global control on the perturbed whiskers. We introduce as in [25] the following integrable Hamiltonian:

$$N(x,y,I;\mu) = \langle \omega, I-a\rangle + bP(x, y+\langle q,a\rangle) + b\langle q, I-a\rangle (y+\langle q,a\rangle) + \frac{b}{2}\langle Q(I-a), I-a\rangle$$

(recall that a and b depend on μ, although this is not made explicit). This Hamiltonian is defined globally in the variable $x \in \mathbb{T}$, and has the same hyperbolic torus $\widetilde{\mathcal{T}}$ and its whiskers $\widetilde{\mathcal{W}}^{\pm}_{\text{loc}}$, and the dynamics on them, as the local normal form $\widetilde{H}$. The only difference is that the parameterization $\tilde{z}(t,\varphi)$ (of the whiskers in $\widetilde{H}$) can now be defined for any $t \in \mathbb{R}$; in this way the local whiskers $\widetilde{\mathcal{W}}^{\pm}_{\text{loc}}$ can be extended to a (unique) global homoclinic whisker of N. Following [25], we are going to take advantage of the fact that the "artificial" Hamiltonian N is integrable.

We assume our starting Hamiltonian H in (2.1) analytic on a complex domain

$$D_{\sigma,s,r,\rho} = \Big\{(x,y,\varphi,I) : |\operatorname{Im} x| \le \sigma, |y| \le s, |I| \le r, |\operatorname{Im}\varphi| \le \rho\Big\},$$

with s big enough in order to contain a neighborhood of the (global) unperturbed whiskers. We denote Υ^t_0, Υ^t, Υ^t_N the time-t flows of the Hamiltonians H_0, H, N respectively.

We choose two points (x^0, y^0), (x^1, y^1) on the positive $(y^0, y^1 > 0)$ homoclinic orbit of the pendulum P, with $x^0 > \pi$ and $x^1 < \pi$, and such that the associated φ-sections $(x^j, y^j - \langle q,a\rangle, \mathbb{T}^n, a)$, $j = 0,1$, are contained in the domain of the normal form $\widetilde{H}$. Consider $T > 0$ such that

$$\Upsilon^T_N\left(x^1, y^1 - \langle q,a\rangle, \mathbb{T}^n, a\right) = \left(x^0, y^0 - \langle q,a\rangle, \mathbb{T}^n, a\right).$$

Comparing the flows of N and H we can define, in a neighborhood Ω of $\Phi\left(x^0, y^0 - \langle q,a\rangle, \mathbb{T}^n, a\right)$, the map

$$\Theta = \Upsilon^T \circ \Phi \circ \Upsilon^{-T}_N \circ \Phi^{-1}.$$

This map is exact symplectic, and takes $\mathcal{W}^{+}_{\text{loc}} \cap \Omega$ (a subset of the local stable whisker) into $\mathcal{W}^-$ (the global unstable whisker). Besides, it is easy to check that the map Θ gives a correspondence between our parameterizations of the whiskers:

$$\Theta\left(z^+(t,\varphi)\right) = z^-(t,\varphi), \tag{4.1}$$

for any t, φ such that $z^+(t,\varphi) \in \Omega$. Therefore, the difference $\Theta - \mathrm{id}$ constitutes a measure for the splitting on $\mathcal{W}^+_{\mathrm{loc}} \cap \Omega$. We stress that a formula like (4.1) cannot be expected outside $\mathcal{W}^+_{\mathrm{loc}}$, since we need the fact that the Hamiltonians $\widetilde{H}$ and N have the same dynamics on the whiskers. Note also that (4.1) tells us that the map Θ does not depend on the points (x^0,y^0), (x^1,y^1) chosen. What does depend strongly on these points is the neighborhood Ω where Θ is defined.

To continue, it is convenient to express the map Θ in the normal form hyperbolic variables $w = (u,v,\varphi,I)$ introduced in Section 3, in which the local whiskers become coordinate planes, and the global unstable whisker can be seen as a graphic over the local stable one. Recall that $\widetilde{G} = \widetilde{H} \circ \Psi$ is our local normal form expressed in the hyperbolic variables. The local whiskers for $\widetilde{G}$ are given by

$$\widehat{\mathcal{W}}^+_{\mathrm{loc}} = \Psi^{-1}\left(\widetilde{\mathcal{W}}^+_{\mathrm{loc}}\right) = \{v = 0, I = a\},$$
$$\widehat{\mathcal{W}}^-_{\mathrm{loc}} = \Psi^{-1}\left(\widetilde{\mathcal{W}}^-_{\mathrm{loc}}\right) = \{u = 0, I = a\}.$$

The exact symplectic map

$$\widehat{\Theta} = (\Phi \circ \Psi)^{-1} \circ \Theta \circ (\Phi \circ \Psi)$$

is defined in the neighborhood $\widehat{\Omega} = (\Phi \circ \Psi)^{-1}(\Omega)$ of $\left(u^0, 0, \mathbb{T}^n, a\right)$, where $u^0 > 0$ is defined by $\left(x^0, y^0\right) = \psi\left(u^0, 0\right)$. It is adequate to define

$$\widehat{\mathcal{W}}^- = \widehat{\Theta}\left(\widehat{\mathcal{W}}^+_{\mathrm{loc}} \cap \widehat{\Omega}\right)$$

as an invariant manifold of $\widetilde{G}$, which is the equivalent in the hyperbolic variables for (a piece of) the global unstable whisker $\mathcal{W}^-$.

In the hyperbolic variables, the problem of measuring the splitting has a simpler formulation, since $\widehat{\mathcal{W}}^-$ can be seen as a graphic over the local whisker $\widehat{\mathcal{W}}^+_{\mathrm{loc}}$. Let us parameterize:

$$\widehat{\mathcal{W}}^+_{\mathrm{loc}} : \ \hat{w}^+(t,\varphi) = \Psi^{-1}\left(\tilde{z}(t,\varphi)\right) = \left(u_0(bt), 0, \varphi - \pi q + \omega t, a\right),$$
$$\widehat{\mathcal{W}}^- : \ \hat{w}^-(t,\varphi) = (\Phi \circ \Psi)^{-1}\left(z^-(t,\varphi)\right),$$

where $u_0(t)$ is defined by $\psi(u_0(t),0) = (x_0(t), y_0(t))$. In components, we will write $\hat{w}^\pm = \left(\hat{u}^\pm, \hat{v}^\pm, \hat{\varphi}^\pm, \hat{I}^\pm\right)$. Then the splitting is given by the difference $\hat{I}^- - \hat{I}^+ = \hat{I}^- - a$. We remark that we do not need to consider the difference $\hat{v}^- - \hat{v}^+ = \hat{v}^-$ because $\hat{I}^-(t,\varphi) = a$ implies that $\hat{v}^-(t,\varphi) = 0$; this is related to the fact that the whiskers are contained in energy levels of $\widetilde{G}$.

The exactness of the symplectic map $\widehat{\Theta}$ and the fact that $\widehat{\Theta} - \mathrm{id} = \mathrm{O}(\mu)$, imply that for μ small enough there exists a *generating function* $\theta\left(u, \check{v}, \varphi, \check{I}\right)$ such that $\widehat{\Theta} : \left(\check{u}, \check{v}, \check{\varphi}, \check{I}\right) \mapsto (u, v, \varphi, I)$ is given by

$$u = \check{u} - \partial_{\check{v}}\theta, \qquad \check{v} = v - \partial_u\theta, \qquad \varphi = \check{\varphi} - \partial_{\check{I}}\theta, \qquad \check{I} = I - \partial_\varphi\theta.$$

We define the *splitting potential* as the following periodic function:

$$\mathcal{L}(\varphi) = \mathcal{L}(t_0, \varphi) = \theta(u_0(bt_0), 0, \varphi - \pi q + \omega t_0, a), \qquad \varphi \in \mathbb{T}^n,$$

with t_0 such that $\hat{w}^+(t_0, \mathbb{T}^n) \subset \widehat{\Omega}$. The generating function θ is determined up to an additive constant, that can be chosen to ensure that $\mathcal{L}$ has zero average. Note that $\mathcal{L}$ depends on μ and on the choice of t_0. The gradient $\mathcal{M}(\varphi) = \partial_\varphi \mathcal{L}(\varphi)$ will be called the (vector) *splitting function,* and the following theorem states that, on a suitable φ-section, the function $\mathcal{M}$ gives the splitting distance.

THEOREM 4.1. *There exist functions $\tau(\varphi) = \mathrm{O}(\mu)$ and $\Xi(\varphi) = \varphi + \mathrm{O}(\mu)$, periodic in φ, such that, for all $\varphi \in \mathbb{T}^n$,*

$$\mathcal{M}(\varphi) = \hat{I}^- \left(t_0 + \tau(\varphi), \Xi(\varphi)\right) - a.$$

For our fixed t_0, note that

$$\hat{I}^-(t_0, \varphi) - a = \partial_\varphi \theta \left(\hat{u}^-(t_0, \varphi), 0, \hat{\varphi}^-(t_0, \varphi), a\right). \tag{4.2}$$

One can think that this should be the most natural election for what we call "splitting function". Nevertheless, it does not come from Formula (4.2) that this is the gradient of a scalar function, although this obstruction has been easily overcome with the change of variables given by the periodic functions τ and Ξ.

An important consequence of Theorem 4.1 is the existence of at least $n+1$ effective intersections between the whiskers $\mathcal{W}^\pm$, giving rise to at least $n+1$ trajectories biasymptotic to the invariant torus $\mathcal{T}$. These intersections are obtained as critical points of a function on $\mathbb{T}^n$, and the minimum number of them comes from Lyusternik–Schnirelman theory (see [15, Sec. 2.12]). This constitutes the main result contained in [25]. Note also that, in nondegenerate cases, the number of intersections becomes at least 2^n, as one deduces from Morse theory.

5. First order approximations for the splitting: Melnikov integrals. We now use Poincaré–Melnikov theory to give a first order approximation (in μ) for the splitting potential introduced in the previous section. We define the (scalar) *Melnikov potential* and the (vector) *Melnikov function* as, respectively, the following periodic functions:

$$L(\varphi) = -\int_{-\infty}^{\infty} \left(H_1 - \overline{H_1} - \{\chi, H_0\}\right)(z_0(t, \varphi))\, \mathrm{d}t, \tag{5.1}$$

$$M(\varphi) = \partial_\varphi L(\varphi) = -\int_{-\infty}^{\infty} \left[\partial_\varphi \left(H_1 - \{\chi, H_0\}\right)\right](z_0(t, \varphi))\, \mathrm{d}t. \tag{5.2}$$

In these formulas, $\chi(x, y, \varphi, I) = \chi(\varphi)$ is the (zero average) function solving the small divisors equation

$$\langle \omega, \partial_\varphi \chi \rangle + \overline{H_1(0, 0, \cdot, 0)} = H_1(0, 0, \cdot, 0). \tag{5.3}$$

It is not difficult to check that the integrals in (5.1, 5.2) are absolutely convergent (in contrast with [29, 50, 57], where conditionally convergent Melnikov integrals are introduced for $n \geq 2$). This absolute convergence is due to the incorporation of the function $\chi(\varphi)$ (introduced by Treschev in [56]), which is closely related to the shift suffered by the perturbed whiskered torus $\mathcal{T}$ with respect to the unperturbed torus $\mathcal{T}_0$ in applying KAM theorem. Indeed, writing the parameterization of the torus $\mathcal{T}$ in components: $z^* = (x^*, y^*, \varphi^*, I^*)$, the following lemma gives a first order approximation for $I^*(\varphi)$.

LEMMA 5.1. *The I-component of $z^*(\varphi)$ satisfies*

$$I^*(\varphi) = \mu\left(\mathrm{const} - \partial_\varphi \chi(\varphi)\right) + \mathrm{O}\left(\mu^2\right).$$

We note that $L(\varphi)$, as defined in (5.1), has zero average ($\overline{L} = 0$), and that $L(\varphi)$ and $M(\varphi)$ do not depend on μ.

We stress that our formula for $L(\varphi)$ is quite compact, and useful in the coupled case ($q \neq 0$), as we show in Section 6. It is also worth remarking that (5.1) it is equivalent to the formula appearing in [13, §4], but much simpler.

We now show some alternative expressions for $L(\varphi)$ and $M(\varphi)$. Using that

$$\int_{t_1}^{t_2} \{\chi, H_0\}\left(z_0(t,\varphi)\right) \mathrm{d}t = \chi\left(z_0(t_2,\varphi)\right) - \chi\left(z_0(t_1,\varphi)\right) \tag{5.4}$$

for any t_1, t_2, and applying (2.7), we get

$$L(\varphi) = \lim_{T\to\infty}\left[-\int_{-T}^{T}\left(H_1 - \overline{H_1}\right)\left(z_0(t,\varphi)\right)\mathrm{d}t \right.$$
$$\left. + \chi(\varphi + \pi q + \omega T) - \chi(\varphi - \pi q - \omega T)\right].$$

Taking a φ-derivative, we obtain

$$M(\varphi) = \lim_{T\to\infty}\left[-\int_{-T}^{T} \partial_\varphi H_1\left(z_0(t,\varphi)\right)\mathrm{d}t \right.$$
$$\left. + \partial_\varphi\chi(\varphi + \pi q + \omega T) - \partial_\varphi\chi(\varphi - \pi q - \omega T)\right].$$

In the uncoupled case $q = 0$, we can provide simpler (and perhaps more classical) expressions. Proceeding like in (5.4) and using that $\langle \omega, \partial_\varphi \chi\rangle = \{\chi, H_0\}(0,0,\cdot,0)$, we have

$$\begin{aligned}\chi(\varphi + \omega t_2) - \chi(\varphi + \omega t_1) &= \int_{t_1}^{t_2} \{\chi, H_0\}(0,0,\varphi+\omega t, 0)\mathrm{d}t \\ &= \int_{t_1}^{t_2} \left(H_1 - \overline{H_1}\right)(0,0,\varphi+\omega t,0)\mathrm{d}t.\end{aligned}$$

Thus, for the special case $q = 0$ we obtain the following expressions (compare with formula (2.15) of [21]):

$$L(\varphi) = -\int_{-\infty}^{\infty} \left[\left(H_1 - \overline{H_1}\right)(z_0(t,\varphi)) - \left(H_1 - \overline{H_1}\right)(0,0,\varphi + \omega t, 0) \right] \mathrm{d}t,$$

$$M(\varphi) = -\int_{-\infty}^{\infty} \left[\partial_\varphi H_1\left(z_0(t,\varphi)\right) - \partial_\varphi H_1\left(0,0,\varphi+\omega t,0\right) \right] \mathrm{d}t.$$

Coming again to the general case, the following standard lemma shows that a first order approximation for the difference $I^-(t_0,\cdot) - I^+(t_0,\cdot)$ is given by the Melnikov function $M(\varphi)$ (we denote $z^\pm = (x^\pm, y^\pm, \varphi^\pm, I^\pm)$ in the parameterizations of the whiskers $\mathcal{W}^\pm$). In fact, an analogous expression is given in [56] for $F_0\left(z^-(t_0,\cdot)\right) - F_0\left(z^+(t_0,\cdot)\right)$, where F_0 is any given first integral of the unperturbed Hamiltonian H_0. Our statement concerns the case $F_0 = I_j$, $j = 1, \dots, n$.

LEMMA 5.2. *For any fixed* $t_0 \in \mathbb{R}$,

$$I^-(t_0,\varphi) - I^+(t_0,\varphi) = \mu M(\varphi) + \mathrm{O}(\mu^2). \tag{5.5}$$

Finally, we see that the splitting potential $\mathcal{L}(\varphi)$ introduced in Theorem 4.1 can be approximated by the Melnikov potential $L(\varphi)$. This requires to show that the approximation (5.5), expressed in the original variables, remains true after changing to the normal form variables in which the splitting potential $\mathcal{L}(\varphi)$ had to be defined.

THEOREM 5.1. *For the splitting function and the splitting potential introduced in Theorem 4.1, one has:*

$$\mathcal{L}(\varphi) = \mu L(\varphi) + \mathrm{O}\left(\mu^2\right), \qquad \mathcal{M}(\varphi) = \mu M(\varphi) + \mathrm{O}\left(\mu^2\right).$$

The proof shows that the Melnikov function $M(\varphi)$ remains valid as a first order approximation for the splitting function $\mathcal{M}(\varphi)$, in spite of the fact that $\mathcal{M}(\varphi)$ was defined in Theorem 4.1 using the normal form variables.

We remark that the first order approximations for $\mathcal{M}(\varphi)$ and $\mathcal{L}(\varphi)$ do not depend on t_0 at first order in μ but they do at higher orders.

Since both the splitting potential and the Melnikov potential are defined on $\mathbb{T}^n$, a direct application of Morse theory implies the existence, for $|\mu|$ small enough, of at least 2^n transverse homoclinic orbits to the whiskered torus $\mathcal{T}$, as long as the Melnikov potential is a Morse function, that is, all its critical points are nondegenerate (a generic property).

6. A computable example in the coupled case. As an example, we consider a perturbation of a coupled integrable Hamiltonian ($q \neq 0$ in (2.2)), with $n+1$ degrees of freedom. In the integrable part, we choose the classical pendulum: $V(x) = \cos x - 1$. In the perturbation, we consider a function only depending on φ:

$$H_1(\varphi) = \sum_{k \in \mathbb{Z}^n} h_k \, e^{i\langle k, \varphi \rangle}.$$

We use the following well-known formulas for the homoclinic trajectory of the standard pendulum:

$$\begin{aligned} x_0(t) &= 4\arctan e^t, \qquad y_0(t) = \frac{2}{\cosh t}, \\ \delta_0(t) &= q(x_0(t)-\pi), \\ z_0(t,\varphi) &= (x_0(t), y_0(t), \varphi+\delta_0(t)+\omega t, 0)\,. \end{aligned}$$

The solution $\chi(\varphi)$ of Equation (5.3) is simply

$$\chi(\varphi) = -\sum_{k\neq 0} \frac{i\,h_k}{\langle k,\omega\rangle}\, e^{i\langle k,\varphi\rangle},$$

and the Melnikov potential is given by its Fourier series

$$L(\varphi) = \sum_{k\neq 0} L_k\, e^{i\langle k,\varphi\rangle},$$

where each coefficient is given by the following integral:

$$L_k = \frac{h_k\ \langle k,q\rangle\, e^{-i\pi\langle k,q\rangle}}{\langle k,\omega\rangle} \int_{-\infty}^{\infty} e^{i\langle k,\omega\rangle t}\, e^{i\langle k,q\rangle x_0(t)}\, y_0(t)\, \mathrm{d}t.$$

Then, for the Melnikov function $M(\varphi) = \sum_{k\neq 0} M_k\, e^{i\langle k,\varphi\rangle}$, it is clear that $M_k = ikL_k$.

For the sake of simplicity, we assume that q is "half-integer" (i.e. $2q \in \mathbb{Z}^n$). In this way, we can easily compute the Melnikov integrals using residue theory since all the singularities of the functions involved are poles. Roughly, we get:

$$|L_k| \sim |h_k|\ \frac{e^{\pm\frac{\pi}{2}\langle k,\omega\rangle}}{|\sinh\left(\pi\,\langle k,\omega\rangle\right)|}\ |\langle k,\omega\rangle|^{|\langle k,2q\rangle|-1}\,.$$

It is very interesting to sketch an analysis of this expression in the case of *fast* frequencies. Thus, we consider $\omega = \omega^*/\sqrt{\varepsilon}$ with ω^* Diophantine, and introduce $\gamma = \gamma^*/\sqrt{\varepsilon}$ in (2.6), for some $\tau \geq n-1$. We also assume exponentially decreasing coefficients for the perturbation, like $|h_k| \sim e^{-|k|\rho}$ for every k (the parameter ρ is the width of analyticity of $H_1(\varphi)$). Then we can make the following estimate: The dominant harmonic $L_k e^{i\langle k,\varphi\rangle}$ in the Fourier series of $L(\varphi)$ is given for the indexes k such that

$$|\langle k,\omega^*\rangle| \sim \gamma^*\,|k|^{-\tau}, \qquad |k| \sim \left(\frac{\pi}{2}\frac{\gamma^*}{\rho\sqrt{\varepsilon}}\right)^{1/(\tau+1)},$$

(assuming that the corresponding coefficients h_k are nonvanishing), and we deduce that the Melnikov potential $L(\varphi)$ is exponentially small in ε: $-\log L = \mathrm{O}\left(\varepsilon^{-1/2(\tau+1)}\right)$, as well as its gradient $M(\varphi)$.

A more precise asymptotic behavior can be obtained for the case of *two* rotators ($n = 2$ in (2.2)), since then one can apply the theory of continued fractions to the ratio of frequencies ω_2^*/ω_1^*. For instance, in the case of a *quadratic* number ω_2^*/ω_1^* like the 'golden mean' $(1+\sqrt{5})/2$, one can find, as in [17], that $\log L \sim -c(\log \varepsilon)\varepsilon^{-1/4}$, where $c(u)$ is a positive periodic function with period $2\log(\omega_2^*/\omega_1^*)$. A direct application of Theorem 5.1 ensures that the splitting distance is as predicted by the Melnikov function, but requires $\mu = \mathrm{o}\left(\exp\left(-c(\log\varepsilon)\varepsilon^{-1/4}\right)\right)$.

A justification of the case $\mu = \varepsilon^p$ for some $p > 0$ is immediate as long as one has a significant refinement of Theorem 5.1:

$$\mathcal{L}(\varphi - t_0\omega^*/\sqrt{\varepsilon}) = \mu L(\varphi - t_0\omega^*/\sqrt{\varepsilon}) + \mathrm{O}\left(\mu^2\varepsilon^{-p}\right),$$

for φ, t_0 on the *complex* strip $|\mathrm{Im}\,\varphi| \leq \rho - \varepsilon^{1/4}$, $|\mathrm{Im}\,t_0| \leq \pi/2 - \varepsilon^{1/4}$. Such kind of result is a byproduct of an extension theorem (see, for instance, [17, 52]), which is currently being researched by the authors.

Acknowledgments. We are indebted to P. Lochak for his critical comments and review, and to L. Niederman for several discussions. One of the authors (A.D.) is very grateful to the hospitality from the Institute for Mathematics and Its Applications in Minneapolis during the preparation of this manuscript.

REFERENCES

[1] V. Arnold, *Proof of a theorem of A.N. Kolmogorov on the invariance of quasi-periodic motions under small perturbations of the Hamiltonian*, Russian Math. Surveys, **18** (1963), pp. 9–36.

[2] ———, *Instability of dynamical systems with several degrees of freedom*, Soviet Math. Dokl., **5** (1964), pp. 581–585.

[3] V. Arnold, V. Kozlov, and A. Neishtadt, *Mathematical aspects of classical and celestial mechanics*, in Dynamical systems III, V. Arnold, ed., Vol. **3** of Encyclopaedia Math. Sci., Springer-Verlag, Berlin–Heidelberg, 1988.

[4] G. Benettin, L. Galgani, A. Giorgilli, and J.-M. Strelcyn, *A proof of Kolmogorov's theorem on invariant tori using canonical transformations defined by the Lie method*, Il Nuovo Cimento B, **79** (1984), pp. 201–223.

[5] P. Bernard, *Perturbation d'un hamiltonien partiellement hyperbolique*, C.R. Acad. Sci. Paris Sér. I Math., **323** (1996), pp. 189–194.

[6] U. Bessi, *An approach to Arnold's diffusion through the calculus of variations*, Nonlinear Anal., **26** (1996), pp. 1115–1135.

[7] ———, *Arnold's example with three rotators*, Nonlinearity, **10** (1997), pp. 763–781.

[8] S. Bolotin, *Homoclinic orbits in invariant tori of Hamiltonian systems*, in Dynamical systems in classical mechanics, V. Kozlov, ed., Vol. **168** of Amer. Math. Soc. Transl. Ser. 2, Amer. Math. Soc., Providence, RI, 1995, pp. 21–90. Adv. Math. Sci., 25.

[9] J.-B. Bost, *Tores invariants des systèmes dynamiques hamiltoniens (d'après Kolmogorov, Arnold, Moser, Rüssmann, Zehnder, Herman, Pöschel, ...)*, Astérisque, **133**, **134** (1986), pp. 113–157.

[10] H. Broer and G. Huitema, *A proof of the isoenergetic KAM-theorem from the 'ordinary' one*, J. Differential Equations, **90** (1991), pp. 52–60.

[11] H. BROER, G. HUITEMA, AND M. SEVRYUK, *Quasi-periodic motions in families of dynamical systems: Order amidst chaos*, Vol. **1645** of Lecture Notes in Math., Springer-Verlag, New York, 1996.

[12] L. CHIERCHIA, *Non-degenerate 'Arnold diffusion'*. mp_arc@math.utexas.edu, Preprint 96-137, 1996.

[13] L. CHIERCHIA AND G. GALLAVOTTI, *Drift and diffusion in phase space*, Ann. Inst. H. Poincaré Phys. Théor., **60** (1994), pp. 1–144.

[14] B. CHIRIKOV, *A universal instability of many-dimensional oscillator systems*, Phys. Rep., **52** (1979), pp. 263–379.

[15] S.-N. CHOW AND J. HALE, *Methods of bifurcation theory*, Vol. **251** of Grundlehren Math. Wiss., Springer-Verlag, New York, 1982.

[16] J. CRESSON, *A λ-lemma for partially hyperbolic tori and the obstruction property*. Preprint, 1998.

[17] A. DELSHAMS, V. GELFREICH, A. JORBA, AND T. SEARA, *Exponentially small splitting of separatrices under fast quasiperiodic forcing*, Comm. Math. Phys., **189** (1997), pp. 35–71.

[18] ———, *Splitting of separatrices for (fast) quasiperiodic forcing*, in Simó [54].

[19] A. DELSHAMS AND P. GUTIÉRREZ, *Effective stability and KAM theory*, J. Differential Equations, **128** (1996), pp. 415–490.

[20] ———, *Splitting potential and Poincaré-Melnikov theory for whiskered tori in Hamiltonian systems*. Preprint, 1998.

[21] A. DELSHAMS AND R. RAMÍREZ-ROS, *Melnikov potential for exact symplectic maps*, Comm. Math. Phys., **190** (1997), pp. 213–245.

[22] A. DELSHAMS AND T. SEARA, *Splitting of separatrices in Hamiltonian systems with one and a half degrees of freedom*, Math. Phys. Electron. J., **3** (1997), pp. 1–40.

[23] R. DOUADY, *Applications du théorème des tores invariants*, PhD thesis, Université Paris VII, 1982.

[24] H. DUMAS, *Ergodization rates for linear flows on the torus*, J. Dynamics Differential Equations, **3** (1991), pp. 593–610.

[25] L. ELIASSON, *Biasymptotic solutions of perturbed integrable Hamiltonian systems*, Bol. Soc. Bras. Mat, **25** (1994), pp. 57–76.

[26] G. GALLAVOTTI, *Twistless KAM tori, quasi flat homoclinic intersections, and other cancellations in the perturbation series of certain completely integrable Hamiltonian systems. A review*, Rev. Math. Phys., **6** (1994), pp. 343–411.

[27] G. GALLAVOTTI, G. GENTILE, AND V. MASTROPIETRO, *Pendulum: Separatrix splitting*. Preprint 97-472, mp_arc@math.utexas.edu, 1997.

[28] S. GRAFF, *On the conservation of hyperbolic invariant tori for Hamiltonian systems*, J. Differential Equations, **15** (1974), pp. 1–69.

[29] P. HOLMES AND J. MARSDEN, *Melnikov's method and Arnold diffusion for perturbations of integrable Hamiltonian systems*, J. Math. Phys., **23** (1982), pp. 669–675.

[30] A. KOLMOGOROV, *The general theory of dynamical systems and classical mechanics*, in Foundations of mechanics, Benjamin/Cummings, Reading, Mass., 2nd ed., 1978, pp. 741–757 (appendix).

[31] V. LAZUTKIN, *Splitting of separatrices for the Chirikov's standard map* (in Russian). Preprint VINITI 6372-84, 1984.

[32] R. LLAVE AND C. WAYNE, *Whiskered and low dimensional tori in nearly integrable Hamiltonian systems*. Preprint, 1989.

[33] P. LOCHAK, *Effective speed of Arnold's diffusion and small denominators*, Phys. Lett. A, **143** (1990), pp. 39–42.

[34] ———, *Canonical perturbation theory via simultaneous approximation*, Russian Math. Surveys, **47** (1992), pp. 57–133.

[35] ———, *Hamiltonian perturbation theory: periodic orbits, resonances and intermittency*, Nonlinearity, **6** (1993), pp. 885–904.

[36] ———, *Arnold diffusion; a compendium of remarks and questions*, in Simó [54].

[37] P. LOCHAK AND A. NEISHTADT, *Estimates of stability time for nearly integrable systems with a quasiconvex Hamiltonian*, Chaos, **2** (1992), pp. 495–499.

[38] J. MARCO, *Transition le long des chaînes de tores invariants pour les systèmes hamiltoniens analytiques*, Ann. Inst. H. Poincaré Phys. Théor., **64** (1996), pp. 205–252.

[39] R. MOECKEL, *Transition tori in the five-body problem*, J. Differential Equations, **129** (1996), pp. 290–314.

[40] A. MORBIDELLI AND A. GIORGILLI, *Superexponential stability of KAM tori*, J. Statist. Phys., **78** (1995), pp. 1607–1617.

[41] J. MOSER, *The analytic invariants of an area-preserving mapping near a hyperbolic fixed point*, Comm. Pure Appl. Math., 9 (1956), pp. 673–692.

[42] ———, *On invariant curves of area-preserving mappings of an annulus*, Nachr. Akad. Wiss. Göttingen Math.-Phys. Kl. II, **2** (1962), pp. 1–20.

[43] A. NEISHTADT, *The separation of motions in systems with rapidly rotating phase*, J. Appl. Math. Mech., **48** (1984), pp. 133–139.

[44] N. NEKHOROSHEV, *An exponential estimate of the time of stability of nearly-integrable Hamiltonian systems*, Russian Math. Surveys, **32** (1977), pp. 1–65.

[45] L. NIEDERMAN, *Dynamic around a chain of simple resonant tori in nearly integrable Hamiltonian systems.* To appear in J. Differential Equations. Preprint 97-142, mp_arc@math.utexas.edu, 1997.

[46] A. PERRY AND S. WIGGINS, *KAM tori are very sticky: Rigorous lower bounds on the time to move away from an invariant Lagrangian torus with linear flow*, Phys. D, **71** (1994), pp. 102–121.

[47] H. POINCARÉ, *Sur le problème des trois corps et les équations de la dynamique*, Acta Math., **13** (1890), pp. 1–271.

[48] J. PÖSCHEL, *Integrability of Hamiltonian systems on Cantor sets*, Comm. Pure Appl. Math., **35** (1982), pp. 653–696.

[49] ———, *Nekhoroshev estimates for quasi-convex Hamiltonian systems*, Math. Z., **213** (1993), pp. 187–216.

[50] C. ROBINSON, *Horseshoes for autonomous Hamiltonian systems using the Melnikov integral*, Ergodic Theory Dynam. Systems, **8** (1988), pp. 395–409.

[51] M. RUDNEV AND S. WIGGINS, *KAM theory near multiplicity one resonant surfaces in perturbations of a-priori stable Hamiltonian systems*, J. Nonlinear Sci., **7** (1997), pp. 177–209.

[52] ———, *Existence of exponentially small separatrix splitting and homoclinic connections between whiskered tori in weakly hyperbolic near-integrable Hamiltonian systems*, Phys. D, **114** (1998), pp. 3–80.

[53] C. SIMÓ, *Averaging under fast quasiperiodic forcing*, in Hamiltonian Mechanics: Integrability and Chaotic Behavior, J. Seimenis, ed., Vol. **331** of NATO ASI Ser. B: Phys., held in Toruń, Poland, 28 June–2 July 1993. Plenum, New York, 1994, pp. 13–34.

[54] ———, ed., *Hamiltonian Systems with Three or More Degrees of Freedom*, NATO Adv. Sci. Inst. Ser. C Math. Phys. Sci., Held in S'Agaró, Spain, 19–30 June 1995. Kluwer Acad. Publ., Dordrecht, Holland, to appear in 1999.

[55] D. TRESCHEV, *The mechanism of destruction of resonance tori of Hamiltonian systems*, Math. USSR Sb., **68** (1991), pp. 181–203.

[56] ———, *Hyperbolic tori and asymptotic surfaces in Hamiltonian systems*, Russian J. Math. Phys., **2** (1994), pp. 93–110.

[57] S. WIGGINS, *Global bifurcations and chaos: Analytical methods*, Vol. **73** of Appl. Math. Sci., Springer, New York, 1990.

[58] Z. XIA, *Arnold diffusion in the elliptic restricted three-body problem*, J. Dynamics Differential Equations, **5** (1993), pp. 219–240.

[59] ———, *Arnold diffusion and oscillatory solutions in the planar three-body problem*, J. Differential Equations, **110** (1994), pp. 289–321.

[60] E. ZEHNDER, *Generalized implicit function theorems with applications to some small divisor problems/II*, Comm. Pure Appl. Math., **29** (1976), pp. 49–111.

GEOMETRIC SINGULAR PERTURBATION THEORY BEYOND NORMAL HYPERBOLICITY

FREDDY DUMORTIER* AND ROBERT ROUSSARIE†

Abstract. Geometric Singular Perturbation theory has traditionally dealt only with perturbation problems near normally hyperbolic manifolds of singularities. In this paper we want to show how blow up techniques can permit enlarging the applicability to non-normally hyperbolic points. We will present the method on well chosen examples in the plane and in 3-space.

1. Introduction. For differential equations $\dot{x} = X(x, \varepsilon)$, with $x \in \mathbb{R}^n$, $\varepsilon \in \mathbb{R}$, in general we can speak about a singular perturbation problem when we study phenomena, for $X(x, \varepsilon)$ with $\varepsilon \sim 0$ and $\varepsilon \neq 0$, which appear near sets that contain continuance of zeroes of $X(x, 0)$. The continuance can be lines or manifolds or more complicated stratified sets. The phenomena under consideration can be periodic (emphasizing the number of periodic orbits, their shape or period) as well as exhibit more complicated dynamical behaviour, especially in higher dimensions. We could also merely try to calculate or estimate transit times near these sets with a view to solving boundary value problems. To make precise what "near" means we prefer - when possible - to work with the Hausdorff metric on compact sets in $\mathbb{R}^n$ (see [Ba]). Let us recall that the Hausdorff distance $d(K_1, K_2)$ between two compact sets $K_1, K_2 \subset \mathbb{R}^n$ is defined to be

$$d(K_1, K_2) = \max\Big\{ \max\{d(x, K_2), x \in K_1\},\ \max\{d(K_1, x), x \in K_2\}\Big\}$$

In this paper we will restrict to smooth differential equations and to simple situations in which $x \in \mathbb{R}^2$ or $x \in \mathbb{R}^3$ while the set of zeroes is 1-dimensional.

The generic "singular" local situation consists of a regular smooth curve of zeroes along which the vector field $X(x, 0)$ is normally hyperbolic. The study of the phase portrait of $X(x, \varepsilon)$ for $\varepsilon > 0$ but small can be made by considering center manifolds along the curve of zeroes, also called the slow manifold. This is the way to deal with such singular perturbation problems in what is called "Geometric Singular Perturbation theory". We can refer to the paper of Fenichel (see [F]) for a precise statement on the existence of center manifolds and the reduction principle to central behaviour in the neighbourhood of compact pieces of normally hyperbolic regular manifolds of zeroes.

---

*Limburgs Universitair Centrum, Universitaire Campus, B-3590 Diepenbeek, BELGIUM (fdumorti@luc.ac.be).

†Université de Bourgogne, Laboratoire de Topologie, CNRS-UMR5584, U.F.R. Sciences et Techniques, 9, avenue Alain Savary, BP400, F-21011-Dijon Cedex, FRANCE (roussari@satie.u-bourgogne.fr).

To keep track, in a useful way, as to how orbits entering these neighbourhoods will leave them again there is the so called "exchange lemma" which one can find in [JK]. By these techniques GSP-theory developed an interesting insight in the way one could create complicated dynamics near relatively simple singular limit situations.

In calculations which require a high degree of differentiability (see e.g. [DR2]) it is interesting to add to the standard techniques of GSP-theory the C^k normal linearization theorem of Takens [T], and its generalization by Bonckaert [B]. In an appendix we show the relevance of these theorems.

In any case neither the standard GSP-theory, nor the theorem of Takens deal with non-normally hyperbolic singular sets.

In [DR1] we presented a technique, which we called "family blow-up" or "global blow-up" to extend GSP-theory to such non-normally hyperbolic situations. In fact in [DR1] we restricted the presentation to the study of a specific singular perturbation problem, namely

$$\begin{cases} \dot{x} = y - \dfrac{x^2}{2} - \dfrac{x^3}{3} \\ \dot{y} = \varepsilon(a - x) \end{cases}$$

coming from Van der Pol's equation $\varepsilon\ddot{x} + (x^2 + x)\dot{x} + x - a = 0$.
A detailed analysis was made, introducing the essential ingredients in a way apt for generalization. We have the impression that it might be interesting to emphasize the basic ideas to a greater extent.
Therefore, in this paper we will concentrate on not only presenting the blow-up technique on the simplest, although interesting, examples possible but we will especially stress the geometric nature of the technique avoiding as much as possible the analytic calculations.
In the same spirit, one can find applications of global blow-up in [Du] in studying the Hopf-bifurcation and in [DS], in treating a singular boundary value problem.
However the first case is not a singular perturbation problem, while the second, although very geometric, is not as simple as some of the examples that we will present here.

The examples we want to treat in $\mathbb{R}^3$ are also interesting in the sense that they will permit us to show how the paper [DRR] on fixed points of planar diffeomorphisms might have some interest in the study of singular perturbation problems in $\mathbb{R}^3$.
Although, for reasons explained before, we will stress examples, this paper can also be seen as the start of a systematic treatment of local non-normally hyperbolic situations, in the sense that we will give a description of the generic cases in $\mathbb{R}^2$ and $\mathbb{R}^3$.

Before doing this let us present *"family blow-up"* or *"global blow-up"*, restricting to families of differential equations $\dot{x} = X(x, \varepsilon)$ depending on n phase variables ($x \in \mathbb{R}^n$) and only 1-parameter ($\varepsilon \in \mathbb{R}$). In [DR1] or [Du]

one can find a more general presentation allowing ε to be a multi-parameter ($\varepsilon \in \mathbb{R}^p$).
Essentially one considers "$\dot{x} = X(x, \varepsilon), \dot{\varepsilon} = 0$" as a singular vector field on $\mathbb{R}^{n+1}$ and one makes a *"quasi-homogeneous" blow up:*

$$x_i = r^{k_i}\overline{x}_i, \qquad \varepsilon = r^{\ell}\overline{\varepsilon} \tag{1}$$

with $\sum_{i=1}^{n} \overline{x}_i^2 + \overline{\varepsilon}^2 = 1$ and k_i, ℓ suitably chosen strictly positive integers. We can restrict the study to $\overline{\varepsilon} \geq 0$. If we denote by $\Phi(\overline{x}_1, \ldots, \overline{x}_n, \overline{\varepsilon})$ the blow up map, then we can consider a C^∞ vector field $\widetilde{X}$ on $E = S^n \times [0, \infty[$, defined by $\Phi_*(\widetilde{X}) = X$. The choice of k_i, ℓ depends on the specific equation as we will make clear in the examples. The simplest choice would be $k_i = \ell = 1$ but it does not always lead to the simplest resulting vector field $\widetilde{X}$. The resulting vector field $\widetilde{X}$ has to be studied for $((\overline{x}_i, \overline{\varepsilon}), r) \in S^n \times [0, r_0[$, for some $r_0 > 0$, in order to provide all the information we need about the phase portraits of X_ε for $\varepsilon > 0$, ε small on a (ε-independent) neighbourhood of 0 in the phase space.
To truly grasp geometrically what happens it might be interesting to start with $n = 2$, so that $x = (x_1, x_2)$, keeping the blown-up vector field 3-dimensional. Often the blow-upleads to a vector field $\widetilde{X}$ which is zero along $S^{n-1} \times \{0\}$ so that one divides by some power of r in order to make the result less degenerate; one introduces then $\overline{X} \doteq \frac{1}{r^m} \cdot \widetilde{X}$, for some well chosen positive integer m.
This division does not alter the phase portrait of $\widetilde{X}$ outside $S^{n-1} \times \{0\}$ and is hence as useful to study the phase portraits of X_ε as $\widetilde{X}$ itself is. One only needs to take this division into account when calculating transit times along the orbits.

In many cases now - as we will see in the examples, and as one can also see in [DR1] - this blow up procedure will permit us to extend the usual "center manifold" approach along the normally-hyperbolic manifolds of zeroes to the non-normally hyperbolic points.

The "spherical" global blow up presented in (1) is very adequate to get a geometric insight. However, as is usual in making precise calculations along a sphere, it is often better to work in "charts".
One specific chart is obtained by changing $(\overline{x}_i, \overline{\varepsilon})$ into $\left(\frac{\overline{x}_i}{\overline{\varepsilon}}, 1\right)$ and restricting $(\overline{\overline{x}}_i) = \left(\frac{\overline{x}_i}{\overline{\varepsilon}_i}\right)$ to some large compact ball in $\mathbb{R}^n$. The ball can be chosen arbitrarily. In fact, in order to keep the notation simple, we will write $\overline{x}_i$ instead of $\overline{\overline{x}}_i$.
In doing this, we see that (1) merely expresses the classical "rescaling", in which the family of vector fields, X_ε is changed into a new family of

vector fields $\overline{X}_r$. The chart is hence called the *"family chart"* or *"family rescaling"*.

To complete the picture, and using the same convention with respect to notation, we have to look at

$$\sum_{i=1}^{n} \overline{x}_i^2 = 1, \qquad \overline{\varepsilon} \in [0, \overline{\varepsilon}_0[$$

for some small $\overline{\varepsilon}_0 > 0$. This we call *"phase directional rescaling"*.

Of course, this phase directional rescaling also leads to a study near an $(n-1)$-sphere, so that further subdivision into charts might be interesting. As such we will often work with so called *"directional blow-ups"* in which, respecting the notational convention, we use: $\overline{\varepsilon} \in [0, \overline{\varepsilon}_0[$, $\overline{x}_{i_0} = \pm 1$ and $(\overline{x}_1, \ldots, \widehat{\overline{x}}_{i_0}, \ldots, \overline{x}_n)$ in some large compact ball in $\mathbb{R}^{n-1}$; we always keep $r \in [0, r_0[$. The union of all the charts will lead to a complete picture near $S^n \times [0, r_0[$.
In the family rescaling, the procedure leads to a new 1-parameter family of vector fields. However the blow-up in the phase direction, of the 1-parameter family of n-dimensional vector fields X_ε, does not lead to a 1-parameter family of n-dimensional vector fields. It leads to an $(n+1)$-dimensional vector field that leaves invariant a foliation defined by a first integral. In fact the function $r^\ell \overline{\varepsilon} = \varepsilon$ stays constant along the orbits. The presence of this invariant foliation is quite useful in the analysis of $\overline{X}$.

2. Treatment of 2-dimensional singular perturbation problems. In this section we give a complete analysis of the generic non-normally hyperbolic singular perturbation problems in the plane.

2.1. Normal crossing of two lines of singularities. A first situation occurs when we have a normal crossing of two regular curves of singularities. In suitably chosen C^∞ coordinates this local situation can be described by

$$\begin{cases} \dot{x} = xy\ f(x,y) \\ \dot{y} = xy\ g(x,y) \end{cases} \tag{2}$$

with f and g both C^∞, and with $f(0,0) \neq 0$ as well as $g(0,0) \neq 0$. It is easy to show (see Appendix 1) that (2) is locally C^∞-equivalent to

$$\begin{cases} \dot{x} = xy \\ \dot{y} = xy \end{cases} \tag{3}$$

As a trivial side remark we can observe that the vector field (2) has a non-degenerate first integral whose level curves are transverse to both lines of singularities.

We can now look to what happens when we add a generic ε-perturbation $\varepsilon Z(x,y)$ with $Z(0,0) \neq 0$. It suffices to consider:

$$\text{(4)} \qquad \begin{cases} \dot{x} = xy + \varepsilon(\sin\theta + f(x,y,\varepsilon)) \\ \dot{y} = xy + \varepsilon(\cos\theta + g(x,y,\varepsilon)) \end{cases}$$

for some $\theta \in [0, 2\pi[$, $f(0,0,0) = g(0,0,0) = 0$, f and g of class C^∞.
The analysis can be done by a global blow up:

$$\text{(5)} \qquad (x,y,\varepsilon) = (r\overline{x}, r\overline{y}, r^2\overline{\varepsilon})$$

with $(\overline{x}, \overline{y}, \overline{\varepsilon}) \in S^2$.
We can first calculate the phase-directional rescaling in the charts $\{\overline{x} = \pm 1\}$ and $\{\overline{y} = \pm 1\}$. We start in $\{\overline{x} = 1\}$ with:

$$(x,y,\varepsilon) = (r, r\overline{y}, r^2\overline{\varepsilon})$$

After a division by r this gives:

$$\text{(6)} \qquad \begin{cases} \dot{r} = r[\overline{\varepsilon}(s + O(r)) + \overline{y}] \\ \dot{\overline{y}} = \overline{\varepsilon}(c + O(r)) - \overline{\varepsilon}\,\overline{y}(s + O(r)) + \overline{y} - \overline{y}^2 \\ \dot{\overline{\varepsilon}} = -2\overline{\varepsilon}[\overline{\varepsilon}(s + O(r)) + \overline{y}] \end{cases}$$

writing $(\sin\theta, \cos\theta)$ as (s,c).

For $\overline{\varepsilon} = 0$ this gives

$$\begin{cases} \dot{r} = r\overline{y} \\ \dot{\overline{y}} = \overline{y}(1 - \overline{y}) \end{cases}$$

presenting a hyperbolic saddle at $(r, \overline{y}) = (0,1)$ and a normally hyperbolic line of singularities at $\{\overline{y} = 0\}$; see Figure 1.
Similar calculations in other charts complete the picture described in Figure 2.
For the family rescaling we use

$$(x,y,\varepsilon) = (r\overline{x}, r\overline{y}, r^2)$$

After a division by r we get:

$$\begin{cases} \dot{\overline{x}} = \overline{x}\,\overline{y} + s + O(r) \\ \dot{\overline{y}} = \overline{x}\,\overline{y} + c + O(r) \end{cases}$$

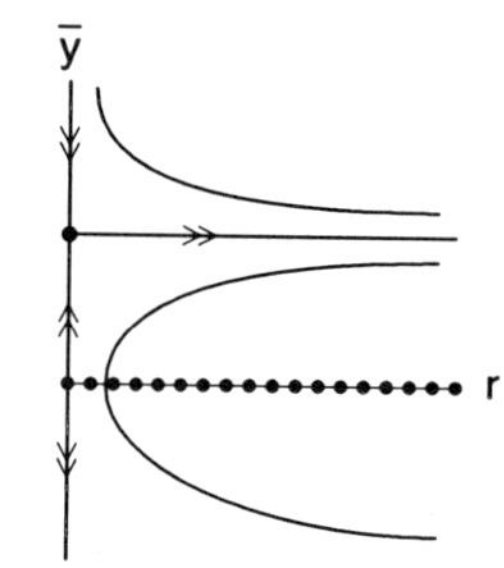

FIG. 1. *Directional blow-up.*

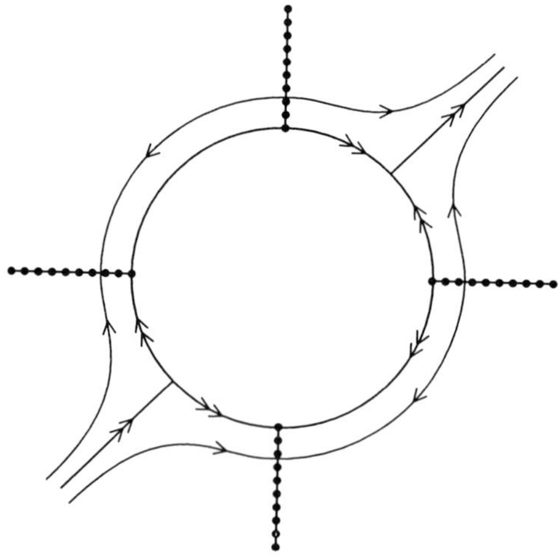

FIG. 2. *Blow-up of the singularity* $xy\left(\dfrac{\partial}{\partial x}+\dfrac{\partial}{\partial y}\right)$.

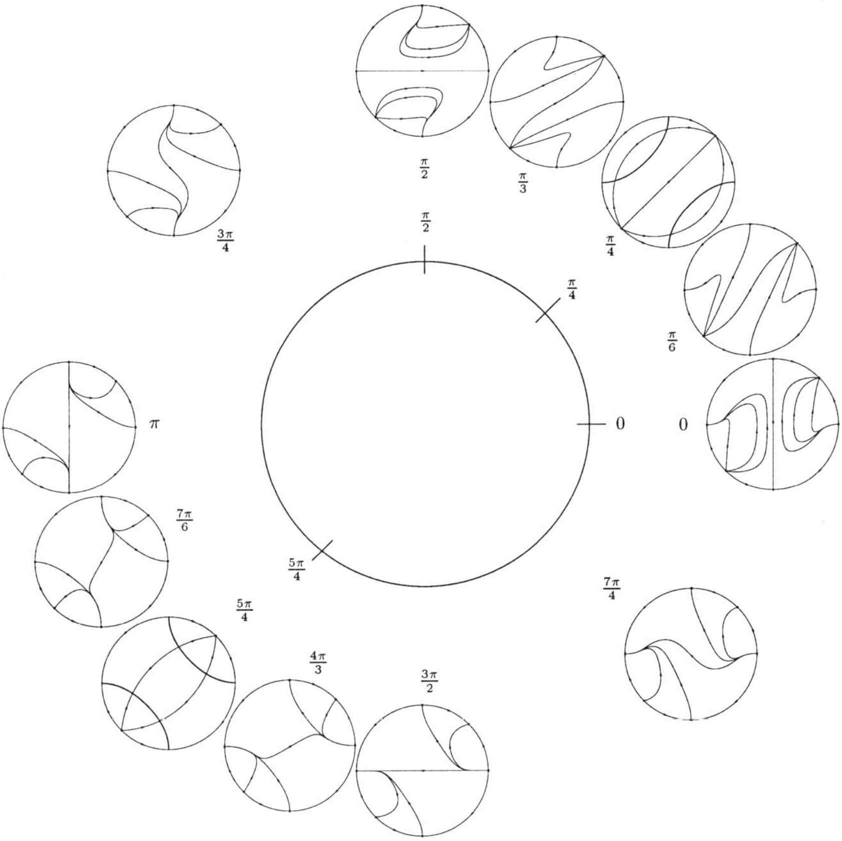

FIG. 3. *Phase portraits of (7) on a Poincaré disc.*

Let us now study the phase portraits of these vector fields for $r = 0$:

$$(7) \qquad \begin{cases} \dot{\overline{x}} = \overline{x}\,\overline{y} + \sin\theta \\ \dot{\overline{y}} = \overline{x}\,\overline{y} + \cos\theta \end{cases}$$

with $\theta \in [0, 2\pi[$. The behaviour near infinity in a Poincaré-compactification can be calculated in the phase-directional rescaling by taking $r = 0$ in (6).

In Figure 3 we show the different phase portraits of (7), drawn on a Poincaré-disc. The different phase portraits are drawn by the programme P4 (see [DH], [H]). Exceptional phase portraits occur at the values $\theta = 0$, $\pi/4$, $\pi/2$, $5\pi/4$; all the others can be considered to be generic. At the values $\theta = \pi/4$ and $\theta = 5\pi/4$ the vector field (4) has two curves of singularities, and its study will be more complicated. At $\theta = 0$ and $\theta = \pi/2$ there are "isolated" connections between normally hyperbolic lines of singularities of opposite type (resp. attracting and repelling). This induces a "canard" phenomenon and requires further analysis. In all the other cases the study of (4) for $\varepsilon \sim 0$ is easy to finish with the information we already have. Before doing this let us remark that the results on (7), summarized in Figure 3, can be proved rigorously by making the necessary calculations at "infinity" (hence by working in the phase-directional rescaling) and by observing, among other easy things, that besides the cases $\cos\theta = \sin\theta$, the others can be changed into $(xy+a)\frac{\partial}{\partial x} + (xy+a+1)\frac{\partial}{\partial y}$ or $(xy+a)\frac{\partial}{\partial x} + (xy+a-1)\frac{\partial}{\partial y}$, both 1-parameter families being rotational (see [P] or [Du]) with respect to a.

Let us now describe the four generic cases, starting with $\theta \in]0, \pi/4[$. The precise phase portrait drawn in Figure 3 is the one for $\theta = \pi/6$. Considering the full (3-dimensional) blow up, described in (5) we get a 3-dimensional phase portrait as given in Figure 4. All the necessary information is contained in the different charts (family rescaling and phase-directional rescaling).

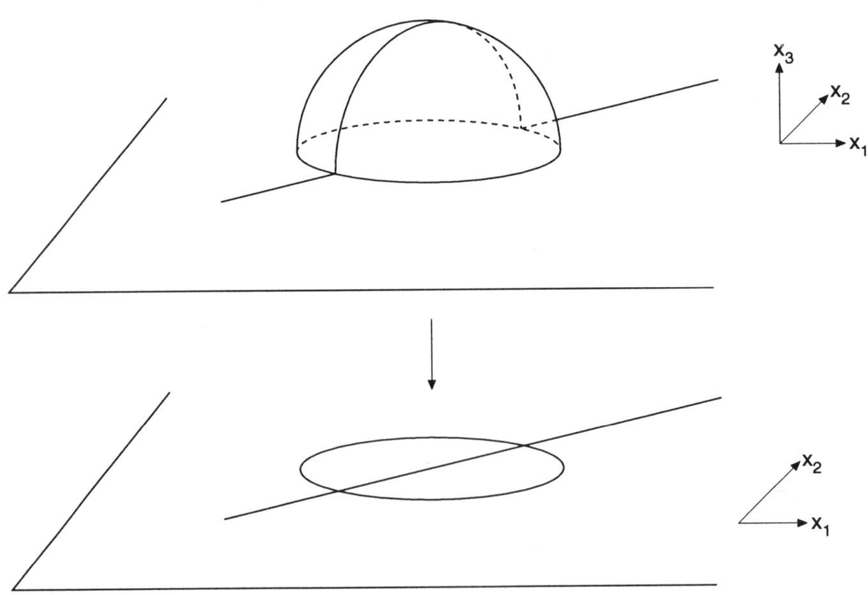

FIG. 4. *Pictorial representation in* $\mathbb{R}^3$.

Figure 4, and many other pictures that we will make, may give the impression of being 2-dimensional. However to understand the picture one has to consider a 3-space with coordinates (x_1, x_2, x_3), like in Figure 4. On the hemisphere $S_+^2 = \{(x_1, x_2, x_3) \mid x_1^2 + x_2^2 + x_3^2 = 1, x_3 \geq 0\}$ one draws the phase portrait of $\overline{X}$ restricted to $\{(\overline{x}, \overline{y}, \overline{\varepsilon}, r) \mid \overline{x}^2 + \overline{y}^2 + \overline{\varepsilon}^2 = 1, \overline{\varepsilon} \geq 0, r = 0\}$, identifying $(x_1, x_2, x_3) = (\overline{x}, \overline{y}, \overline{\varepsilon})$; on $P_+^2 = \{(x_1, x_2, x_3) \mid x_3 = 0, x_1^2 + x_2^2 \geq 1\}$ one draws the phase portrait of $\overline{X} \mid \{\overline{\varepsilon} = 0\}$ identifying $(x_1, x_2) = ((1 + r)\overline{x}_1, (1 + r)\overline{x}_2)$. Both identifications lead to the same vector field on $S_+^2 \cap P_+^2 = S^2 \cap \{x_3 = 0\}$.

Finally one uses the projection $\pi_{1,2} \mid S_+^2 \cup P_+^2$ with $\pi_{1,2}(x_1, x_2, x_3) = (x_1, x_2)$
Figure 5 represents a 3-dimensional limiting phase portrait for $\varepsilon = 0$. We divide the set of singularities into 4 halflines V^u, V^s, H^u and H^s as in Figure 5, and denote the unstable (resp. stable) manifold of p_1 (resp. p_2) by W^u (resp. W^s). It is easy to see, using e.g. the techniques described in [DR1], that for $\varepsilon > 0$, but small, the phase portrait will be as given in Figure 6.
Orbits starting in the neighbourhood (on center manifolds) of V^s will end up near W^u, and similarly for orbits starting near H^u. There is clearly no possibility of having a "canard phenomenon" in the sense that orbits can travel from center manifolds of one of the singular halflines to another one.

Something similar happens for $\theta \in]\pi/4, \pi/2[$. In Figure 7 we show the limiting system and in Figure 8 the phase portrait for $\varepsilon > 0$ but small.

The situation becomes much more interesting in the two other cases. In Figure 9 we show the limiting system for $\theta \in]\pi/2, 5\pi/4[$. We see that starting near V^u, on center manifolds of V^u, it is now possible to end up near W^u, near H^u, and near V^s. There is hence a strong scattering. There will clearly be only a very thin region of initial values near V^u that will give birth to orbits that will stay near H^u along an a priori given segment. There is clearly also a possibility to move from small neighbourhoods of V^u to small neighbourhoods of H^s, eventually ending up near V^s. Phase portraits for $\varepsilon > 0$, ε small, appear as in Figure 10. All points near H^s end up near V^s.

Similar phenomena happen for $\theta \in]5\pi/4, 2\pi[$ as one can see in Figures 11 and 12 for, respectively, the limiting system and the genuine phase portraits for $\varepsilon > 0$, ε small. There is now a large scattering, in positive time, for orbits starting near H^u.

2.2. Generic turning point. For the next generic non-normally hyperbolic situation we can restrict to the case in which we have exactly one regular curve of singularities, which we can consider to be the x-axis. The vector field then, locally, has the expression :

$$\begin{cases} \dot{x} = y\ f(x, y) \\ \dot{y} = xy\ g(x, y) \end{cases} \tag{8}$$

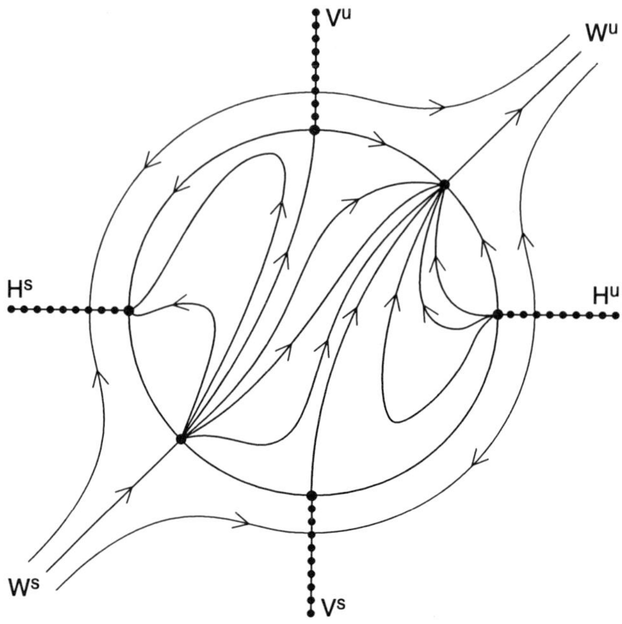

FIG. 5. *Limiting system for blow-up of (4) for* $\theta \in]0, \pi/4[$.

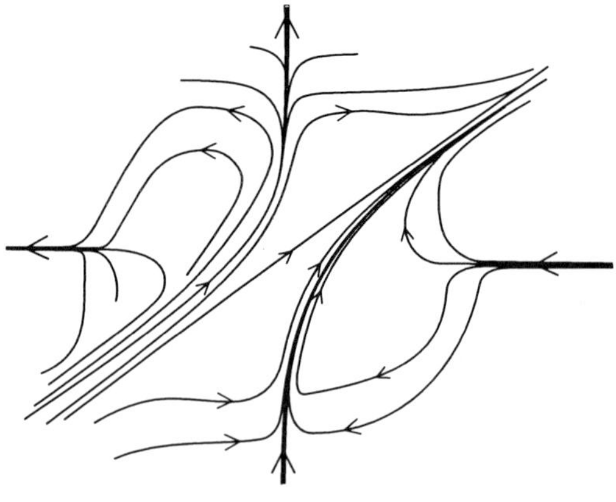

FIG. 6. *Phase portrait of (4) for* $\theta \in]0, \pi/4[$ *and* $\varepsilon > 0$ *small.*

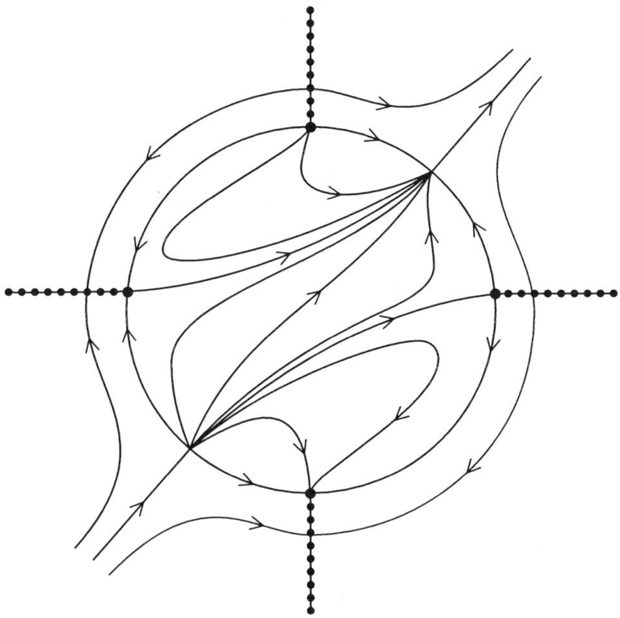

FIG. 7. *Limiting system for blow up of (4) for* $\theta \in]\pi/4, \pi/2[$.

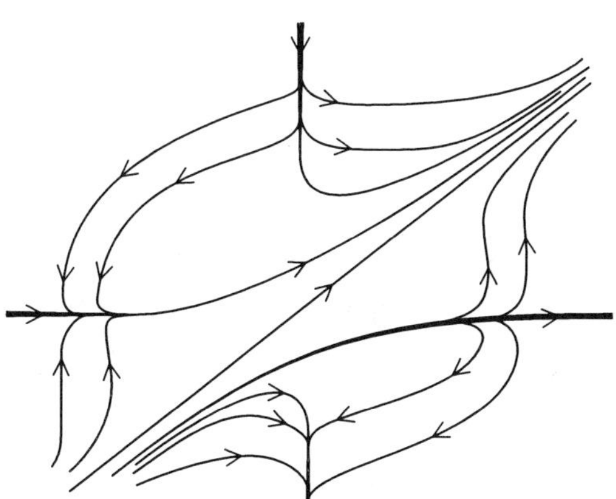

FIG. 8. *Phase portrait of (4) for* $\theta \in]\pi/4, \pi/2[$ *and* $\varepsilon > 0$ *small.*

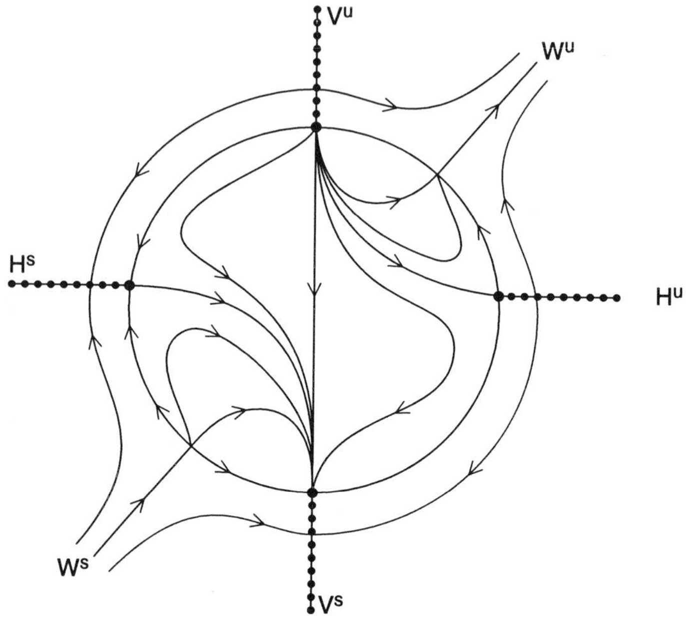

FIG. 9. *Limiting system for blow up of (4) for* $\theta \in]\pi/2, 5\pi/4[$.

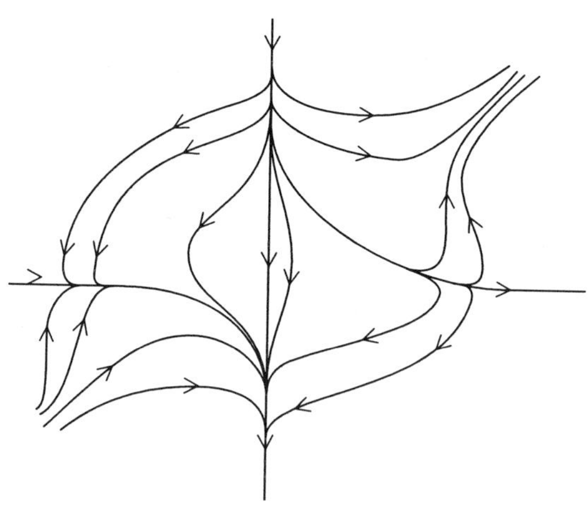

FIG. 10. *Phase portrait of (4) for* $\theta \in]\pi/2, 5\pi/4[$ *and* $\varepsilon > 0$ *small.*

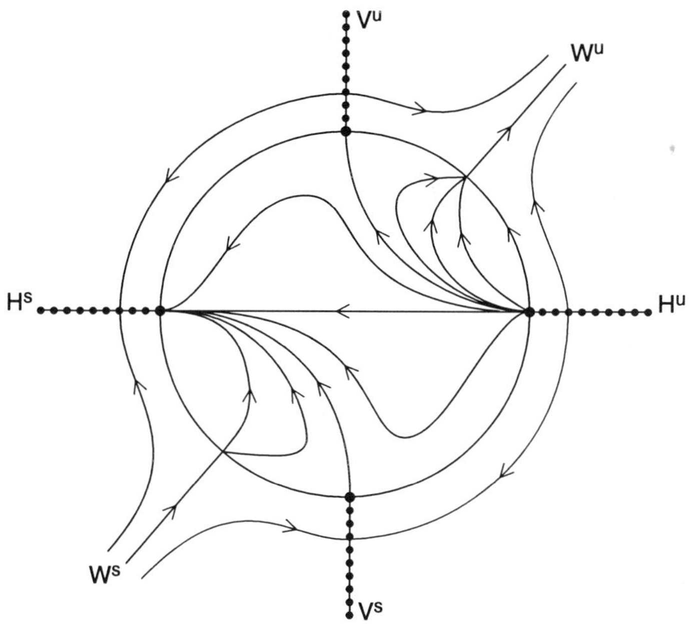

FIG. 11. *Limiting system for blow up of (4) for* $\theta \in]5\pi/4, 2\pi[$.

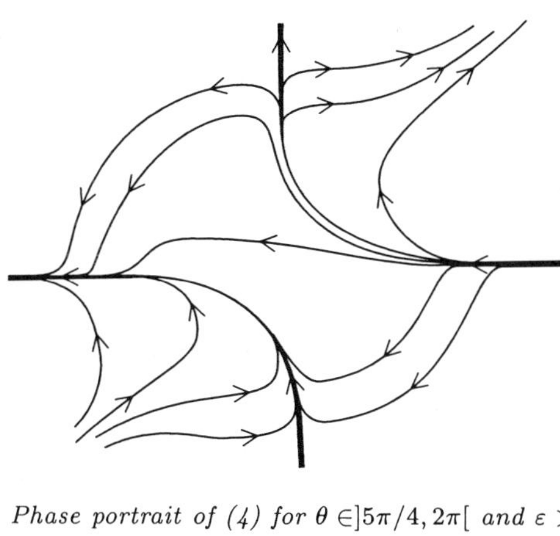

FIG. 12. *Phase portrait of (4) for* $\theta \in]5\pi/4, 2\pi[$ *and* $\varepsilon > 0$ *small.*

and generically we can suppose that $f(0,0) \neq 0$ as well as $g(0,0) \neq 0$, inducing a nilpotent 1-jet at $(0,0)$. The singularity is accordingly called *nilpotent*. It is again easy to show (see Appendix 1) that (8) is locally C^∞ equivalent to

$$X : \begin{cases} \dot{x} = y \\ \dot{y} = xy \end{cases} \tag{9}$$

and we can again observe that X has a regular first integral. We also see that the Y-orbit of $(0,0)$ has a quadratic tangency with the line of singularities of X, where $Y = \frac{\partial}{\partial x} + x\frac{\partial}{\partial y}$. Let us now consider what happens when we add a generic ε-perturbation to (9), i.e.

$$\begin{cases} \dot{x} = y + \varepsilon(a + o(1)) \\ \dot{y} = xy + \varepsilon(b + o(1)) \end{cases} \tag{10}$$

keeping $b \neq 0$. Although the study has been given in [DR1] we, nevertheless, include it in this paper since it is a nice simple example to see how the blow-up procedure works. The analysis can be made by a global blow-up

$$(x, y, \varepsilon) = (r\overline{x}, r^2\overline{y}, r^3\overline{\varepsilon}) \tag{11}$$

with $(\overline{x}, \overline{y}, \overline{\varepsilon}) \in S^2$, and requires a division by r. The phase-directional rescaling for $\overline{\varepsilon} = 0$, or in other words the blow-up of the singularity (9) gives the picture in Figure 13, with two hyperbolic saddles and two normally hyperbolic lines of singularities.
The family rescaling, at $r = 0$, renders the vector fields:

$$\begin{cases} \dot{\overline{x}} = \overline{y} \\ \dot{\overline{y}} = \overline{x}\,\overline{y} + b \end{cases} \tag{12}$$

Generically we can take $b \neq 0$, and a linear equivalence permits us to restrict to $b = \pm 1$. The study of these fields , preferably on a $(1,2)$-Lyapunov-Poincaré disc (cfr. [Du] or [DH]), is very simple and leads to the 3-dimensional phase portraits as presented in Figure 14.
To show how the change occurs from $b < 0$ to $b > 0$, we add the picture for $b = 0$, although it is not a generic one and it requires more elaboration In the generic cases $b \neq 0$, it is again easy to see that for $\varepsilon > 0$, but small, the phase portraits will be as shown in Figure 15.
Certainly, there is no "canard" phenomenon possible, in the sense that orbits starting near the attracting normally hyperbolic line of singularities either stay there (in case $b < 0$) or escape to "regular" movements (in case $b > 0$).

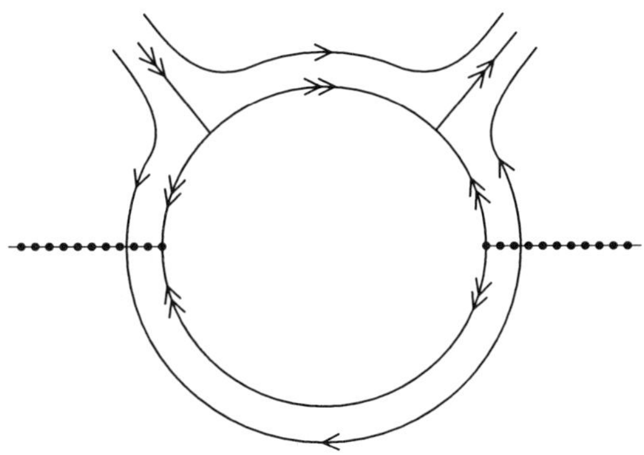

FIG. 13. *Blow-up of the singularity* $y\left((1+o(1))\dfrac{\partial}{\partial x}+x(1+o(1))\dfrac{\partial}{\partial y}\right)$.

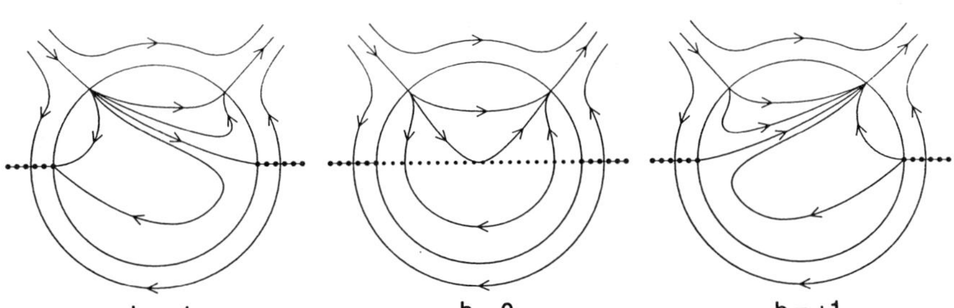

FIG. 14. *Limiting system for the blow-up of (10).*

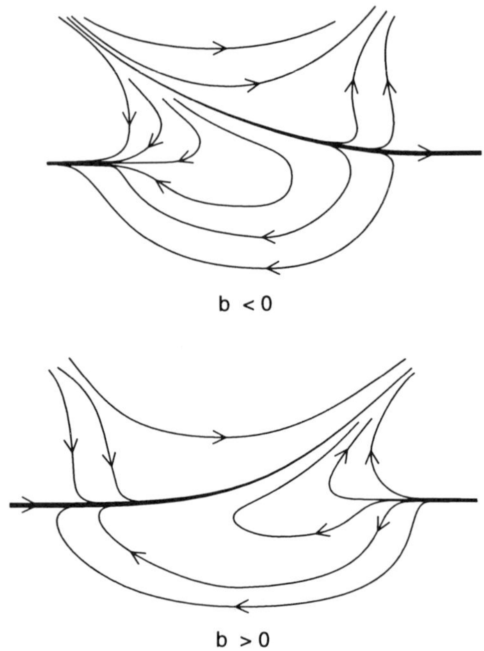

FIG. 15. *Phase portraits of (10) for $\varepsilon > 0$ small.*

There can be a canard phenomenon for $b = 0$, as has been shown in [DR1], when considering the perturbation $\varepsilon\left((a + o(1))\dfrac{\partial}{\partial x} + (b + x + o(2))\dfrac{\partial}{\partial y}\right)$ with b a parameter staying close to 0.

Exactly as in the study of (10), blow-up leads to a "desingularization" of the family in the sense that it leads to singularities that are hyperbolic or partially hyperbolic but no longer nilpotent. In fact such "desingularization" for "nilpotent" singularities is possible in general, as has been shown in the thesis of D. Panazzolo (cfr. [Pa]).

2.3. Generic pitchfork. Besides the "generic" singular situations it might however also be interesting to look at less generic ones that can easily be encountered in the presence of extra constraints.
As such, symmetry reasons can impose the consideration of a less generic situation of normal crossing. This occurs, for example, in studies of vector fields that are $\mathbb{Z}_2$-equivariant with respect to the mapping $(x, y) \mapsto (x, -y)$. In that case in the general expression (2), namely

$$\begin{cases} \dot{x} = xy(a + o(1)) \\ \dot{y} = xy(b + o(1)), \end{cases}$$

we have to put $a = 0$, implying that the regular vector field $Y = (a + o(1))\frac{\partial}{\partial x} + (b+o(1))\frac{\partial}{\partial y}$ is tangent to the y-axis. Generically, we can require this contact to be quadratic, which implies that

$$\left\{ \begin{array}{l} \dot{x} = xy(cy + o(|x,y|)) \\ \dot{y} = xy(b + o(1)) \end{array} \right. \tag{13}$$

with $bc \neq 0$. Since after division by xy we obtain a regular vector field, we again have a regular first integral. In fact (13) represents, up to diffeomorphism, the most generic expression of a "pitchfork bifurcation". Up to linear conjugacy we can always take $b = 1$ and $c = \pm 1$; after time reversal we can even take $c = 1$. So let us continue by considering

$$\left\{ \begin{array}{l} \dot{x} = xy(y + o(|x,y|)) \\ \dot{y} = xy(1 + o(1)) \end{array} \right.$$

The most generic singular perturbation problem, respecting the chosen $\mathbb{Z}_2$-equivariance, is given by

$$\left\{ \begin{array}{l} \dot{x} = xy(y + o(|x,y|)) + \varepsilon(d + f(x,y)) \\ \dot{y} = xy(1 + o(1)) + \varepsilon g(x,y) \end{array} \right. \tag{14}$$

with $d \neq 0$, where both f and g are $O(|x,y|)$, and because of the $\mathbb{Z}_2$-equivariance g needs to be divisible by y. We can, of course, restrict to $d = \pm 1$, and start with $d = 1$.

Let us now use the blow-up

$$(x, y, \varepsilon) = (r^2\overline{x}, r\overline{y}, r^4\overline{\varepsilon})$$

For the family rescaling ($\overline{\varepsilon} = 1$) we obtain:

$$\left\{ \begin{array}{l} r^2\dot{\overline{x}} = r^4\overline{x}\,\overline{y}(\overline{y} + O(r)) + r^4(1 + O(r)) \\ r\dot{\overline{y}} = r^3\overline{x}\,\overline{y}(1 - O(r)) + O(r^4) \end{array} \right.$$

After division by r^2, and for $r = 0$, this gives:

$$\left\{ \begin{array}{l} \dot{\overline{x}} = \overline{x}\,\overline{y}^2 + 1 \\ \dot{\overline{y}} = \overline{x}\,\overline{y} \end{array} \right.$$

The global blow-up gives the picture as in Figure 16.
Since the y-component of (14) is divisible by y we necessarily obtain that $\{\overline{y} = 0\}$ represents a center manifold both at p_1 and p_2. Other center manifolds near p_1 and p_2 will be infinitely tangent to $\{\overline{y} = 0\}$.
Phase portraits of (14), for $\varepsilon > 0$ but small, appear as in Figure 17.

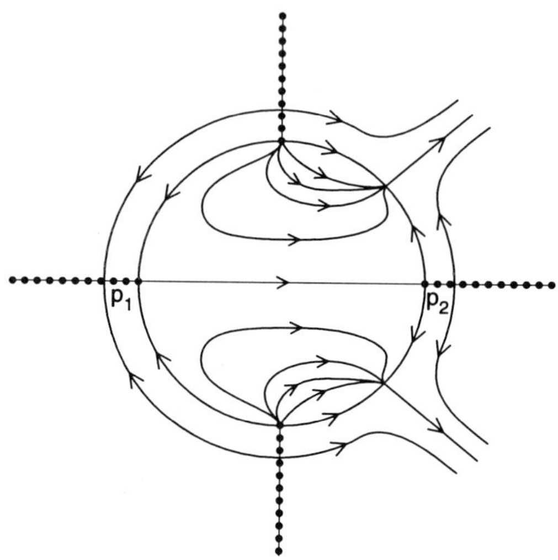

FIG. 16. *Limiting system for blow-up of (14) in case $d = 1$.*

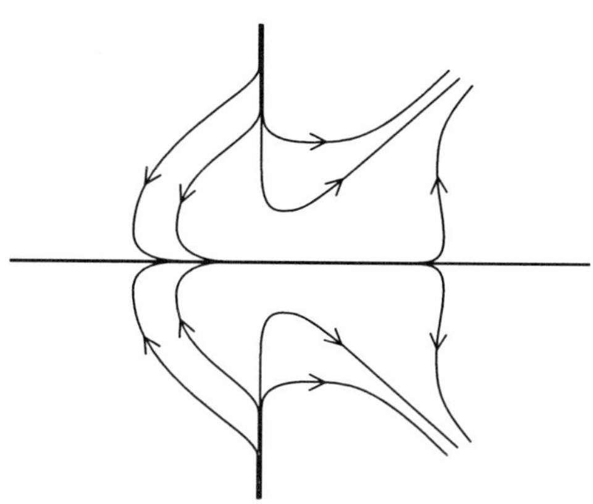

FIG. 17. *Phase portrait of (14) for $d > 0$ and $\varepsilon > 0$ small.*

There remains to consider the case $d = -1$.

$$\begin{cases} \dot{x} = xy(y + o(|x, y|)) + \varepsilon(-1 + f(x, y)) \\ \dot{y} = xy(1 + o(1)) + \varepsilon g(x, y) \end{cases}$$

where both f and g are $O(|x, y|)$, and g is divisible by y. We again blow up by $(x, y, \varepsilon) = (r^2\overline{x}, r\overline{y}, r^4\overline{\varepsilon})$.
For the family rescaling ($\overline{\varepsilon} = 1$) we obtain:

$$\begin{cases} r^2\dot{\overline{x}} = r^4\overline{x}\,\overline{y}(\overline{y} + O(r)) + r^4(-1 + O(r)) \\ r\dot{\overline{y}} = r^3\overline{x}\,\overline{y}(1 + O(r)) + O(r^4) \end{cases}$$

After division by r^2, and for $r = 0$, this gives

$$\begin{cases} \dot{\overline{x}} = \overline{x}\,\overline{y}^2 - 1 \\ \dot{\overline{y}} = \overline{x}\,\overline{y} \end{cases}$$

The global blow-up gives the picture as in Figure 18.

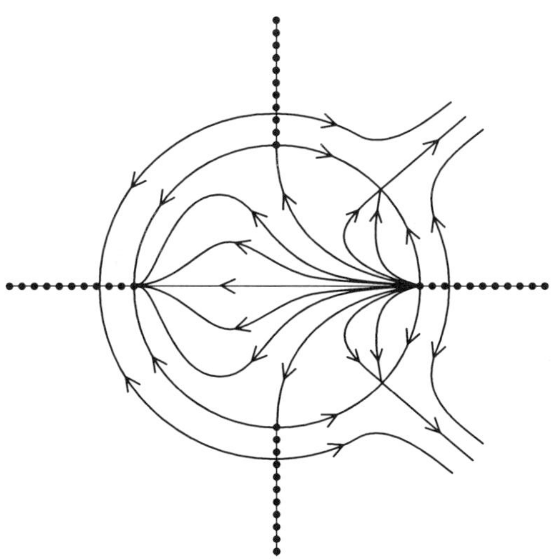

FIG. 18. *Limiting system for blow-up of (14) in case $d = -1$.*

Phase portraits of (14), for $\varepsilon > 0$ but small, appear as in Figure 19.
In case $d = 1$, we can ask up to which position orbits starting near the x-axis will stay close to the x-axis. More precisely we take initial values at, for instance, $(-x_0, 1)$, with $x_0 > 0$, for different $\varepsilon > 0$; we follow the

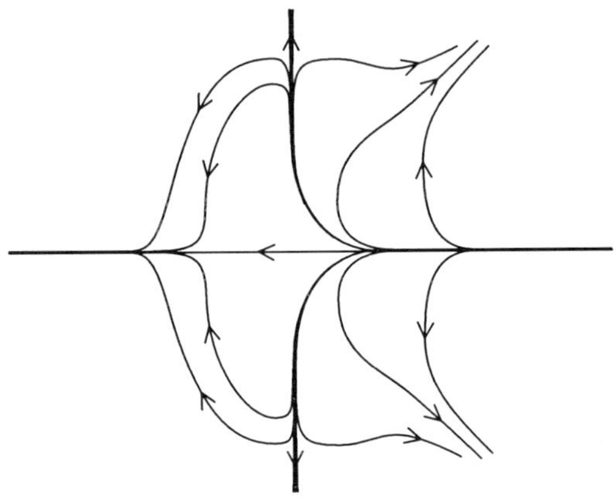

FIG. 19. *Phase portrait of (14) for d < 0 and ε > 0 small.*

different orbits until they hit the line $\{y = 1\}$ again and try to determine whether these pieces of orbits Γ_ε have a (Hausdorff-) limit. In fact, with the techniques developed in [DR1] one can prove that the Hausdorff limit exists and is equal to the union $\{x = -x_0, y \in [0,1]\} \cup \{x = x_1, y \in [0,1]\} \cup \{y = 0, x \in [-x_0, x_1]\}$ for some $x_1 > 0$. To calculate the dependence $x_1 = x_1(x_0)$ there is a procedure in which the divergence plays an essential role. The derivative of the transition map of the flow of X_ε along Γ_ε is given by

$$\exp\left(-\frac{1}{\varepsilon}\int_{\Gamma_\varepsilon} \text{div } X_\varepsilon dt\right) \tag{15}$$

We therefore write (14) as:

$$X_\varepsilon : \begin{cases} \dot{x} = xy(y + o(|x,y|) + \varepsilon(1 + f(x,y)) \\ \dot{y} = xy(1 + \alpha(x,y)) + \varepsilon g(x,y) \end{cases} \tag{16}$$

The divergence $\text{div}X_\varepsilon$ of this field is equal to

$$y(y + o(|x,y|)) + x\left(1 + \alpha(x,y) + y\frac{\partial \alpha}{\partial y}\right) + \varepsilon\left(\frac{\partial f}{\partial x} + \frac{\partial g}{\partial y}\right) \tag{17}$$

and at the limit $\varepsilon = 0$ we have to consider this expression along the slow manifold (line of singularities) $\{y = 0\}$; this gives:

$$x(1+\alpha(x,0))+\varepsilon\left(\frac{\partial f}{\partial x}(x,0)+\frac{\partial g}{\partial y}(x,0)\right) \tag{18}$$

Similarly along $\{y=0\}$ we have to use

$$dt=\frac{dx}{\varepsilon(1+f(x,0))} \tag{19}$$

It follows now from the expression (18) that the function $x_1(x_0)$ is a solution of

$$\int_{-x_0}^{x_1(x_0)} \frac{x(1+\alpha(x,0))}{1+f(x,0)}dx=0 \tag{20}$$

This fact had already been obtained in [L] by means of Non-Standard Analysis.
By implicit differentiation it is clear that

$$\frac{dx_1}{dx_0}(x_0)>0$$

for $x_0>0$, so that $x_1=x_1(x_0)$ is a strictly monotone function.

3. Generic phase portraits along curves of singularities in $\mathbb{R}^3$. To show that the techniques are not limited to the plane we will treat some generic situations of curves of singularities in $\mathbb{R}^3$. Since transverse crossings between two lines are not possible, we can restrict to a single regular curve in describing the generic non-normally hyperbolic situation. We position the curve of singularities at $\{x=y=0\}$, and put the non-normally hyperbolic point at $\{x=y=z=0\}$. There are now two possibilities.

3.1. The partially hyperbolic case. At $(0,0,0)$ we have as 1-jet

$$\begin{pmatrix} \lambda & 0 & 0 \\ 0 & 0 & 0 \\ 0 & 1 & 0 \end{pmatrix}$$

with $\lambda \neq 0$. Since one eigenvalue is hyperbolic, we can reduce the study to a 2-dimensional center manifold, ending up with the generic cases described in the previous paragraph. Of course center manifolds are not necessarily C^∞ but they are as differentiable as needed to apply the methods presented in paragraph 2.
Now, up to C^r-conjugacy, we can straighten an invariant center manifold N and consider one of the planar, generic, non-normally hyperbolic situations. Exactly as in the previous chapter we can now add a perturbation εY with Y regular. The situation with Y tangent to N gives the results described in the previous chapter. In the 3-dimensional case, however, it is more generic to take $Y(p)\in \mathbb{R}^3\backslash T_pN$ at each $p\in N$. We do not wish to work this out in this paper, but prefer to focus on the second possibility.

3.2. The Hopf-zero case. At $(0,0,0)$ we have as 1-jet

$$\begin{pmatrix} 0 & -1 & 0 \\ 1 & 0 & 0 \\ 0 & 0 & 0 \end{pmatrix}$$

and along the line of singularities we can suppose that the trace of the 1-jet undergoes a generic change from negative to positive, i.e. there is a generic loss of stability. The situation has also been studied in [FLA1] by means of different techniques. We can also refer to [FLA2] for a description of an application where the situation arises naturally. We recall from paragraph 2 that in $\mathbb{R}^2$ the most generic cases always have a regular first integral, meaning that we could in fact start the perturbation from a 1-parameter family of 1-dimensional vector fields. As we will see in this chapter there is generically no regular first integral in this 3-dimensional case.

Let us now first uncover the topology of the generic situation. It will enable us to use again the blow-up technique and to focus on some results on germs of planar diffeomorphisms, described in [DRR] and useful in the study of 3-dimensional singular perturbation problems.

A less generic but interesting situation occurs when we require a regular first integral, meaning that we start with a 1-parameter family of planar vector fields. The most generic situation in this case is known to be a Hopf bifurcation (of codimension 1). We refer to [Du] for a study of the Hopf-bifurcation by means of global blow up.
We now begin the description of the generic singular Hopf-zero singularity.

As already mentioned, we suppose that the z-axis consists of singularities, and that the isolated, non-normally hyperbolic point is situated at the origin. As the 1-jet at $(0,0,0)$ we have an infinitesimal rotation around the z-axis, i.e.

$$-y\frac{\partial}{\partial x} + x\frac{\partial}{\partial y}.$$

We restrict to $I = \{x = y = 0, -z_0 \leq z \leq z_0\}$, for some $z_0 > 0$, along which the eigenvalues of X are 0, $a(z) \pm ib(z)$ with $b(z) > 0$. By an analytic change of coordinates we bring the 1-jet of X along I to the expression:

$$b(z)\left(x\frac{\partial}{\partial y} + y\frac{\partial}{\partial x}\right) + a(z)\left(x\frac{\partial}{\partial x} + y\frac{\partial}{\partial y}\right)$$

Using a C^∞ change of coordinates whose 1-jet is the identity along I we can now bring the ∞-jet along I to an expression that is invariant under rotations around the z-axis. This construction works exactly like the usual normal form (cfr. [T]) and has been elaborated in full detail in an appendix of [DK]. Let us recall the exact statement.

THEOREM 3.1 (see [DK]). *Let X be a smooth vector field on $\mathbb{R}^3$ defined in a neighbourhood of the line segment I with all the conditions as described above.*
Then there exists a smooth diffeomorphism φ defined in a neighbourhood of I with the 1-jet being the identity at each point of I, such that $\varphi_* X$ has an expression:

$$\begin{aligned}&\big(b(z)+f(x^2+y^2,z)\big)\left(x\frac{\partial}{\partial y}-y\frac{\partial}{\partial x}\right)\\ +\ &\big(a(z)+g(x^2+y^2,z)\big)\left(x\frac{\partial}{\partial x}+y\frac{\partial}{\partial y}\right)+h(x^2+y^2,z)\frac{\partial}{\partial z}+Z(x,y,z)\end{aligned}$$

with $f,g,h,Z\in C^\infty$, $f(0,z)=g(0,z)=h(0,z)=0$ and $j_\infty Z(0,0,z)=0$ for all $z\in I$.

We can hence suppose that:

$$X=X^N+X_\infty=f(\rho^2,z)\frac{\partial}{\partial\theta}+g(\rho^2,\theta)\rho\frac{\partial}{\partial\rho}+\rho^2 h(\rho^2,z)\frac{\partial}{\partial z}+X_\infty \tag{21}$$

where f,g,h are C^∞ functions, and X_∞ is flat along the z-axis Oz where $X_\infty=f_\infty\dfrac{\partial}{\partial\theta}+g_\infty\rho\dfrac{\partial}{\partial\rho}+\rho^2 h_\infty\dfrac{\partial}{\partial z}$ with $f_\infty,g_\infty,h_\infty$ flat functions in ρ.

In the formula (21), ρ and θ are polar coordinates defined as $x=\rho\cos\theta$, $y=\rho\sin\theta$. The expression X^N is the principal part of X and its particular form is an expression of the rotational invariance of the formal normal form.

Our first hypothesis was: $f(0,0)=1$. We can restrict the domain of study W to satisfy $F=f(\rho^2,z)+f_\infty(\rho,z,\theta)\neq 0$ everywhere in W. Then, being interested only in the phase portrait of X, we can replace X by $\dfrac{1}{F}X$ which is C^∞-equivalent to X:

$$\begin{aligned}\frac{1}{F}X&=\frac{\partial}{\partial\theta}+\big(G(\rho^2,z)+G_\infty(\rho,z,\theta)\big)\rho\frac{\partial}{\partial\rho}\\ &\quad+\rho^2\big(H(\rho^2,z)+H_\infty(\rho,z,\theta)\big)\frac{\partial}{\partial z}\end{aligned} \tag{22}$$

where G,H are C^∞ functions and G_∞, H_∞ are flat along Oz.
Let us write $G(\rho^2,z)=g_0(z)+O(\rho^2)$, with $g_0(0)=0$. As a generic hypothesis on the eigenvalues as functions of z, we can consider

$$g_0'(0)\neq 0\,. \tag{H_1}$$

Up to the change $z\to -z$ one can even assume that

$$g_0'(0)<0\,. \tag{H_1'}$$

This implies that points go from repelling to attracting when z passes through $0\in Oz$.

Next, let us write $H(\rho^2, z) = h_0(z) + O(\rho^2)$. We pose as a second generic condition:

$$(H_2) \qquad h_0(0) \neq 0\,.$$

We can now begin the study of X under the generic conditions (H_1'), (H_2). Consider the half-plane $\Pi = \{x \geq 0, y = 0\}$ parametrized by (x, z) with $x \geq 0$. If V is a neighbourhood of 0 in Π, each trajectory through $(z, x) \in V\backslash Oz$ $(x > 0)$ returns to Π at time 2π. This defines a Poincaré return map for $X, P\colon V \to \Pi$ taking $P(0, z) \equiv z$. The vector field $\dfrac{X}{F}$ is the suspension of P. It suffices then to study P.

We can introduce the so-called reduced vector field on Π (see [T]):

$$X_R(x, z) = G(x^2, z)x\frac{\partial}{\partial x} + x^2 H(x^2, z)\frac{\partial}{\partial z} \tag{23}$$

We denote by $\varphi(t, m)$, with $m = (x, z) \in \Pi$, the flow of X_R. Now, because X is a flat perturbation of $X^N = \dfrac{\partial}{\partial\theta} + X_R(\rho, z)$, we have :

$$P(z, x) = P_0(z, x) + P_\infty(z, x) \tag{24}$$

with $P_0(z, x) = \varphi(2\pi, (z, x))$ and $P_\infty(z, x) \in C^\infty$ and flat along $\{x = 0\}$.

First, we may notice that it is very easy to describe the map P_0, inducing the phase portrait of $X^N = \dfrac{\partial}{\partial\theta} + X_R(\rho, z)$.
In fact $X_R = xY_R$ with:

$$Y_R = G(x^2, z)\frac{\partial}{\partial x} + H(x^2, z)x\frac{\partial}{\partial z} \tag{25}$$

We can write $g_0'(0) = -\alpha$ and $h_0(0) = \beta$, with $\alpha > 0$ and $\beta \neq 0$. One has $G(x^2, z) = -\alpha z + O(|x|^2)$ and $H(x^2, z) = \beta + o(1)$.
If $\beta > 0$, the linearization of Y_R at $(0, 0)$ has complex eigenvalues. The origin is a focus-center point of Y_R, considered as a vector field on $\mathbb{R}^2$, and the orbits turn around 0. As a consequence, for any $m \in V\backslash Oz$ the X_R-orbit has an α-limit point on Oz_- and an ω-limit point on Oz_+ (see Figure 20).
If $\beta < 0$, Y_R has two real eigenvalues and the origin is a saddle point. The separatrices are transverse to the z-axis and the phase portrait of X_R easily follows (see Figure 21).
We will refer to the case $\beta > 0$ as the *elliptic case* and to the case $\beta < 0$ as the *hyperbolic case.*

The orbits of P_0 "follow" the orbits of X_R because P_0 is the time-2π mapping of X_R, and the dynamics of P_0 is clear. To obtain the dynamics of P, the idea is to compare it with the dynamics of P_0. The problem is non-trivial because the flat difference between P_0 and P could produce secular qualitative differences for an infinite number of iterations.

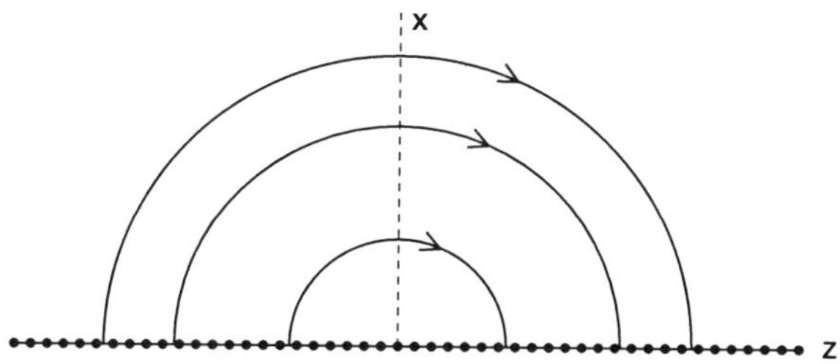

FIG. 20. *Phase portrait of X_R for $\beta > 0$.*

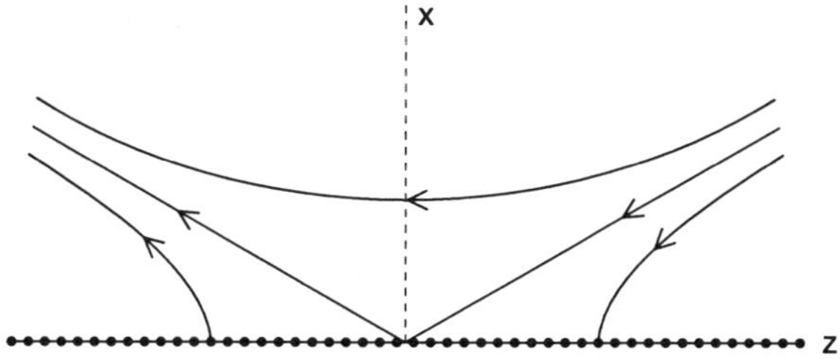

FIG. 21. *Phase portrait of X_R for $\beta < 0$.*

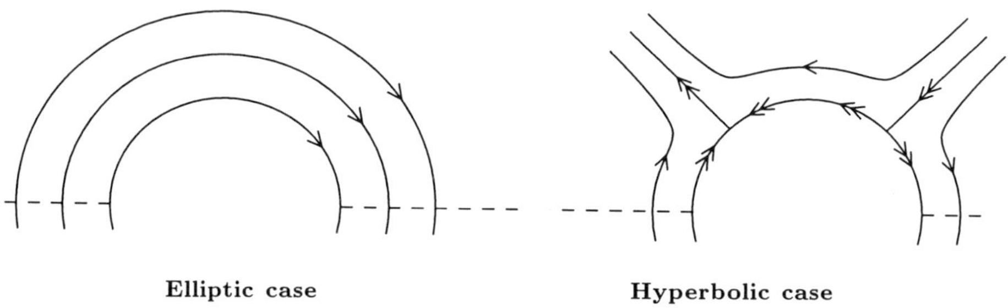

FIG. 22. *Phase portrait of* $\widetilde{X}_R$.

To prove that it is not the case, and that indeed the dynamics of P is very similar to that of P_0, we perform a blow-up of the origin of coordinates, using standard polar coordinates :

$$\Phi(\varphi, u) \begin{cases} x = u \cos\varphi \\ z = u \sin\varphi \end{cases}$$

with $\Pi = \left\{(u, \varphi) \mid u \geq 0, -\frac{\pi}{2} \leq \varphi \leq \frac{\pi}{2}\right\}$.
First, let us consider the blowing-up of X_R. One finds :

$$\Phi_* X_R = \widehat{X}_R : \begin{cases} \dot{\varphi} = uc(\alpha s^2 + \beta c^2 + O(u)) \\ \dot{u} = u^2 c^2((\beta - \alpha)s + O(u)) \end{cases} \tag{26}$$

where $c = \cos\varphi$, $s = \sin\varphi$.

The vector field $\widehat{X}_R$ is equal to $uc\widetilde{X}_R$ where $\widetilde{X}_R$ has just *elementary singularities* located on the critical locus $\{u = 0\}$. In fact, in the elliptic case, $\widetilde{X}_R$ has no singularities at all. In the hyperbolic case, $\widetilde{X}_R$ has two singular points $(\pm\varphi_0, 0)$ with $\varphi_0 = \arctan(-\beta/\alpha)^{1/2}$.
The tangential eigenvalue is non zero because $\pm\varphi_0$ are simple roots of $\{\alpha s^2 + \beta c^2 = 0\}$. At these points, the radial eigenvalue is equal to $\pm(\beta - \alpha)s(\varphi_0)c^2(\varphi_0) \neq 0$. Moreover, looking at the signs of the eigenvalues, we verify that these two points are hyperbolic saddles. The phase portrait of $\widetilde{X}_R$ is presented in the Figure 22 .

Now $\widehat{X}_R$ is the product of $\widetilde{X}_R$ with the function uc which is zero on $\widehat{\Sigma}$, the result of blowing-up the z-axis: $\widehat{\Sigma} = \{u = 0\} \cup \{c = 0\} \subset \widehat{\Pi} = \left\{(u, \varphi) \mid u \geq 0, -\frac{\pi}{2} \leq \varphi \leq \frac{\pi}{2}\right\}$.

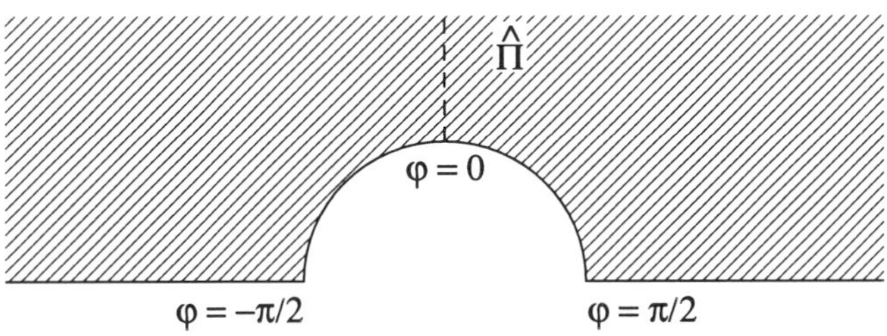

FIG. 23. *Picture of* $\widehat{\Sigma}$.

Recall now that P_0 is the time-2π mapping of X_R and that P is a flat perturbation of P_0 along Oz.
If we denote by $\widehat{P}_0$, $\widehat{P}$ the lifts of P_0, P in $\widehat{\Pi}$ (with $\Phi \circ \widehat{P}_0 = P_0 \circ \Phi, \Phi \circ \widehat{P} = P \circ \Phi$), it follows that $\widehat{P}$ is a *flat perturbation of* $\widehat{P}_0$ *along* $\widehat{\Sigma}$. $\widehat{P}_0$ itself is the time-2π map of the vector field $\widehat{X}_R$. We can apply to $\widehat{P}$ the main result of [DRR], the so called "Fundamental Theorem" (cfr. p.20 of [DRR]) $\widehat{P}$ verifies precisely the conditions of this theorem at each point of $\widehat{\Sigma}$. For a detailed and precise description of the conditions we refer to [DRR] (p.18 and 19), but roughly it means that we can find local coordinates (u, v) in which $\widehat{\Sigma} = \{u = 0\}$ (respectively $\widehat{\Sigma} = \{uv = 0\}$), $\widehat{P} - \widehat{P}_0$ is ∞-flat along $\widehat{\Sigma}$, $\widehat{P}_0$ is the time 1-mapping of some C^∞ vector field $u^k Y$ (resp. $u^k v^\ell Y$) with Y having at most elementary singularities. Under these conditions the Fundamental Theorem states that each point of $\widehat{\Sigma}$ has a neighbourhood W_p in which $\widehat{P}$ and $\widehat{P}_0$ are C^∞ conjugate by a diffeomorphism $g = Id + \nu$ where ν is flat along $W_p \cap \widehat{\Sigma}$.

Let us examine the global consequences of this in the two cases. We begin with the hyperbolic case, which is the simplest.

3.2.1. The hyperbolic case. It is clear that $\widehat{P}_0$ and $\widehat{P}$ are topologically conjugate. In fact [DRR] gives, in this case, a more precise result: there is no obstruction to gluing together the local conjugacies to form a global, C^∞ conjugacy $\widehat{g}$ of $\widehat{P}_0$ with $\widehat{P}$, infinitely close to the identity along $\widehat{\Sigma}$. Due to this fact, $\widehat{g}$ induces a C^∞ conjugacy of P_0 with P, which is infinitely close to the identity along the z-axis, and finally we can lift this conjugacy between return maps to a C^∞ equivalence of X with X^N, which is infinitely near the identity along the z-axis of $\mathbb{R}^3$. It follows that X inherits all qualitative properties of the normal form. It has a smooth invariant cone at 0 on which the orbits tend to 0 on the side $z > 0$ and escape from 0 on the side $z < 0$.

The inside of the cone is filled with invariant disks (obtained by the saturation of the orbits of X_R) which are the stable and unstable invariant manifolds of the generic part of the line of singularities Oz. Figure 24 gives the picture of the phase portrait in the hyperbolic case.

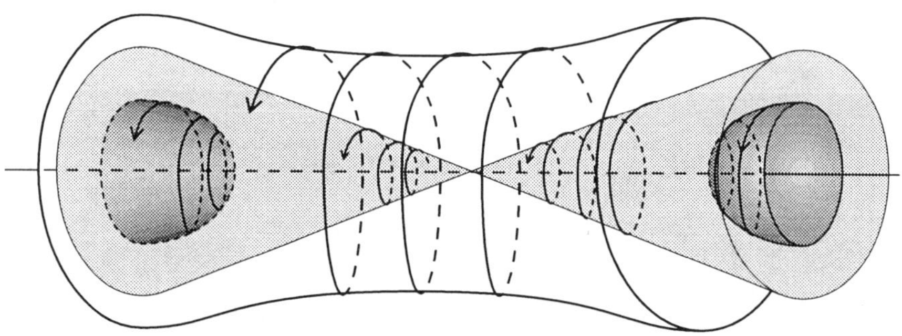

FIG. 24. *Phase portrait of X in the hyperbolic case.*

3.2.2. The elliptic case. Exactly as in the hyperbolic case we can glue together the local conjugacies given by the Fundamental Theorem, in a way as described in [DRR]. This is however not possible on a full neighbourhood of $\{u = 0, -\pi/2 \leq \varphi \leq \pi/2\}$ inside V, but only on neighbourhoods $\widehat{W}_-$ and $\widehat{W}_+$ of respectively $\{u = 0, -\pi/2 \leq \varphi \leq 0\}$ and $\{u = 0, 0 \leq \varphi \leq \pi/2\}$, as represented in Figure 25.

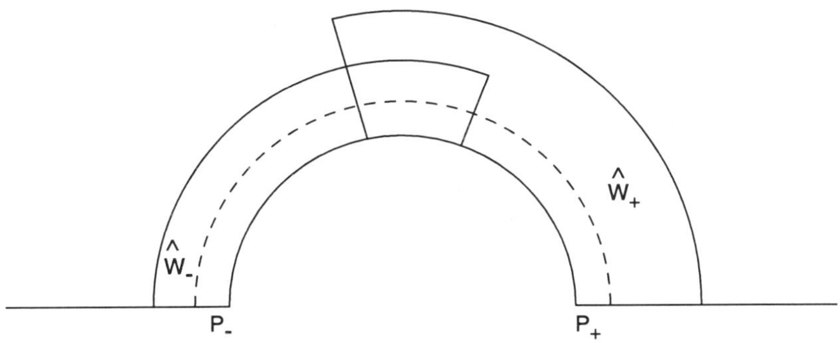

FIG. 25. *Subdivision of blown-up situation.*

Using these conjugacies, we see that $\widehat{P}$ is embedded as time-2π mapping of some vector field $\widehat{X}_-$ on $\widehat{W}_-$ and as time-2π mapping of a second vector field $\widehat{X}_+$ on $\widehat{W}_+$, which are both conjugate to $\widehat{X}_R$ on their respective domain. In general, one cannot choose $\widehat{X}_-$ equal to $\widehat{X}_+$ in $\widehat{W}_- \cap \widehat{W}_+$ and this difference is the obstruction to embedding $\widehat{P}$ as a time-2π mapping of the flow of a vector field (conjugate to $\widehat{X}_R$) and hence to conjugating $\widehat{P}$ with $\widehat{P}_0$.

Nevertheless, it easily follows from the existence of the two embeddings that in the domain $\widehat{W} = \widehat{W}_+ \cup \widehat{W}_-$, for each point $m \in \widehat{W} \backslash \widehat{\Sigma}$ the orbit $\{\widehat{P}^n(m)\}$ converges to a point $\alpha(m) \in \widehat{O}z_- = \{u > 0, \varphi = -\pi/2\}$ for $n \to -\infty$ and to a point $\omega(m) \in \widehat{O}z_+ = \{u > 0, \varphi = \pi/2\}$ for $n \to +\infty$. If we take a point $\alpha \in \widehat{O}z_-$, the set $W^u(\alpha) = \{m \in W \mid \widehat{P}^n(m) \to \alpha \text{ for } n \to -\infty\}$ is a smooth 1-dimensional manifold; it is C^∞ diffeomorphic to an orbit of $\widetilde{X}_R = (1/uc)\widehat{X}_R$. $W^u(\alpha)$ is the unstable manifold of α, and the $\{W^u(\alpha)\}_\alpha$ form a C^∞ foliation of $W \backslash \widehat{O}z_+$.

In the same way, one has a foliation $\{W^s(\omega)\}_\omega$ formed by the stable manifolds of the points $\omega \in \widehat{O}z_+$. In general the two foliations do not coincide in W, as it would be the case for $\widehat{P}_0$.

The non-coincidence of $\widehat{X}_+$ and $\widehat{X}_-$ gives rise to functional moduli which are obviously obstructions to conjugating $\widehat{P}$ with $\widehat{P}_0$ or, to embedding $\widehat{P}$ in a flow. With each $\alpha \in \widehat{O}z_-$ one can associate $\ell^u(\alpha) \subset \widehat{O}z_+$, the closed interval of ω-limit points of W^u_α. In the same way, one has a mapping $\ell^s(\omega)$ of $\widehat{O}z_+$ into the family of closed intervals of $\widehat{O}z_-$. Let us notice that the functions $\alpha \to |\ell^u(\alpha)|$, $\omega \to |\ell^s(\omega)|$ are C^∞ and flat at 0.

A nice problem consists in trying to describe the precise functions ℓ^s and ℓ^u that one can get when working with analytic vector fields X, or with polynomial vector fields of a certain degree.

We can now describe the phase portrait of X itself: there exists a neighbourhood W of $0 \in \mathbb{R}^3$ in which each orbit through a point $m \in W \backslash Oz$ has for α-limit a point $\alpha(m) \in Oz_-$ and for ω-limit a point $\omega(m) \in Oz_+$. As such, each orbit outside Oz is a heteroclinic orbit spiraling around Oz from a point of Oz_- to a point of Oz_+.

Moreover, each $\alpha \in W \cap Oz_-$ has a 2-dimensional unstable manifold W^u_α, and the W^u_α foliate a neighbourhood of 0 in $W \backslash Oz_+$. In the same way one has a foliation W^s_ω, $\omega \in Oz_+$. The two foliations do not coincide unless X is equivalent to its normal form. The above functional moduli are clearly defined for X and can be interpreted as topological obstructions for X to being topologically equivalent to its normal form.

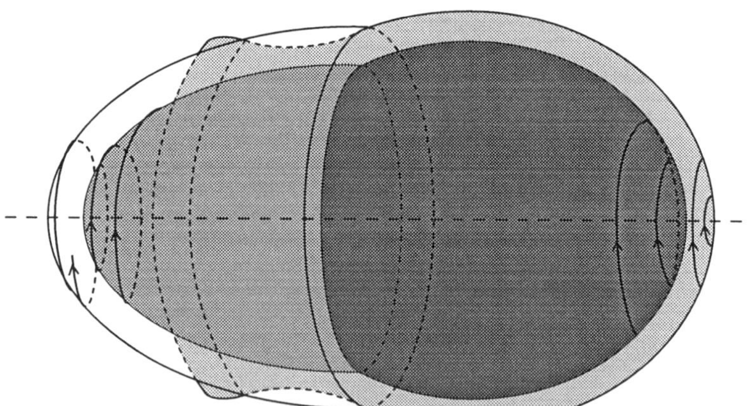

FIG. 26. *The stable and unstable foliations in the elliptic case.*

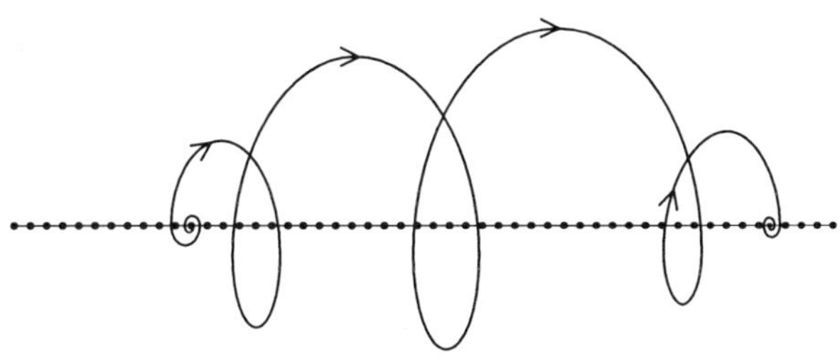

FIG. 27. *Phase portrait of X in the elliptic case.*

APPENDIX

Appendix 1: Simple normal forms. We first show that a C^∞ vector field as in (2)

$$X : \begin{cases} \dot{x} = xy\ f(x,y) \\ \dot{y} = xy\ g(x,y) \end{cases} \tag{27}$$

with $f(0,0) \cdot g(0,0) \neq 0$ is C^∞-equivalent to

$$\begin{cases} \dot{x} = xy \\ \dot{y} = xy \end{cases} \tag{28}$$

Let $\overline{X}$ be $f(x,y)\dfrac{\partial}{\partial x} + g(x,y)\dfrac{\partial}{\partial y}$.
By the flow box theorem, we know there exists a C^∞ change of coordinates, respecting the y-axis,

$$\varphi(\overline{x},\overline{y}) = (\overline{x}\varphi_1(\overline{x},\overline{y}), \varphi_2(\overline{x},\overline{y}))$$

with the property $\varphi_* \left(\dfrac{\partial}{\partial \overline{x}}\right) = \overline{X}$, $\varphi_1(0,0) \neq 0$ and $\varphi_2(0,\overline{y}) = \overline{y}$.

Clearly we have $X = \varphi_* \left(\overline{x}\varphi_1(\overline{x},\overline{y}) \cdot \varphi_2(\overline{x},\overline{y}) \frac{\partial}{\partial \overline{x}} \right)$.
Now $\{\varphi_2(\overline{x},\overline{y}) = 0\}$ can be written as $\{\overline{x} = \overline{y}\beta(\overline{y})\}$ with β of class $C^\infty$and $\beta(0) \neq 0$. We introduce the new coordinates $(s,\overline{y})$ with $(\overline{x},\overline{y}) = (\beta(\overline{y})s,\overline{y})$, in which the vector field

$$\begin{cases} \dot{\overline{x}} = \overline{x}\varphi_1(\overline{x},\overline{y}) \cdot \varphi_2(\overline{x},\overline{y}) \\ \dot{\overline{y}} = 0 \end{cases}$$

can be written as

$$\begin{cases} \dot{s} = s(s-\overline{y}) \cdot \gamma(s,\overline{y}) \\ \dot{\overline{y}} = 0 \end{cases}$$

with γ of class C^∞ and $\gamma(0,0) \neq 0$ because of the genericity conditions on (27). If we now change to the coordinates (s,z) with $z = s - \overline{y}$ we get

$$\begin{cases} \dot{s} = sz\gamma(s,s-z) \\ \dot{z} = sz\gamma(s,s-z) \end{cases}$$

proving our claim. We may observe that at each step we can change C^∞ into C^ω.

Similarly we can show that a C^∞ vector field as in (8)

$$X : \begin{cases} \dot{x} = y\ f(x,y) \\ \dot{y} = xy\ g(x,y) \end{cases} \tag{29}$$

with $f(0,0) \cdot g(0,0) \neq 0$ is C^∞ equivalent to

$$\begin{cases} \dot{x} = y \\ \dot{y} = xy \end{cases} \tag{30}$$

Let $\overline{X}$ denote $f(x,y)\frac{\partial}{\partial x} + xg(x,y)\frac{\partial}{\partial y}$. We know the existence of a C^∞ change of coordinates, being the identity on the y-axis:

$$\varphi(\overline{x},\overline{y}) = (\overline{x}\varphi_1(\overline{x},\overline{y}), \overline{y} + \overline{x}\varphi_2(\overline{x},\overline{y}))$$

with the property $\varphi_* \left(\frac{\partial}{\partial \overline{x}} \right) = \overline{X}$. Clearly we have

$$X = \varphi_* \left((\overline{y} + \overline{x}\varphi_2(\overline{x},\overline{y})) \frac{\partial}{\partial \overline{x}} \right)$$

A straightforward calculation shows that $\varphi_2(0,0) = 0$, $\frac{\partial \varphi_2}{\partial \overline{x}}(0,0) \neq 0$ and $\frac{\partial \varphi_2}{\partial y}(0,0) = 0$. The curve of zeroes is now situated along some C^∞ curve

$\overline{y} = g(\overline{x})$ with $g(0) = g'(0) = 0$ and $g''(0) \neq 0$. With a C^∞ change of coordinates of the form $\overline{y} = Y$, $\overline{x} = \alpha(X)$ we can change this curve of zeroes into $Y = X^2$. The expression of the vector field is now

$$\begin{aligned}\dot{X} &= Y(1 + G(X,Y)) \\ \dot{Y} &= -2XY(1 + G(X,Y))\end{aligned}$$

for some C^∞ function G with $G(0,0) = 0$, and this vector field is clearly C^∞ equivalent to (30). Again we can replace C^∞ by C^ω everywhere.

Appendix 2. As we have mentioned in the introduction, in the normally hyperbolic situation some refinements of classical GSP-theory can also be helpful in treating problems that require a high degree of differentiability. This is, for instance, the case in [DR1] and certainly in [DR2]. We consider the *C^k-normal linearization theorem of F. Takens.* We refer to [T] for a precise description of a quite general statement. We also refer to [B] for an adaptation of the theorem, taking care of certain constraints given by invariant foliations. In order to avoid the technical complexity linked to resonances among eigenvalues let us only consider the simple case of a line of singularities in $\mathbb{R}^2$. Up to a C^k change of coordinates, with k any a priori given positive number, we can suppose that the family of vector fields has the expression

$$\alpha(x,\varepsilon)\left(\pm y\frac{\partial}{\partial y} + f(x,\varepsilon)\frac{\partial}{\partial x}\right).$$

where α and f are both C^k and α is strictly positive. As such, after division by α, i.e. up to a C^k-equivalence, we can suppose that the normal form expression is

$$\pm y\frac{\partial}{\partial y} + f(x,\varepsilon)\frac{\partial}{\partial x}$$

In fact f is divisible by ε, and if we suppose the ε-perturbation to be generic in the sense that $\dfrac{\partial f}{\partial \varepsilon}(0,0) \neq 0$, then locally near $(0,0)$ we can prove the normal form for C^k-equivalence to be

$$\pm y\frac{\partial}{\partial y} + \varepsilon\frac{\partial}{\partial x}$$

which is an extremely easy expression to work with. For a typical use of this result we refer to [DR1] and [DR2]. Let us however recall a number of results. We simplify the presentation by supposing that the regular curve of normally hyperbolic zeroes is situated along the y-axis, which can always been achieved by a C^∞ change of coordinates.

We consider a C^∞ family of vector fields $X(x,y,\varepsilon) = X_\varepsilon(x,y)$. Let us not specify the domain on which we work, supposing that the whole y-axis consists of singularities at which the vector field X_0 is normally attracting with respect to $\{x = 0\}$, i.e. $X(0, y, 0) = 0$ and $\frac{\partial X_1}{\partial x}(0, y, 0) < 0$; we also suppose that $\frac{\partial X}{\partial \varepsilon}(0, y, 0) < 0$. Let W_y^s denote the stable manifold of the point $(0, y, 0)$.

We choose any $y_0 \in I\!R$, $p \in W_{y_0}^s \backslash \{(0, y_0, 0)\}$, $C_p = \{(x(\varepsilon), y(\varepsilon)\}$ a regular C^∞ curve transversally cutting $\{\varepsilon = 0\}$ at p. We also choose a regular C^∞ surface $\sum_p$, with $C_p \subset \sum_p$, transversally cutting $W_{y_0}^s$ at p.

For any $q = (x, y, \varepsilon) \in I\!R^3$ with $\varepsilon > 0$, and any $T > 0$ let $\varphi_\varepsilon(q, T)$ denote the position of the X_ε-orbit with initial condition q at some t, and let $O(q, T) = \{\varphi(q, t) \mid t \in [0, T]\}$. Consider also $\sum_{y_1}$ to be a regular C^∞ surface transversaly cutting the y-axis at $(0, y_1, 0)$ with $y_1 < y_0$. By means of the normal form of Takens, the following results can be obtained:

1) If we choose $q \in \sum_p \backslash \{\varepsilon = 0\}$ sufficiently close to p, then there exists some $T_{y_1}^q \in I\!R$ such that $\varphi(q, T_{y_1}^q) \in \sum_{y_1}$, and for $q \to p$, the Hausdorff limit of $O(q, T_{y_1}^q)$ will be equal to the union of $\{(0, y, 0) \mid y_1 \le y \le y_0\}$ and the piece of $W_{y_0}^s$ between y_0 and p.

2) For C_p sufficiently small, the C^∞ manifold $M(C_p, y_1) = \cup\{O(q, T_{y_1}^q) \mid q \in C_p \backslash \{p\}\}$ has a closure $\overline{M(C_p, y_1)}$ which is a C^∞ center manifold along the y-axis for $y_1 \le y < y_0$. For the rest it contains the piece of $W_{y_0}^s$ between y_0 and p.

3) For $\sum_p$ sufficiently small, the mapping $F : \sum_p \backslash \{\varepsilon = 0\} \to \sum_{y_1}$, $F(q) = \varphi(q, T_{y_1}^q)$ is a C^∞ mapping, which extends in a C^∞ way to $\sum_p$ by defining $F(q) = (0, y_1, 0) \in \sum_{y_1}$ for $q \in \sum_p \cap \{\varepsilon = 0\}$. The transition map from $\sum_p$ to $\sum_{y_1}$ is hence a C^∞ map seen in any regular C^∞ parametrization on respectively $\sum_p$ and $\sum_{y_1}$. It is of course not a diffeomorphism (except on $\sum_p \backslash \{\varepsilon = 0\}$) but, on the contrary, it is flat along $\{\varepsilon = 0\}$.

REFERENCES

[B] P. Bonckaert, *Partially hyperbolic fixed points with constraints*, Trans. Amer. Math. Soc. **348**, 1996, 997–1011.

[Ba] M. Barnsley, *Fractals everywhere*, Academic Press, San Diego, 1988.

[BCD] E. Benoit, J.L Callot, F. Diener, and M. Diener, *Chasse au canard*, Collect. Math. **31, 32** (1–3), 1981, 37–119.

[D] M. Diener, *Etude générique des canards, Thèse*, Univ. de Strasbourg, 1981.

[DH] F. Dumortier and C. Herssens, *Tracing phase portraits of planar polynomial vector fields with detailed analysis of the singularities*, Qualitative theory of dynamical systems.

[DK] F. Dumortier and H. Kokubu, *Chaotic dynamics in $Z\!\!Z_2$-equivariant unfoldings of codimension 3 singularities of vector fields in $I\!R^3$*, Ergodic Theory and Dynamical Systems.

[DR1] F. Dumortier and R. Roussarie, *Canard cycles and center manifolds*, Memoirs of the AMS, **121**, 1996.

[DR2] F. Dumortier and R. Roussarie, *Multiple canard cycles*, Preprint.

[DRR] F. Dumortier, P. Rodrigues, and R. Roussarie , *Germs of diffeomorphisms in the plane*, Lect. Notes Math., **902**, 1981, Springer-Verlag.

[DS] F. DUMORTIER AND B. SMITS, *Transition time analysis in singularly perturbed boundary value problems*, Trans. Amer. Math. Soc. **347**, 10, 1995, 4129–4145.

[Du] F. DUMORTIER, *Techniques in the theory of local bifurcations, blow-up, normal forms, nilpotent bifurcations, singular perturbations, in "Bifurcations of periodic orbits and vector fields"*, D. Schlomiuk, edit., NATO ASI-series C408, 19–73, 1993.

[F] N. FENICHEL, *Geometric Singular Perturbation Theory for Ordinary Differential Equations, Jr. of Differential Equations*, **31**, 53–98, 1979.

[FLA1] B. FIEDLER, S. LIEBSCHER, AND J.C. ALEXANDER , *Generic Hopf bifurcation from lines of equilibria without parameters : I. Theory*, Preprint, FU Berlin, 1998.

[FLA2] B. FIEDLER, S. LIEBSCHER, AND J.C. ALEXANDER, *Generic Hopf bifurcation from lines of equilibria without parameters*, II, Preprint, FU Berlin, 1998.

[H] C. HERSSENS, *Drawing and classifying phase portraits of planar polynomial vector fields*, Ph.D.-thesis, LUC, 1998.

[JK] C.K.R.T. JONES, N. KOPELL, *Tracking invariant manifolds with differential forms in singularly perturbed systems*, Jr. of Differential Equations **108**, 1994, 64–88.

[L] C. LOBRY, *Dynamic Bifurcations in Dynamic Bifurcations*, Ed. E. Benoît, Lect. Notes, Math., **1493**, 1991, Springer-Verlag.

[Pa] D. PANAZZOLO, *Desingularisation des singularités nilpotentes dans les familles analytiques*, Ph.D.-thesis, Univ. de Bourgogne, 1997.

[P] L. PERKO, *Differential Equations and Dynamical Systems*, 2nd ed., Texts in Applied Mathematics, **7**, Springer-Verlag, Berlin-Heidelberg - New York, 1996.

[T] F. TAKENS, *Partially hyperbolic fixed points*, Topology 10, 133–147, 1971.

A PRIMER ON THE EXCHANGE LEMMA FOR FAST-SLOW SYSTEMS

TASSO J. KAPER* AND CHRISTOPHER K.R.T. JONES†

Abstract. In this primer, we give a brief overview of the Exchange Lemma for fast-slow systems of ordinary differential equations. This Lemma has proven to be a useful tool for establishing the existence of homoclinic and heteroclinic orbits, especially those with many components or jumps, in a variety of traveling wave problems and perturbed near-integrable Hamiltonian systems. It has also been applied to models in which periodic orbits and solutions of boundary value problems are sought, including singularly perturbed two-point boundary value problems. The Exchange Lemma applies to fast-slow systems that have normally hyperbolic invariant manifolds (that are usually center manifolds). It enables one to track the dynamics of invariant manifolds and their tangent planes while orbits on them are in the neighborhood of a normally hyperbolic invariant manifold. The end result is a closeness estimate in the C^1 topology of the tracked manifold to a certain submanifold of the normally hyperbolic's local unstable manifold. We review the general version of the Exchange Lemma due to Tin [24] that treats problems in which there is both fast and slow evolution on the center manifolds. The main normal form used in the neighborhoods of the invariant manifolds is obtained from the persistence theory for normally hyperbolic invariant manifolds due to Fenichel [5, 6, 7]. The works of Jones and Tin [11, 15, 24] form the basis for this work, and [15, 24] contain full presentations of all of the results stated here.

1. Introduction.

1.1. Homoclinic and heteroclinic orbits.

Homoclinic and heteroclinic orbits connecting hyperbolic critical points, hyperbolic periodic orbits, or other hyperbolic invariant sets play central roles in the analysis of dynamical systems. For example, in the context of partial differential equations, travelling or standing waves will satisfy ordinary differential equations (ODEs), and in the phase space of these ODEs, such waves will be manifest as homoclinic or heteroclinic orbits connecting the critical points that represent the asymptotic states at $\pm\infty$. Another example arises in the context of integrable systems, where homoclinic and heteroclinic orbits occur at energy levels at which energy surfaces change in topological type. Indeed, they form the boundary of regions with a fixed set of action-angle variables.

This role of homoclinic and heteroclinic orbits as the boundary of a consistent type of behavior is also important in other problems. For instance, in the theory of Lagrangian transport, both types of orbit occur as the impermeable boundary between regions of ostensibly distinct motion, particularly in physical spaces that are restricted to 2-dimensions. Under perturbations that break this restriction, such as time-dependent

*Department of Mathematics & Center for BioDynamics, Boston University, 111 Cummington Street, Boston, MA 02215.

†Division of Applied Mathematics & Lefschetz Center for Dynamical Systems, Brown University, 182 George Street, Providence, RI 02912.

or 3-dimensional effects, complex dynamics can result from the complicated intersections that result between stable and unstable manifolds in 3-dimensions.

In perturbed integrable systems and more general dissipative systems, homoclinic and heteroclinic orbits are also often associated with the presence of chaotic dynamics in the form of an embedded horseshoe. Chaotic dynamics depend on two essential features: hyperbolic effects giving complementary directions of expansion and compression and recurrence of motion. Homoclinic orbits and loops of heteroclinic orbits to a saddle point or higher-dimensional hyperbolic invariant set provide, by their very nature, both of these effects.

1.2. Transverse intersections of invariant manifolds. Given the central role of homoclinic and heteroclinic orbits in so many systems, the question then is how to isolate them in particular cases. One particular perspective that has been successful on this fundamental problem comes from invariant manifold theory in dynamical systems and may be viewed as an outgrowth of what is traditionally known as a shooting method. The basic idea is simple: start with the unstable manifold of the desired state at $-\infty$ and try to see if it 'hits' the stable manifold of the corresponding state at $+\infty$. This technique is most readily visualized in the search for a homoclinic orbit, as such an orbit requires the variation of a parameter. This can be seen simply from the dimensions of the stable and unstable manifolds. These must add up, if the asymptotic critical point is hyperbolic, to the full dimension of the phase space. If there is a homoclinic orbit, it will be a 1-dimensional curve lying in the intersection of these two manifolds. By an elementary dimension count, this will require at least one parameter, in other words it is a codimension one phenomenon. The 'aiming' of the 'shooting' is then achieved through the tuning of the parameter to cause the desired intersection.

More generally, homoclinic and heteroclinic orbits may consist of several distinct segments. Traveling wave solutions in bistable reaction-diffusion equations, such as in the FitzHugh-Nagumo model [14] for example, are pulse-like homoclinic orbits in the associated ODE, consisting of active and quiescent states connected to each other via front and back segments that are jumps from one state to the other. The critical point represents the backward asymptotic state and lies on one locally invariant manifold representing the quiescent phase. The front lies in the local unstable manifold of this manifold and connects it to the vicinity of another locally invariant manifold. Immediately following this up jump, the homoclinic solution has a nontrivial segment near (but not on) that second invariant manifold. This middle segment represents the plateau of the pulse in a time trace, *e.g.*, the active phase in the case of action potentials modeled by the FitzHugh-Nagumo system. Finally, at the end of this middle segment, the homoclinic orbit leaves the vicinity of the second invariant manifold, returning to the first one along its stable manifold. In summary, the homoclinic solution in these types of bistable systems lies in the transverse intersection of the stable and unstable manifolds of the critical point, but it consists of several distinct segments, and the unstable manifold must be tracked over substantial portions of the phase space in order to find the full homoclinic orbit.

Similar types of homoclinic orbits with distinct segments near – and between – normally hyperbolic locally invariant manifolds occur in many other dynamical systems models. One may consider the general framework of an ODE that has a sequence $\{\mathcal{I}_i\}_{i=1}^{I+1}$ of locally invariant manifolds for some $I \geq 1$ with $\mathcal{I}_{I+1} = \mathcal{I}_1$ and the following type of heteroclinic transition chain. For each $i = 1, \ldots, I$, let $\mathcal{I}_i$ have local stable and unstable manifolds, $W^S_{\text{loc}}(\mathcal{I}_i)$ and $W^U_{\text{loc}}(\mathcal{I}_i)$, and assume that pairwise $W^U_{\text{loc}}(\mathcal{I}_i)$ and $W^S_{\text{loc}}(\mathcal{I}_{i+1})$ transversely intersect. Finally, assume that the flows on the manifolds $\mathcal{I}_i$ are of a suitable type so that a closed heteroclinic transition chain exists and allow for the possibility that $\mathcal{I}_{j_1} = \mathcal{I}_{j_2}$ for any pairs j_1, j_2.

Within the framework of this type of heteroclinic transition chain, the critical point representing the biasymptotic state for the desired homoclinic orbit lies on $\mathcal{I}_1$ ($\equiv \mathcal{I}_{I+1}$), and the entire orbit lies in the transverse intersection of $W^U(\mathcal{I}_1)$ and $W^S(\mathcal{I}_1)$. However, in between the semi-infinite-in-time 'tails' near $\mathcal{I}_1$, the homoclinic orbit (and nearby solutions on $W^U(\mathcal{I}_1)$) has segments of nontrivial duration near each of the intervening locally invariant manifolds $\mathcal{I}_2, \ldots, \mathcal{I}_I$. Orbits on $W^U(\mathcal{I}_1)$ that are near (but not in) the transverse intersection of $W^U_{\text{loc}}(\mathcal{I}_1)$ and $W^S_{\text{loc}}(\mathcal{I}_2)$ will spend significant times near $\mathcal{I}_2$ and then veer away from $\mathcal{I}_2$ close to $W^U_{\text{loc}}(\mathcal{I}_2)$. In turn, given that $W^U_{\text{loc}}(\mathcal{I}_2)$ and $W^S_{\text{loc}}(\mathcal{I}_3)$ intersect each other transversely, there will be a transverse intersection of $W^U(\mathcal{I}_1)$ and $W^S_{\text{loc}}(\mathcal{I}_3)$, as well, because a portion of $W^U(\mathcal{I}_1)$ is close to $W^U_{\text{loc}}(\mathcal{I}_2)$. Proceeding iteratively, we find that there are orbits on $W^U(\mathcal{I}_1)$ near (but not in) the intersection of $W^U(\mathcal{I}_1)$ and $W^S_{\text{loc}}(\mathcal{I}_3)$ that spend significant times near $\mathcal{I}_3$, and then veer away from $\mathcal{I}_3$ close to its local unstable manifold, and so forth. One may think that the sequence of pairwise transverse intersections constitutes a 'backbone' or sequence of allowable transitions (such as the fronts and backs in bistable systems), and that the tracking of the critical point's unstable manifold draps the flesh over the backbone. The resulting homoclinic orbit is a locally unique orbit along the flesh.

We remark that these homoclinic orbits are also referred to as multi-bump homoclinic orbits, because of the slow segments in between successive near-heteroclinic excursions. Moreover, each of the individual heteroclinic (or homoclinic) orbits in the transition chain may be multiple pulse orbits as long as they are in the transverse intersections of the appropriate invariant manifolds. In this way, one can also treat problems with Silnikov-type multipulse homoclinic orbits that consist of a sequence of near-homoclinic excursions with no slow segments in between.

1.3. Boundary value problems and problems with periodic orbits. Invariant manifold theory from dynamical systems has also been useful for singularly perturbed two-point boundary value problems. This may be illustrated as follows on an elementary linear, singularly perturbed, second-order equation for $y(x)$ on $[0, 1]$, with Dirichlet conditions $y(0) = A$ and $y(1) = B$, that has a locally unique solution. One can naturally form a pair of 2-dimensional manifolds in the 3-dimensional $x - y - y'$ space, as follows. Take a collection of initial conditions $(y(0) = A, y'(0))$ with $y'(0)$ varying over a suitable interval. Flow these initial conditions forward (in the 'time' x) to some suitable point x_0, thus obtaining a 2-dimensional manifold, $\mathcal{L}$. A second 2-dimensional manifold, $\mathcal{R}$, is obtained by flowing the points $(y(1) = B, y'(1))$ backward in x also to x_0, where also here $y'(1)$ is chosen over an appropriate interval. Then, the desired solution with a boundary layer to the problem is found by showing that, in a suitably transformed coordinate system that is naturally suggested by the leading order outer and inner equations, the manifolds $\mathcal{L}$ and $\mathcal{R}$ transversely intersect each other at x_0, and indeed the locally unique solution lies in this intersection. The intersection (which a priori need not be transverse) is precisely what one would find using a topological or numerical shooting procedure, and the transversality of the intersection (obtained via tracking the manifolds and their tangent planes) guarantees the local uniqueness and enables one to conclude persistence. See [9] for details. Moreover, this procedure can be extended naturally to other singularly perturbed boundary value problems, see [9, 16, 25], especially to problems in which the solutions consist of several inner and outer segments. Geometrically, one then tracks the manifolds of solutions satisfying the boundary conditions around in the extended phase space near the slow manifolds corresponding to the outer solutions.

Also, periodic orbits consisting of fast-slow segments, with multiple fast pulses, arise in applications. Typically, one first finds a heteroclinic transition chain of the type discussed above in the phase space, and then the desired periodic orbit lies close to this 'backbone,' *i.e.*, close to, but not on, the intersections of the center manifolds' stable and unstable manifolds. See [21, 22, 23, 4] for some examples, and see also section 4.5.

1.4. Tracking invariant manifolds and the Exchange Lemma. In all of the problems mentioned above, considerable work is required to achieve the needed control for tracking the (global) unstable manifold around in phase space. When the tracked manifolds are in the neighborhoods of other invariant manifolds (such as when the unstable manifold of the critical point in the FitzHugh-Nagumo system is in the vicinity of the second locally invariant manifold), control may be achieved by obtaining good estimates of their positions and of their tangent planes. Moreover, control over the tangent planes is essential as we have seen in the above problems so that one can guarantee that the tracked manifold transversely intersects the local stable manifold of the 'next' locally invariant manifold in the transition chain. An especially useful estimate is one that gives, at a point on an exit face of a prescribed neighborhood of the 'other' invariant manifold, closeness in the C^1 topology of the tracked manifold to an appropriate submanifold of the 'other' invariant manifold's local unstable manifold. The appropriate submanifold will have the same dimension as the tracked manifold, which of course imposes some condition on the class of manifolds that can be tracked. Nevertheless, this condition is naturally satisfied in the case of transverse homoclinic orbits, and we specify it in Section 4.

The theory of the Exchange Lemma reviewed here provides precisely this type of useful estimate. It applies to a general class of tracked invariant manifolds, including the unstable manifolds of critical points, the unstable manifolds of periodic orbits or of other hyperbolic invariant sets. The tracking of these manifolds is carried out in the vicinities of normally hyperbolic locally invariant manifolds. Broadly speaking, the main version of this theory states that, if a tracked manifold transversely intersects the local stable manifold of a normally hyperbolic locally invariant manifold, then it will be C^1 exponentially close to a certain submanifold of that manifold's local unstable manifold on exit from the neighborhood, see Figure 1. Moreover, the argument of the exponential factor is known explicitly in terms of the rates of exponential growth along the manifold's local stable and unstable manifolds, the rates of growth along the manifold itself, and the time orbits on the tracked manifold spend in the neighborhood. A precise statement is given in Theorem 4.2.

The Exchange Lemma theory was first developed for singularly perturbed systems with strictly slow dynamics on the normally hyperbolic, locally invariant manifold in [14], [13], and [12]. In [14], the singularly perturbed FitzHugh-Nagumo model was analyzed. The unstable manifold of the critical point was tracked, and the orbit homoclinic to it was constructed as described above in Section 1.2. The tracking of manifolds whose dimension is one bigger than that of the local unstable manifolds to which the closeness estimates refer in singularly-perturbed systems is carried out in [13]. A general theorem is also given there for multi-pulse homoclinic orbits that exist due to the presence of a heteroclinic transition chain of the type described above. An improved formulation of the normal form was developed in [12] and used to sharpen the tracking estimates. Applications to near-integrable systems where the sharpened closeness estimates are needed are given in [12, 17].

Tin [24] presents the most general version of the Exchange Lemma. See also [15]. Consider a normally hyperbolic invariant manifold of dimension ℓ that has an unstable manifold of dimension $k + \ell$ for arbitrary positive integers k and ℓ. Tin's Exchange Lemma enables one to track manifolds whose dimensions are any integer in $[k + 1, k + \ell]$. Moreover, Tin's work extends the class of problems to which the Exchange Lemma can be applied to include systems in which there is both fast and slow dynamics on the normally hyperbolic locally invariant manifold, whereas the earlier works had only allowed for slow dynamics. Finally, it substantially weakens the requirement imposed in [14, 13, 12] that orbits on the tracked manifold spend asymptotically long times ($\mathcal{O}(1/\varepsilon)$) near the manifold; it also yields closeness results when these times are long but $\mathcal{O}(1)$.

The Exchange Lemma has found application to many systems. A nonexhaustive list includes [1, 2, 4, 8, 11, 15, 17, 19, 23, 25]. We refer the reader to [24] for a more detailed history.

The name Exchange stems from the fact that in the FitzHugh-Nagumo example information about the wave speed on entry into the neighborhood of the second slow manifold, which represents the active phase, is exchanged for information about the slow variable on exit from that neighborhood. In particular in this example, there is a transverse intersection of $W^U_{\text{loc}}(\mathcal{I}_1)$ and $W^S_{\text{loc}}(\mathcal{I}_2)$ at an isolated value of the wavespeed, and the up jump lies close to the heteroclinic orbit in this intersection. Now, in that neighborhood, all of the tangent vectors turn so that they are nearly parallel to the fast unstable direction. As a result, the components of the tangent vectors in the direction of the wave speed are nearly indistinguishable. However, new information is obtained in exchange, namely the value of the slow variable at which the down jump (near the transverse intersection of $W^U(\mathcal{I}_2)$ and $W^S(\mathcal{I}_1)$) occurs is the information gained. A similar exchange of significant tangent directions occurs in general systems, during the passage near a center-slow manifold.

The brunt of the work associated to the Exchange Lemma concerns analyzing and estimating the dynamics of the tangent planes to the tracked manifolds. The tangent space to an invariant manifold is spanned by evolving vectors that obey the tangent equation, often called the equation of first variations (or variational equations). Any chosen initial set of vectors that forms a basis will remain a basis of the tangent space at future times because of the uniqueness of solutions. Following a full set of basis vectors may not always be judicious, however. In higher dimensions it means keeping track of an excess of information where a tangent space should really be described more efficiently. More serious is the phenomena that arises in multiple time scale problems that the chosen spanning vectors may become almost linearly dependent. Indeed, fast directions will tend to dominate, and two vectors initially chosen as independent may become indistinguishable as they are 'pushed out' in strongly unstable fast directions, as mentioned above. The resolution is to follow the entire tangent space as encoded in its exterior basis, *i.e.*, in terms of the appropriate differential forms, see Section 4.4. Calculating the equations on the forms is a very natural and elementary, algebraic procedure using Plücker coordinates, as is done in [24].

1.5. Relation between the Exchange and Lambda Lemmas. In low dimensional spaces, such as 2-dimensional spaces, control over segments of the global unstable manifold is possible because of the elementary topology of the phase space. A prototypical example is the 'Lambda' (or Inclination) Lemma for diffeomorphisms of the plane, see for example [18]. In its simplest form, the λ-Lemma applies to diffeomorphisms ϕ of the plane to itself that have a saddle fixed point, x_0, and for which the 1-dimensional local stable and unstable manifolds of x_0 intersect transversely at a point p. The λ-Lemma states that, under a sufficiently large number N of iterations of ϕ, the image of a segment of $W^U(x_0)$ containing p lies in a prescribed neighborhood of x_0, and it gets pushed C^1 close to $W^U_{\mathrm{loc}}(x_0)$. This Lemma is important for a number of fundamental results in dynamical systems theory, including for establishing the existence of hyperbolic invariant sets known as Smale horseshoes. Moreover, there exist generalizations to, among others, planar diffeomorphisms with two distinct saddle fixed points and transverse heteroclinic orbits.

Viewed from the perspective of the more-widely known λ-Lemma theory, the Exchange Lemma theory offers a similar type of C^1 closeness result for certain tracked manifolds, but it applies to a much more general situation. The planar diffeomorphism is replaced by the flow map of a high-dimensional system of ODEs. The saddle fixed point is generalized to an arbitrary finite-dimensional normally hyperbolic, locally invariant manifold, and there can be nontrivial dynamics on this manifold. Moreover, the stable and unstable manifolds of this manifold are generally also of higher dimension. Finally, the closeness of the tracked manifold is not to the entire local unstable manifold, but rather is, in general, only to a particular submanifold of the local unstable manifold. The tracking theory in these more general situations is considerably more difficult to establish.

1.6. Fenichel theory. The mathematical underpinning of the persistence theory, the uniform asymptotic rate conditions, and the general singular perturbation theory used below is provided by the theorems of Fenichel, see [5, 6, 7]. In these works, Fenichel develops a theory of invariant manifolds that survive perturbations due to characteristics of normal hyperbolicity, that is to say the directions normal to the manifold are more strongly expanding and/or compressing than the flow on the manifold itself. [7] gives the most general results for singularly perturbed systems in which there is both fast and slow motion on the normally hyperbolic, locally invariant manifolds. See Sections 2 and 3.

For singularly perturbed problems with strictly slow center manifolds, [11] and [20] each present new proofs of the Fenichel theory. Also, [11] contains a rich trove of examples, and [16] presents a somewhat more elementary introduction (with many examples, but no proofs). Finally, similar theorems for the persistence and foliations of compact, boundaryless invariant manifolds were established independently in [10].

1.7. Organization of this primer. This rest of this primer is organized as follows. In Section 2, we state the general class of ODEs to which the Exchange Lemma may be applied, and we state some basic results from the Fenichel persistence theory. Then, in Section 3, we describe the derivation of the Fenichel normal form for these ODEs in neighborhoods of normally hyperbolic invariant manifolds. This normal form plays a central role in the theory of the Exchange Lemma. In section 4, we present the tracking results that constitute the theory of the Exchange Lemma due to Tin [24], see also [15]. In particular, we carefully lay out the hypotheses on the tracked manifolds, prove an elementary C^0 tracking result, and then present the main Exchange Lemma that gives the desired C^1 tracking. Also in section 4, we give a brief overview of the proof of the Exchange Lemma and a brief discussion of extensions involving less restrictive hypotheses on the tracked manifolds.

2. The class of systems under study.

This primer concerns the systems of ODEs

$$\mathbf{x}' = \mathbf{f}(\mathbf{x}, \varepsilon), \tag{2.1}$$

where $\mathbf{x} \in \mathbb{R}^N$, $\mathbf{f}$ is a C^r function of $\mathbf{x}$ and ε, $r \geq 2$, $\varepsilon \in [0, \varepsilon_0]$ for some $\varepsilon_0 > 0$. The prime denotes differentiation with respect to the independent variable τ.

Throughout this work, we assume that for every $\varepsilon \in [0, \varepsilon_0]$ the system (2.1) has an ℓ-dimensional, C^r smooth, normally hyperbolic center manifold $\mathcal{S}_\varepsilon$, possibly with boundary. For concreteness, we assume that the manifolds $W^U_{\text{loc}}(\mathcal{S}_\varepsilon)$ and $W^S_{\text{loc}}(\mathcal{S}_\varepsilon)$ are of dimension $k + \ell$ and $m + \ell$, respectively, where k and m are arbitrary nonnegative integers such that $k + m + \ell = N$. We also assume that the foliations by invariant families of fast stable and unstable fibers of these local stable and unstable manifolds are C^r smooth in $\mathbf{x}$ and C^{r-1} smooth in ε. As we discuss in the remainder of this section, one may directly verify these assumptions by using the Fenichel theory.

Systems of the form (2.1) that possess one or more families of normally hyperbolic, locally invariant manifolds arise ubiquitously. A common scenario is that for the unperturbed vector field, (2.1) with $\varepsilon = 0$, there exists a manifold $\mathcal{S}_0$ that is a C^r, properly embedded, compact, normally hyperbolic, Riemannian submanifold of $\mathbb{R}^N$, possibly with boundary, $\partial\mathcal{S}_0$. One may refer to $\bar{\mathcal{S}}_0 = \mathcal{S}_0 \cup \partial\mathcal{S}_0$ as a critical manifold. Moreover, in various circumstances, the critical manifold is overflowing invariant, inflowing invariant, or invariant, and then $\bar{\mathcal{S}}_0$ persists as a C^r, locally invariant (or even invariant) manifold $\mathcal{S}_\varepsilon$ for $\varepsilon \in (0, \varepsilon_0]$.

Normal hyperbolicity is the essential ingredient that guarantees this persistence. The growth and contraction in directions normal to $\mathcal{S}_0$ must be exponential and at a rate that (in absolute value) is greater than the rate of growth in the tangential directions. The examples in Section I of

[5], especially the example in Section I.3, provide excellent illustrations of normal hyperbolicity and of what happens when this assumption breaks down. Generalized Lyapunov type numbers measure the normal and tangential growth rates.

Specifically, consider a critical manifold with boundary, $\bar{\mathcal{S}}_0 = \mathcal{S}_0 \cup \partial\mathcal{S}_0$, that is overflowing invariant, *i.e.*, $\bar{\mathcal{S}}_0$ is invariant under the backward flow of the system and the vector field is everywhere transverse to the boundary $\partial\mathcal{S}_0$. Also, for the tangent spaces, let $T\mathbb{R}^N|_{\bar{\mathcal{S}}_0}$ have a continuous subbundle N^+ that contains $T\bar{\mathcal{S}}_0$ and that is overflowing invariant with respect to the equation of first variations on $T\mathbb{R}^N$. Then, let $I \subseteq N^+$ be any continuous subbundle such that $N^+ = T\bar{\mathcal{S}}_0 \oplus I$, and let $J \subseteq T\mathbb{R}^N|_{\bar{\mathcal{S}}_0}$ be any continuous subbundle such that $T\mathbb{R}^N|_{\bar{\mathcal{S}}_0} = N^+ \oplus J$. Note that by definition,

$$T\mathbb{R}^N|_{\bar{\mathcal{S}}_0} = T\bar{\mathcal{S}}_0 \oplus I \oplus J,$$

and hence it makes sense to define the projection operators π_I and π_J. Moreover, this decomposition of $T\mathbb{R}^N|_{\bar{\mathcal{S}}_0}$ naturally leads one to study, for any point $p \in \bar{\mathcal{S}}_0$, the images of $\mathbf{u}_0 \in T_p\bar{\mathcal{S}}_0$, $\mathbf{v}_0 \in I_p$, $\mathbf{w}_0 \in J_p$, under the flow map F^τ for time $\tau \leq 0$:

$$\mathbf{u}^\tau = DF^\tau(p)\mathbf{u}_0, \qquad \mathbf{v}^\tau = \pi_I DF^\tau(p)\mathbf{v}_0, \qquad \mathbf{w}^\tau = \pi_J DF^\tau(p)\mathbf{w}_0.$$

The generalized Lyapunov type numbers are now defined as follows:

$$\begin{aligned} \lambda^+(p) &= \inf\left\{\lambda \in \mathbb{R}^+ : \lambda^\tau|\mathbf{v}^\tau| \to 0 \text{ as } \tau \to -\infty \text{ for all } \mathbf{v}_0 \in I_p\right\} \\ \nu^+(p) &= \inf\left\{\nu \in \mathbb{R}^+ : \frac{\nu^\tau}{|\mathbf{w}^\tau|} \to 0 \text{ as } \tau \to -\infty \text{ for all } \mathbf{w}_0 \in J_p\right\}. \end{aligned} \tag{2.2}$$

And, if $\nu^+(p) < 1$, then we also define

$$\begin{aligned} \sigma^+(p) = \inf\Big\{\sigma \in \mathbb{R}^+ : \frac{|\mathbf{u}^\tau|}{|\mathbf{w}^\tau|^\sigma} &\to 0 \text{ as } \tau \to -\infty \\ &\text{for all } \mathbf{u}_0 \in T_p\bar{\mathcal{S}}_0, \mathbf{w}_0 \in J_p\Big\}. \end{aligned} \tag{2.3}$$

Fenichel proves that these type numbers are independent of the choice of I, J and the Riemannian metric $|\cdot|$, see [5], especially Section II.1, and also Section V. Since ν^τ grows as $\tau \to -\infty$ for $0 < \nu < 1$ and since there are normal vectors $\mathbf{w}^\tau$ that grow in backward time, we want $\nu^+ < 1$ so that there is exponential growth in the normal directions. The other type number σ^+ measures the relative rates of the exponential growth in the normal and tangential directions. The more rapidly the normal vectors contract exponentially relative to the tangential vectors, the smaller σ^+ is and the smoother the manifold is. In many problems, σ^+ is zero, as occurs for instance when there is no exponential growth in the tangential directions anywhere on $\mathcal{M}$.

The persistence of overflowing invariant manifolds is established in:

Fenichel Unstable Manifold Theorem [5] (see also [24]). Let $\bar{\mathcal{S}}_0$ be a C^r overflowing invariant manifold under the flow of (2.1) with $\varepsilon = 0$. Let N^+ be a C^0 subbundle of $T\mathbb{R}^N|_{\bar{\mathcal{S}}_0}$ containing $T\bar{\mathcal{S}}_0$ that is overflowing invariant under the equations of first variations. Then, if $\lambda^+(p) < 1$, $\nu^+(p) < 1$, and $\sigma^+(p) < 1/r$ for all $p \in \bar{\mathcal{S}}_0$, there exists a C^r family of C^r overflowing invariant manifolds, W_ε^U, parametrized by $\varepsilon \in [0, \varepsilon_0]$ where ε_0 is sufficiently small, such that W_0^U contains $\bar{\mathcal{S}}_0$ and $TW_0^U|_{\bar{\mathcal{S}}_0} = N^+$.

Now, for inflowing invariant manifolds $\bar{\mathcal{S}}_0$, a similar theorem exists giving a C^r family of C^r stable manifolds $W_\varepsilon^S(\mathcal{S}_\varepsilon)$. Next, if $\mathcal{S}_0$ is boundaryless, then it is trivially both overflowing and inflowing invariant (due to compactness), and hence its persistence is guaranteed along with the persistence of its stable and unstable manifolds.

In addition, in a given application with a locally invariant critical manifold that has nonempty boundary, it may not be the case that $\bar{\mathcal{S}}_0$ is either overflowing or inflowing invariant. Nevertheless, in these cases one can still apply the above persistence theory by using bump functions. One modifies the dynamics outside of $\bar{\mathcal{S}}_0$ in a C^∞ manner so that there are slightly larger manifolds that are overflowing and inflowing invariant and $\bar{\mathcal{S}}_0 \subseteq \bar{\mathcal{S}}_{\mathbf{O}} \subseteq \bar{\mathcal{S}}_{\mathbf{I}}$. The Fenichel theory can then be applied to each of the overflowing and inflowing invariant manifolds $\bar{\mathcal{S}}_{\mathbf{O}}$ and $\bar{\mathcal{S}}_{\mathbf{I}}$, respectively, and the desired result about persistence of the original manifold $\bar{\mathcal{S}}_0$ and its stable and unstable manifolds is obtained simply by restriction. Of course, the choice of modification fixes the resultant persistent manifold $\mathcal{S}_\varepsilon$, eliminating the nonuniqueness that is generally the case.

Finally, Fenichel establishes what has become to be called an invariant family theorem (see section 2.1.1 of [24]) that describes the fiber structure on the unstable (and stable) manifolds. We will not present any of this here, referring instead the reader to [24], except to remark that the smoothness of the families of invariant fibers is C^r in $\mathbf{x}$ but only C^{r-1} in ε, so that our assumption is as stated above.

3. The Fenichel normal form.

In this section, we state the Fenichel normal form [13, 12, 24] for a system (2.1) in which the center manifold $\mathcal{S}_0$ is diffeomorphic to a contractable open set U in $\mathbb{R}^\ell$. A straightforward extension is at hand afterward to the more general case of compact manifolds $\bar{\mathcal{S}}_0$ that are not necessarily contractable simply by covering $\bar{\mathcal{S}}_0$ with finitely many contractable open submanifolds and by then patching the results from the Fenichel 'charts' for the individual submanifolds together smoothly.

3.1. The geometrically-induced Fenichel coordinates.

The point of departure is the case of $\varepsilon = 0$ in (2.1). For each point $p \in \mathcal{S}_0$, let

$$E_p^+ = T_p W_{\text{loc}}^U(\mathcal{S}_0) \quad \text{and} \quad E_p^- = T_p W_{\text{loc}}^S(\mathcal{S}_0). \tag{3.1}$$

The linear vector bundles $E^+ = \bigcup_{p \in \mathcal{S}_0} E_p^+$ and $E^- = \bigcup_{p \in \mathcal{S}_0} E_p^-$ are of rank k and m, respectively, and tangent to $W^U_{\text{loc}}(\mathcal{S}_0)$ and $W^S_{\text{loc}}(\mathcal{S}_0)$, respectively, at points on $\mathcal{S}_0$. Moreover, they are trivial since U is contractable to a point, and they are locally invariant. Finally, $T\mathbb{R}^n|_{\mathcal{S}_0} = T\mathcal{S}_0 \oplus E^+ \oplus E^-$. This geometric decomposition of the linearized dynamics near $\mathcal{S}_0$ permits introduction of a coordinate system $(\mathbf{a}, \mathbf{b}, \mathbf{y})$ in $\mathbb{R}^k \times \mathbb{R}^m \times \mathbb{R}^\ell$ such that

$$\begin{aligned} \mathcal{S}_0 &= \{(\mathbf{a}, \mathbf{b}, \mathbf{y}) | \mathbf{a}, \mathbf{b} = \mathbf{0}\} \\ E^+ &= \{(\mathbf{a}, \mathbf{b}, \mathbf{y}) | \mathbf{b} = \mathbf{0}\} \\ E^- &= \{(\mathbf{a}, \mathbf{b}, \mathbf{y}) | \mathbf{a} = \mathbf{0}\}. \end{aligned} \tag{3.2}$$

The Fenichel theory [5, 7], in particular the Unstable Manifold Theorem cited above and the analogous stable manifold theorem, establishes the existence of functions $\mathbf{h}^+(\mathbf{a}, \mathbf{y}, \varepsilon)$ and $\mathbf{h}^-(\mathbf{b}, \mathbf{y}, \varepsilon)$ such that

$$\begin{aligned} W^U_{\text{loc}}(\mathcal{S}_\varepsilon) &= \{(\mathbf{a}, \mathbf{b}, \mathbf{y}) | \mathbf{b} = \mathbf{h}^+(\mathbf{a}, \mathbf{y}, \varepsilon)\} \\ W^S_{\text{loc}}(\mathcal{S}_\varepsilon) &= \{(\mathbf{a}, \mathbf{b}, \mathbf{y}) | \mathbf{a} = \mathbf{h}^-(\mathbf{b}, \mathbf{y}, \varepsilon)\}. \end{aligned} \tag{3.3}$$

Moreover, since $W^U_{\text{loc}}(\mathcal{S}_\varepsilon)$ and $W^S_{\text{loc}}(\mathcal{S}_\varepsilon)$ intersect locally precisely in $\mathcal{S}_\varepsilon$ for each $\varepsilon \in (0, \varepsilon_0]$, the functions $\mathbf{h}^+$ and $\mathbf{h}^-$ vanish on $\mathcal{S}_0$ when $\varepsilon = 0$. Also, the first derivatives with respect to $\mathbf{a}, \mathbf{b}$, and $\mathbf{y}$, of these functions vanish on $\mathcal{S}_0$, as well, due to the tangency properties stated above. Finally, the C^r $\mathcal{O}(\varepsilon)$ closeness of the manifolds $W^U_{\text{loc}}(\mathcal{S}_\varepsilon)$ and $W^S_{\text{loc}}(\mathcal{S}_\varepsilon)$ to their unperturbed counterparts $W^U_{\text{loc}}(\mathcal{S}_0)$ and $W^S_{\text{loc}}(\mathcal{S}_0)$ implies $\mathbf{h}^\pm(\cdot, \mathbf{y}, \varepsilon) = \mathbf{h}^\pm(\cdot, \mathbf{y}, 0) + \mathcal{O}(\varepsilon)$.

The Fenichel coordinates for system (2.1) with $\varepsilon \in (0, \varepsilon_0]$ are then derived by straightening out the manifolds $W^U_{\text{loc}}(\mathcal{S}_\varepsilon)$ and $W^S_{\text{loc}}(\mathcal{S}_\varepsilon)$ and also the individual unstable and stable fibers. We outline the procedure here and refer the reader to Section 2.1.2 of [24] for details. Let $\mathbf{a}_1 = \mathbf{a}, \mathbf{b}_1 = \mathbf{b} - \mathbf{h}^+(\mathbf{a}, \mathbf{y}, \varepsilon)$, and $\mathbf{y}_1 = \mathbf{y}$. This invertible coordinate change straightens out $W^U_{\text{loc}}(\mathcal{S}_\varepsilon)$, mapping it to the set $\{(\mathbf{a}_1, \mathbf{b}_1, \mathbf{y}_1) | \mathbf{b}_1 = \mathbf{0}\}$. Next, noting that this first coordinate change maps $W^S_{\text{loc}}(\mathcal{S}_\varepsilon)$ to the set $\{(\mathbf{a}_1, \mathbf{b}_1, \mathbf{y}_1) | \mathbf{a}_1 = \mathbf{h}^-(\mathbf{h}^+(\mathbf{a}_1, \mathbf{y}_1, \varepsilon) + \mathbf{b}_1, \mathbf{y}_1, \varepsilon)\}$, one makes a second invertible coordinate change to straighten $W^S_{\text{loc}}(\mathcal{S}_\varepsilon)$ out, as well. Namely, let $\mathbf{a}_2 = \mathbf{a}_1 - \mathbf{h}^-(\mathbf{h}^+(\mathbf{a}_1, \mathbf{y}_1, \varepsilon) + \mathbf{b}_1, \mathbf{y}_1, \varepsilon)$, $\mathbf{b}_2 = \mathbf{b}_1$, and $\mathbf{y}_2 = \mathbf{y}_1$, so that the manifolds $W^S_{\text{loc}}(\mathcal{S}_\varepsilon)$ and $W^U_{\text{loc}}(\mathcal{S}_\varepsilon)$ are given by the sets $\{\mathbf{a}_2 = \mathbf{0}\}$ and $\{\mathbf{b}_2 = \mathbf{0}\}$, respectively, and the set $\{\mathbf{a}_2, \mathbf{b}_2 = \mathbf{0}\}$ is precisely $\mathcal{S}_\varepsilon$.

To straighten out the individual unstable fibers, we note that $W^U_{\text{loc}}(\mathcal{S}_\varepsilon)$ is coordinatized by $\mathbf{a}_2$ and $\mathbf{y}_2$. Also, letting π^+ denote the projection along an unstable fiber of a point on it to the base point of that fiber, we see that $\pi^+(\mathbf{0}, \mathbf{y}_2, \varepsilon) = \mathbf{y}_2$. Hence, $\pi^+(\mathbf{a}_2, \mathbf{y}_2, \varepsilon) = \mathbf{y}_2 + \Pi^+(\mathbf{a}_2, \mathbf{y}_2, \varepsilon)\mathbf{a}_2$, for some $\ell \times k$ matrix Π^+. Thus, the individual unstable fibers are straightened out by choosing the new coordinates

$$\mathbf{a}_3 = \mathbf{a}_2, \quad \mathbf{b}_3 = \mathbf{b}_2, \quad \mathbf{y}_3 = \mathbf{y}_2 + \Pi^+(\mathbf{a}_2, \mathbf{y}_2, \varepsilon)\mathbf{a}_2.$$

In a completely similar fashion, the projection π^- can be decomposed as $\pi^-(\mathbf{b}_3, \mathbf{y}_3, \varepsilon) = \mathbf{y}_3 + \Pi^-(\mathbf{b}_3, \mathbf{y}_3, \varepsilon)\mathbf{b}_3$, and hence the stable fibers are straightened out in the (last) new coordinates:

$$\mathbf{a}_4 = \mathbf{a}_3, \quad \mathbf{b}_4 = \mathbf{b}_3, \quad \mathbf{y}_4 = \mathbf{y}_3 + \Pi^-(\mathbf{b}_3, \mathbf{y}_3, \varepsilon)\mathbf{b}_3.$$

Summarizing, the sets $\{\mathbf{a}_4 = \mathbf{0}\}$ and $\{\mathbf{b}_4 = \mathbf{0}\}$ are the manifolds $W^S_{\text{loc}}(\mathcal{S}_\varepsilon)$ and $W^U_{\text{loc}}(\mathcal{S}_\varepsilon)$, respectively, and the stable and unstable fiberings are $(\mathbf{b}_4, \mathbf{y}_4) \to \mathbf{y}_4$ and $(\mathbf{a}_4, \mathbf{y}_4) \to \mathbf{y}_4$, respectively. Finally, for convenience, we reset $\mathbf{a} = \mathbf{a}_4$, $\mathbf{b} = \mathbf{b}_4$, and $\mathbf{y} = \mathbf{y}_4$, and this final set of coordinates is referred to as the Fenichel coordinates.

3.2. The structure of the Fenichel normal form. A useful expression for the normal form for (2.1) in the neighborhood of $\mathcal{S}_\varepsilon$ is immediately at hand in terms of the Fenichel coordinates $(\mathbf{a}, \mathbf{b}, \mathbf{y})$ introduced in Section 3.1. The local invariance of the manifolds $\mathcal{S}_\varepsilon$, $W^U_{\text{loc}}(\mathcal{S}_\varepsilon)$ and $W^S_{\text{loc}}(\mathcal{S}_\varepsilon)$ implies that the sets $\{\mathbf{a} = \mathbf{0}\}$ and $\{\mathbf{b} = \mathbf{0}\}$ must also be invariant (for the appropriate time intervals) in the ordinary differential equations for $(\mathbf{a}, \mathbf{b}, \mathbf{y})$. In addition, since the individual fibers have also been straightened out, the $\mathbf{y}$ component of the vector field must be independent of $\mathbf{a}$ on the $\{\mathbf{b} = \mathbf{0}\}$ hyperplane and independent of $\mathbf{b}$ on the $\{\mathbf{a} = \mathbf{0}\}$ hyperplane. Therefore, by these two observations, system (2.1) has the following general structure in the neighborhood of $\mathcal{S}_\varepsilon$:

$$\begin{aligned} \mathbf{a}' &= \Lambda(\mathbf{a}, \mathbf{b}, \mathbf{y}, \varepsilon)\mathbf{a} \\ \mathbf{b}' &= \Gamma(\mathbf{a}, \mathbf{b}, \mathbf{y}, \varepsilon)\mathbf{b} \\ \mathbf{y}' &= \mathbf{h}(\mathbf{y}, \varepsilon) + \mathbf{H}(\mathbf{a}, \mathbf{b}, \mathbf{y}, \varepsilon) \otimes \mathbf{a} \otimes \mathbf{b}. \end{aligned} \tag{3.4}$$

Here, Λ and Γ are C^r smooth, $k \times k$ and $m \times m$ dimensional matrix-valued functions, respectively, $\mathbf{h}$ is a C^r smooth function in $\mathbb{R}^\ell$, and $\mathbf{H}$ is a C^r smooth rank three tensor, for all $r < \infty$, with $\otimes$ denoting the tensor product. In component notation, the $\mathbf{y}$ component of the vector field in (3.4) is expressed as: $\mathbf{y}_i' = \mathbf{h}_i(\mathbf{y}, \varepsilon) + \sum_{u=1}^k \sum_{s=1}^\ell H_{ius}\mathbf{a}_u\mathbf{b}_s$.

3.3. A refined Fenichel normal form for systems with center variables of both fast and slow types. For the purposes of tracking manifolds while keeping the scope of applicability of the tracking results analyzed here as large as possible, we make one final refinement of the variables and a concomitant assumption on the vector field. In particular, we follow [24] and explicitly decompose the center variable $\mathbf{y}$ into slow and fast components: $\mathbf{y} = (\mathbf{w}, \theta)$ where $\mathbf{w} \in \mathbb{R}^{\ell_w}$ and $\theta \in \mathbb{R}^{\ell_\theta}$, with $0 \le \ell_w, \ell_\theta \le \ell$ such that $\ell_w + \ell_\theta = \ell$, and where $U = U_\mathbf{w} \times U_\theta$. Thus, we take (3.4) to have the form:

$$\begin{aligned} \mathbf{a}' &= \Lambda(\mathbf{a}, \mathbf{b}, \mathbf{y}, \varepsilon)\mathbf{a} \\ \mathbf{b}' &= \Gamma(\mathbf{a}, \mathbf{b}, \mathbf{y}, \varepsilon)\mathbf{b} \end{aligned}$$

(3.5)

$$\mathbf{w}' = \varepsilon \mathbf{h}_1(\mathbf{y}, \varepsilon) + \mathbf{H}_1(\mathbf{a}, \mathbf{b}, \mathbf{y}, \varepsilon) \otimes \mathbf{a} \otimes \mathbf{b}$$
$$\theta' = \mathbf{h}_2(\mathbf{y}, \varepsilon) + \mathbf{H}_2(\mathbf{a}, \mathbf{b}, \mathbf{y}, \varepsilon) \otimes \mathbf{a} \otimes \mathbf{b}.$$

With this explicit distinction between slow and fast center variables, we assume that

(H1) $\Lambda(\mathbf{0}, \mathbf{0}, \mathbf{y}, 0)$ and $\Gamma(\mathbf{0}, \mathbf{0}, \mathbf{y}, 0)$ are independent of the fast center variable θ, as is $D_{\mathbf{y}}\mathbf{h}_2(\mathbf{y}, 0)$.

The assumption (H1) about Λ and Γ guarantees that the hyperbolic dynamics in the directions normal to $\mathcal{S}_\varepsilon$ is driven purely by the slow center variable $\mathbf{w}$. The same assumption for $D_{\mathbf{y}}\mathbf{h}_2(\mathbf{y}, 0)$ implies that, on $\mathcal{S}_\varepsilon$, the dependence on θ in the fast center component of the vector field is simple. In particular, the dependence on θ can only lead to algebraic (and not exponential) growth in time.

System (3.5) will be referred to as the Fenichel normal form. We shall exploit it throughout the analysis of the tracking results. In particular, for fixed Δ, ε_0, and U, we focus on the dynamics of (3.5) in the neighborhood

$$\mathcal{B} = \{(\mathbf{a}, \mathbf{b}, \mathbf{y}) | |\mathbf{a}|, |\mathbf{b}| < \Delta, \mathbf{y} \in U, \varepsilon \in [0, \varepsilon_0]\} \tag{3.6}$$

of a slow manifold $\mathcal{S}_\varepsilon$.

In the analysis of (3.4) and (3.5), the following spectral bounds will be useful in making the estimates on the dynamics in Section 4:

$$\begin{aligned} \lambda_0 &= \inf \bigcup_{\mathbf{y} \in \bar{U}} \operatorname{Re}\left(\operatorname{spec} \Lambda(\mathbf{0}, \mathbf{0}, \mathbf{y}, 0)\right), \\ \gamma_0 &= \sup \bigcup_{\mathbf{y} \in \bar{U}} \operatorname{Re}\left(\operatorname{spec} \Gamma(\mathbf{0}, \mathbf{0}, \mathbf{y}, 0)\right), \\ \hat{\alpha}_0 &= \sup \bigcup_{\mathbf{y} \in \bar{U}} \operatorname{Re}\left(\operatorname{spec} D_{\mathbf{y}}\mathbf{h}(\mathbf{y}, 0)\right). \end{aligned} \tag{3.7}$$

Here, the bounds exist and satisfy $\lambda_0 > 0$ and $\gamma_0 < 0$, since $\mathcal{S}_0$ is normally hyperbolic and $W^U_{\text{loc}}(\mathcal{S}_0)$ and $W^S_{\text{loc}}(\mathcal{S}_0)$ are of dimensions k and m, respectively, as assumed in Section 2.

4. Tracking invariant manifolds near center manifolds.

Through out this section, we assume that we are given a C^r smooth family $\{\mathcal{M}_\varepsilon\}_{\varepsilon \in [0, \varepsilon_0]}$ of C^r locally invariant manifolds, each of which is of dimension $k + \sigma$, where $1 \leq \sigma \leq \ell$. These will be the manifolds that we track while orbits on them are near a center manifold $\mathcal{S}_\varepsilon$. The overall goal of the tracking theory is to identify which submanifold of $W^U_{\text{loc}}(\mathcal{S}_\varepsilon)$ a tracked manifold $\mathcal{M}_\varepsilon$ is topologically close to on exit from a neighborhood $\mathcal{B}$ and then to show that $\mathcal{M}_\varepsilon$ is actually differentiably close to this submanifold, *i.e.*, close in the C^1 topology.

4.1. Basic hypotheses on tracked manifolds.

The situation that arises most commonly in applications is that in which, on entry into a

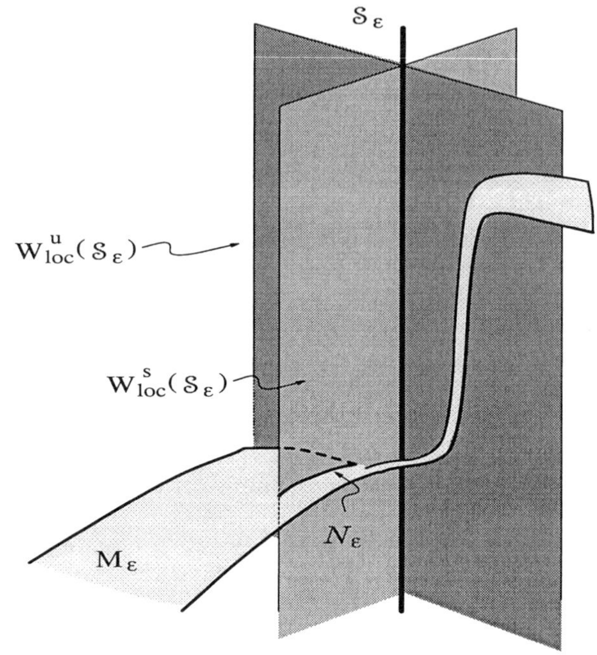

FIG. 1. *A schematic illustration of the slow manifold $\mathcal{S}_\varepsilon$, its local stable and unstable manifolds, and the tracked manifold $\mathcal{M}_\varepsilon$, with the intersection $\mathcal{N}_\varepsilon$ of $\mathcal{M}_\varepsilon$ and $W^S_{\rm loc}(\mathcal{S}_\varepsilon)$.*

neighborhood $\mathcal{B}$, a tracked manifold $\mathcal{M}_\varepsilon$ intersects the local stable manifold $W^S_{\rm loc}(\mathcal{S}_\varepsilon)$ transversely, as illustrated in Figure 1. This is also the main situation we study here. We remark on how to extend the results to other situations, for example to those in which the tracked manifolds are assumed instead to be differentiably close to $W^S_{\rm loc}(\mathcal{S}_\varepsilon)$ on entry (and do not intersect $W^S_{\rm loc}(\mathcal{S}_\varepsilon)$ nearby) in Section 4.5.

The first hypothesis is:

(T1) For each $\varepsilon \in (0, \varepsilon_0]$ there exists a point $q_\varepsilon \in \mathcal{M}_\varepsilon \cap \partial\bar{\mathcal{B}}$ such that q_ε is smooth in ε; and, at the point q_ε, the tracked manifold $\mathcal{M}_\varepsilon$ transversely intersects $W^S_{\rm loc}(\mathcal{S}_\varepsilon)$.

Hypothesis (T1) is the central hypothesis we have mentioned above stating that, at entry to a neighborhood $\mathcal{B}$, the tracked manifold transversely intersects the local stable manifold of $\mathcal{S}_\varepsilon$. It also implies that, for the critical case $\varepsilon = 0$, $\mathcal{M}_0 \cap W^S_{\rm loc}(\mathcal{S}_0)$ in the point q_0. However, this intersection need not be transverse, and often is not, as in Hamiltonian systems, for example.

Next, consider the ball V_{r_0} of radius $r_0 > 0$ about the point q_0 in the phase space of the augmented system (3.5) with the equation $\varepsilon' = 0$

appended. Let

$$\mathcal{Q}_\varepsilon = \mathcal{M}_\varepsilon \cap V_{r_0}.$$

The second hypothesis is a time estimate that concerns the length of time trajectories through $\mathcal{Q}_\varepsilon$ spend in $\mathcal{B}$. In particular, for an arbitrary point $q \in \mathcal{B}$ (and hence also for any $q \in \mathcal{Q}_\varepsilon$), let $\tau_q^{\mathcal{B}}$ be defined as $\inf\{\tau > 0 | q \cdot \tau \notin \mathcal{B}\}$. Then, we fix $\delta_0 > 0$ and assume:

(TE) For each $\varepsilon \in (0, \varepsilon_0]$ and for each $\delta \in [0, \delta_0]$, there exists a subset $\mathcal{Q}_\varepsilon^\delta \subset \mathcal{Q}_\varepsilon$ such that $\inf_{q \in \mathcal{Q}_\varepsilon^\delta} \tau_q^{\mathcal{B}}$ is monotonically increasing as δ decreases, and such that $\lim_{\delta \to 0} \inf_{q \in \mathcal{Q}_\varepsilon^\delta} \tau_q^{\mathcal{B}} = \infty$.

The infinite limit condition in (TE) implies that $\mathcal{Q}_\varepsilon^0 \subset W_{\text{loc}}^S(\mathcal{S}_\varepsilon)$, since only the points on $W_{\text{loc}}^S(\mathcal{S}_\varepsilon)$ never leave the complement of $\mathcal{S}_\varepsilon$ in $\mathcal{B}$ in forward time. Moreover, the geometric interpretation naturally suggests that there is at least one method of selecting the sets $\mathcal{Q}_\varepsilon^\delta$, namely to choose $\delta = \sup_{q \in \mathcal{Q}_\varepsilon^\delta} |\mathbf{a}(q)|$, so that δ measures the maximum distance from points in $\mathcal{Q}_\varepsilon^\delta$ to $W_{\text{loc}}^S(\mathcal{S}_\varepsilon)$.

Before stating the third (and last) hypothesis, it is necessary to identify the set

$$\mathcal{N}_\varepsilon = W_{\text{loc}}^S(\mathcal{S}_\varepsilon) \cap \mathcal{Q}_\varepsilon,$$

and we recall that this intersection is transverse by assumption (T1). See Figure 1. For each $\varepsilon \in (0, \varepsilon_0]$, the set $\mathcal{N}_\varepsilon$ is a σ–dimensional submanifold of $W_{\text{loc}}^S(\mathcal{S}_\varepsilon)$, since $\dim(\mathcal{M}_\varepsilon) = \dim(\mathcal{Q}_\varepsilon) = k + \sigma$, $\dim(W_{\text{loc}}^S(\mathcal{S}_\varepsilon)) = m + \ell$, and the dimension of the ambient space is $k + m + \ell$. In addition, $\mathcal{N}_\varepsilon$ varies smoothly in ε. The base points of the stable fibers in $\mathcal{M}_\varepsilon$ are given by the projection π^- onto $\mathcal{S}_\varepsilon$ along the stable fibers. The final hypothesis is:

(T2) The linear map

$$\left(\pi^- |_{\mathcal{N}_\varepsilon}\right)_* : \; T_{q_\varepsilon}\mathcal{N}_\varepsilon \longrightarrow T_{\pi^-(q_\varepsilon)}\mathcal{S}_\varepsilon$$

is injective.

This last hypothesis is central to the tracking theory: it assures that the manifold

$$\mathcal{J}_\varepsilon = \pi^-(\mathcal{N}_\varepsilon)$$

is a σ–dimensional submanifold of $\mathcal{S}_\varepsilon$, a fact that will be important in the tracking below. In particular, it will be by flowing points in $\mathcal{J}_\varepsilon$ forward along $\mathcal{S}_\varepsilon$ that we will identify the submanifold of $W_{\text{loc}}^U(\mathcal{S}_\varepsilon)$ to which $\mathcal{M}_\varepsilon$ is close to exit of $\mathcal{B}$.

Remark. The assumptions (T1), (TE) and (T2) made here are taken from the assumptions $(T1)_{FD}$, $(TE)_{FD}$, and $(T2)_{FD}$ in Chapter 3.2 of [24].

4.2. An elementary topological tracking result. With the geometric framework for the manifolds $\mathcal{M}_\varepsilon$ established in the previous subsection, we can now track these manifolds while orbits through the entry points $q_\varepsilon \in \mathcal{M}_\varepsilon \cap \partial\mathcal{B}$ spend time intervals of length $\tau_q^{\mathcal{B}}$ in $\mathcal{B}$. The topological (*i.e.*, C^0) tracking is carried out here, and the main C^1 tracking result, namely the Exchange Lemma, is given in the next subsection.

Recall that the manifold $\mathcal{J}_\varepsilon = \pi^-(\mathcal{N}_\varepsilon)$ is the set of base points of the stable fibers in $\mathcal{N}_\varepsilon$. The orbits of these base points on $\mathcal{S}_\varepsilon$ 'shadow' the orbits through the points in $\mathcal{N}_\varepsilon$:

$$\pi^-(q\cdot\tau) = \left(\pi^- q\right)\cdot\tau \qquad \text{for } q\in\mathcal{N}_\varepsilon \text{ and } \tau\in[0,\tau_q^{\mathcal{B}}].$$

The C^0 tracking result may now be stated as:

Proposition 4.1. (Proposition 3.11 in [24]) Assume that $\lambda_0 > \hat{\alpha}_0$. Fix a positive $\kappa < \min(\lambda_0 - \hat{\alpha}_0, -\gamma_0)$. There exist $\Delta_0, \varepsilon_0 > 0$ such that for each $\Delta \in (0,\Delta_0)$ and $\varepsilon\in(0,\varepsilon_0)$ there exists a $\delta_0 > 0$ such that for any $0<\delta<\delta_0$ and for any $q\in\mathcal{Q}_\varepsilon^\delta$ it is the case that

$$\mathcal{M}_\varepsilon\cdot\tau_q^{\mathcal{B}} \text{ is } C^0\ \mathcal{O}(e^{-\kappa\tau_q^{\mathcal{B}}}) \text{ close to } W^U_{\rm loc}(\mathcal{S}_\varepsilon)|_{\mathcal{J}_\varepsilon\cdot\tau_q^{\mathcal{B}}} \text{ at } q\cdot\tau_q^{\mathcal{B}}.$$

Proposition 4.1 may be interpreted geometrically as follows. The base points in $\mathcal{J}_\varepsilon$ evolve along $\mathcal{S}_\varepsilon$, and $\mathcal{J}_\varepsilon\cdot\tau_q^{\mathcal{B}}$ denotes the union of their images after the times $\tau_q^{\mathcal{B}}$. Taking the union of the unstable fibers with base points in $\mathcal{J}_\varepsilon\cdot\tau_q^{\mathcal{B}}$, one obtains the submanifold $W^U_{\rm loc}(\mathcal{S}_\varepsilon)|_{\mathcal{J}_\varepsilon\cdot\tau_q^{\mathcal{B}}}$ of $W^U_{\rm loc}(\mathcal{S}_\varepsilon)$. Proposition 4.1 then states that the tracked manifold $\mathcal{M}_\varepsilon$ exits $\mathcal{B}$ topologically close to this submanifold. In other words, the orbits of the base points on $\mathcal{S}_\varepsilon$ guide the orbits of points on $\mathcal{M}_\varepsilon$ to the right exit place. Moreover, the uniform control of the exponential behavior of the normal components $\mathbf{a}$ and $\mathbf{b}$ over the domain U makes it possible to get the exponential estimate. The constant κ is chosen based on the spectral gap $\lambda_0 - \hat{\alpha}_0 > 0$, which implies that the exponential rate of expansion in the directions normal to $\mathcal{S}_\varepsilon$ are stronger than those in the tangential directions. It is the guaranteed minimum of exponential closeness in general systems, and may be sharpened in some particular cases.

Proof of Proposition 4.1: The proof may be carried out directly and without loss of generality on system (3.4). Fix arbitrary $0<\lambda<\lambda_0$, $\gamma_0<\gamma<0$, and $\hat{\alpha}>\hat{\alpha}_0$ such that $\kappa<|\gamma|$ and $\lambda>\kappa+\hat{\alpha}$. From the structure of the $\mathbf{a}$ and $\mathbf{b}$ components of the vector field (3.4), it follows immediately that there exist exponential bounds on the growth and decay of $\mathbf{a}$ and $\mathbf{b}$, respectively. In particular, there exist positive constants Δ_0, ε_0, K_a, and K_b such that

$$|\mathbf{a}(q\cdot\tau)| \le \frac{\Delta}{K_a}e^{-\lambda(\tau_q^{\mathcal{B}}-\tau)} \qquad \text{and} \qquad |\mathbf{b}(q\cdot\tau)| \le \Delta K_b e^{\gamma\tau} \tag{4.1}$$

for all $\tau\in[0,\tau_q^{\mathcal{B}}]$, $\varepsilon\in(0,\varepsilon_0)$, $\Delta\in(0,\Delta_0)$, and $q\in\mathcal{Q}_\varepsilon^\delta$.

With these bounds on the normal components $\mathbf{a}$ and $\mathbf{b}$ in hand, it remains to establish control over the center variable $\mathbf{y}$. For each point $q \in \mathcal{Q}_\varepsilon^\delta$, when ε_0 and r_0 are small enough, one can pick a point $x(q) \in \mathcal{N}_\varepsilon$ smoothly in q such that $|x(q) - q| \leq C \cdot \operatorname{dist}(q, W^S_{\text{loc}}(\mathcal{S}_\varepsilon))$ for some $C = C(\varepsilon) > 0$. We compare the evolution of the $\mathbf{y}$ component of $q \cdot \tau$ (denoted $\mathbf{y}(\tau)$) to that of the $\mathbf{y}$ component of $x(q)$ (denoted $\mathbf{y}^x(\tau)$). The closeness property of $x(q)$ and estimate (4.1) with $\tau = 0$ directly imply

$$|\mathbf{y}(0) - \mathbf{y}^x(0)| \leq \frac{C\Delta}{K_a} e^{-\lambda \tau_q^{\mathcal{B}}}.$$

Then, for $\tau > 0$, the difference $\mathbf{y}(\tau) - \mathbf{y}^x(\tau)$ satisfies:

$$(\mathbf{y} - \mathbf{y}^x)' = \mathbf{h}(\mathbf{y}, \varepsilon) - \mathbf{h}(\mathbf{y}^x, \varepsilon) + \mathbf{H}(\mathbf{a}, \mathbf{b}, \mathbf{y}, \varepsilon) \otimes \mathbf{a} \otimes \mathbf{b}. \tag{4.2}$$

If one Taylor expands $\mathbf{h}(\mathbf{y}^x, \varepsilon)$ about $\mathbf{y}$, uses natural bounds $K_D|\mathbf{y} - \mathbf{y}^x|$ and $K_N|\mathbf{y} - \mathbf{y}^x|$ on the linear and nonlinear terms from this expansion, and bounds the remaining term $|\mathbf{Hab}| \leq K_H|\mathbf{a}||\mathbf{b}|$ for some $K_H > 0$, then Gronwall's inequality applied to (4.2) yields:

$$\begin{aligned}
|\mathbf{y}(\tau) - \mathbf{y}^x(\tau)| \leq {} & K_D \frac{C\Delta}{K_a} e^{-\lambda \tau_q^{\mathcal{B}} + \hat{\alpha}\tau} \\
& + K_D \int_0^\tau e^{\hat{\alpha}(\tau - t)} \left[\frac{K_N C^2(\varepsilon)\Delta^2}{K_a^2} e^{-2\lambda\tau_q^{\mathcal{B}} + 2(\lambda - \kappa)t} \right. \\
& \left. \qquad + \frac{K_H \Delta^2 K_b}{K_a} e^{-\lambda(\tau_q^{\mathcal{B}} - t) + \gamma t} \right] dt.
\end{aligned}$$

Then, carrying out the integrals and discarding the contributions from the lower limits of integration (since upper bounds are sought), we find:

$$\begin{aligned}
|\mathbf{y}(\tau) - \mathbf{y}^x(\tau)| \leq {} & K_D e^{-\lambda\tau_q^{\mathcal{B}}} e^{(\lambda - \kappa)\tau} \cdot \\
& \cdot \left[\frac{C\Delta}{K_a} e^{(\hat{\alpha} - \lambda + \kappa)\tau} + \frac{K_N C^2(\varepsilon)\Delta^2}{K_a^2(2\lambda - 2\kappa - \hat{\alpha})} e^{-\lambda\tau_q^{\mathcal{B}} + (\lambda - \kappa)\tau} \right. \\
& \left. + \frac{K_H \Delta^2 K_b}{K_a(\lambda + \gamma - \hat{\alpha})} e^{(\gamma + \kappa)\tau} \right].
\end{aligned}$$

Finally, recalling that it was assumed that $0 < \kappa < \lambda - \hat{\alpha}$ and $\kappa < |\gamma|$, so that the exponential factors in the first and third terms are smaller than one and so that the exponential in the second term may be rewritten in terms of $e^{-\kappa\tau_q^{\mathcal{B}}}$ and a factor smaller than one, we find:

$$\begin{aligned}
|\mathbf{y}(\tau) - \mathbf{y}^x(\tau)| \leq {} & K_D e^{-\lambda\tau_q^{\mathcal{B}}} e^{(\lambda - \kappa)\tau} \left[\frac{C\Delta}{K_a} + \frac{K_N C^2 \Delta^2}{K_a^2(2\lambda - 2\kappa - \hat{\alpha})} e^{-\kappa\tau_q^{\mathcal{B}}} \right. \\
& \left. + \frac{K_H \Delta^2 K_b}{K_a(\lambda + \gamma - \hat{\alpha})} \right].
\end{aligned} \tag{4.3}$$

This last estimate can now be used to complete the proof of the desired result, as follows. In particular, we claim that for $\delta_0 > 0$ sufficiently small, if $\delta \in [0, \delta_0]$, $q \in \mathcal{Q}_\varepsilon^\delta$ and $\tau \in [0, \tau]$, then

$$|\mathbf{y}(\tau) - \mathbf{y}^x(\tau)| \leq \bar{C}(\varepsilon) e^{-\lambda \tau_q^{\mathcal{B}}} e^{(\lambda - \kappa)\tau}, \tag{4.4}$$

where $\bar{C}(\varepsilon) = K_D \left[\frac{C\Delta}{K_a} + 1 + \frac{K_H \Delta^2 K_b}{K_a(\lambda + \gamma - \tilde{\alpha})} \right]$. Indeed, we know this holds at $\tau = 0$. Moreover, we may assume, by continuity, that there exists a τ' such that (4.4) holds for $\tau \in [0, \tau']$. Then, the estimate (4.3) actually shows that the inequality in (4.4) is strict at τ', because the time estimate (TE) implies that we can make δ_0 small enough so that $\frac{K_N C^2 \Delta^2}{K_a^2(2\lambda - 2\kappa - \tilde{\alpha})} e^{-\kappa \tau_q^{\mathcal{B}}} < 1$. Hence, the τ interval on which this inequality holds can be enlarged, and then the same estimates can again be used to show that the inequality is strict also on the enlarged interval. Indeed, this procedure may be carried out finitely often so that the whole interval $[0, \tau_q^{\mathcal{B}}]$ is covered. Therefore, we have

$$|\mathbf{y}(\tau_q^{\mathcal{B}}) - \mathbf{y}^x(\tau_q^{\mathcal{B}})| \leq \bar{C}(\varepsilon) e^{-\kappa \tau_q^{\mathcal{B}}},$$

and the proof of Proposition 4.1 is complete. □

An immediate corollary is that under the conditions of Proposition 4.1, for every $q \in \mathcal{Q}_\varepsilon^\delta$, there exists a point $r(q)$ on the good submanifold $W_{\mathrm{loc}}^U(\mathcal{S}_\varepsilon)|_{\mathcal{J}_\varepsilon \cdot \tau_q^{\mathcal{B}}}$ that is exponentially close to the image $(q \cdot \tau_q^{\mathcal{B}})$ of q on exit of $\mathcal{B}$:

$$|r(q) - q \cdot \tau_q^{\mathcal{B}}| \leq K(\Delta, \varepsilon) e^{-\kappa \tau_q^{\mathcal{B}}}$$

for some positive constant $K(\Delta, \varepsilon)$ that is independent of q. Using the notation of Proposition 4.1, we may in fact determine such a point $r(q)$ explicitly as $r(q) = \big(\mathbf{a}(q \cdot \tau_q^{\mathcal{B}}), \mathbf{0}, \mathbf{y}(x(q) \cdot \tau_q^{\mathcal{B}})\big)$. In this way, each point $q \cdot \tau_q^{\mathcal{B}}$ has a partner point on the correct submanifold of $W_{\mathrm{loc}}^U(\mathcal{S}_\varepsilon)$, and the existence of this partner point may be exploited if necessary in the further tracking of $\mathcal{M}_\varepsilon$.

4.3. The Exchange Lemma for systems (2.1) and (3.5). We now state the Exchange Lemma for systems (2.1) and (3.5). This Lemma establishes the desired result, Theorem 4.2 below, namely that the C^0 closeness of the two manifolds $\mathcal{M}_\varepsilon$ and $W_{\mathrm{loc}}^U(\mathcal{S}_\varepsilon)|_{\mathcal{J}_\varepsilon \cdot \tau_q^{\mathcal{B}}}$ given by Proposition 4.1 is actually C^1, with a similar exponential estimate.

Since we work with differential forms in order to track the tangent planes, we will make use of the notion of induced derivation from the linear algebra. Briefly, let $\mathcal{V}$ denote a q-dimensional vector space, and consider a set of p vectors $v_1, \ldots, v_p \in \mathcal{V}$, where $1 \leq p \leq q$. Let $\mathcal{A} : \mathcal{V} \to \mathcal{V}$ be a linear map. Then, the induced derivation is the following map on p-forms:

$$\mathcal{A}^{(p)}(v_1 \wedge \cdots \wedge v_p) = \Sigma_{i=1}^p v_1 \wedge \cdots \wedge \mathcal{A} v_i \wedge \cdots \wedge v_p.$$

Observe that $\text{Tr}\mathcal{A} = \mathcal{A}^{(q)}$. Standard algebra texts give the following example (see also [24]): Consider $\mathcal{V} = \mathbb{R}^3$ with the standard basis $\{e_1, e_2, e_3\}$ and use $a_{11}, \ldots, a_{33}$ to denote the 3×3 matrix $\mathcal{A}$ in terms of this basis. Then, using the ordered basis $\{e_1 \wedge e_2, e_2 \wedge e_3, e_3 \wedge e_1\}$ for the space of 2-forms, the matrix representation of $\mathcal{A}^{(2)}$ is:

$$\Big\{\{a_{11}+a_{22}, -a_{13}, -a_{23}\}, \{-a_{31}, a_{22}+a_{33}, -a_{21}\}, \{-a_{32}, -a_{12}, a_{33}+a_{11}\}\Big\}.$$

The first column of $\mathcal{A}^{(2)}$, for example, is calculated from $\mathcal{A}e_1 \wedge e_2 + e_1 \wedge \mathcal{A}e_2 = (-a_{31}, 0, a_{11})^T + (0, -a_{32}, a_{22})^T$, and the first entry in this column is given by the third component (the e_3 component) of this sum, because $e_1 \wedge e_2 = e_3$, the second entry by the component in the direction of $e_1 = e_2 \wedge e_3$, and so forth.

We use this notation for the $\ell \times \ell$ matrix $D_{\mathbf{y}}\mathbf{h}(\mathbf{y}, 0)$. In particular, we need the additional spectral bounds, which exist due to the earlier assumptions and condition (H1):

$$\tag{4.5} \begin{aligned} \alpha_0 &= \sup \bigcup_{\mathbf{y}\in\bar{U}} \text{Re}\left(\text{spec } (D_{\mathbf{y}}\mathbf{h}(\mathbf{y}, 0))^{(\sigma)}\right), \\ \beta_0 &= \inf \bigcup_{\mathbf{y}\in\bar{U}} \text{Re}\left(\text{spec } (D_{\mathbf{y}}\mathbf{h}(\mathbf{y}, 0))^{(\sigma)}\right). \end{aligned}$$

Also, we set $\rho_0 = \hat{\alpha}_0$ if $\sigma < \ell$, and $\rho_0 = |\gamma_0|$ if $\sigma = \ell$. These spectral bounds enable us to identify the spectral gap conditions needed for control over the evolution of the tangent planes, with α_0 playing the analogous role for the tracking of tangent planes that $\hat{\alpha}_0$ plays for the C^0 tracking.

Theorem 4.2. (Theorem 3.12 in [24].) Fix an integer $1 \leq \sigma \leq \ell$. Assume that

$$\begin{aligned} \lambda_0 &> \max(\hat{\alpha}_0, \rho_0 + \alpha_0 - \beta_0, \alpha_0 - \beta_0), \\ |\gamma_0| &> \alpha_0 - \beta_0. \end{aligned}$$

Let $\mathcal{M}_\varepsilon$ be a manifold of the type assumed in Section 4.1, and let $\kappa > 0$ be chosen so that

$$\kappa < \min\Big(\lambda_0 - \hat{\alpha}_0, |\gamma_0|, \lambda_0 - (\rho_0 + \alpha_0 - \beta_0), \lambda_0 - (\alpha_0 - \beta_0), |\gamma_0| - (\alpha_0 - \beta_0)\Big).$$

Then, there exist $\Delta_0, \varepsilon_0 > 0$ such that, whenever $\Delta \in (0, \Delta_0)$ and hypotheses (T1), (TE) and (T2) are satisfied, for each $\varepsilon \in (0, \varepsilon_0)$ there exists a $\delta_0 = \delta_0(\varepsilon) > 0$ so that for $\delta \in [0, \delta_0]$ and $q \in \mathcal{Q}_\varepsilon^\delta$

$$\mathcal{M}_\varepsilon \cdot \tau_q^{\mathcal{B}} \text{ is } C^1 \ \mathcal{O}(e^{-\kappa \tau_q^{\mathcal{B}}}) \text{ close to } W^U_{\text{loc}}(\mathcal{S}_\varepsilon)|_{\mathcal{J}_\varepsilon \cdot \tau_q^{\mathcal{B}}} \text{ at } q \cdot \tau_q^{\mathcal{B}}.$$

Remark. The assumptions on λ_0 and $|\gamma_0|$ are made for the following reasons. First, $\lambda_0 > \hat{\alpha}_0$ implies that there exists a spectral gap between the minimum rate of exponential expansion in the fast normal directions and the maximum rate of exponential expansion in the directions tangential to $\mathcal{S}_\varepsilon$, as is needed for the estimates on the unstable directions. Next, $\lambda_0 > \alpha_0 - \beta_0$ guarantees that there is gap between the minimum rate of exponential expansion in the fast normal directions and the total possible rate of growth along tangential directions in both forward and backward time.

4.4. Brief overview of the proof of Theorem 4.2. The main goal of the proof is to track the tangent planes to the manifold $\mathcal{M}_\varepsilon$. We begin with the equations of first variations of (2.1) that govern the evolution of individual tangent vectors:

$$\begin{aligned} d\mathbf{a}' &= \Lambda d\mathbf{a} + D_\mathbf{z}\Lambda \otimes d\mathbf{z} \otimes \mathbf{a} \\ d\mathbf{b}' &= \Gamma d\mathbf{b} + D_\mathbf{z}\Gamma \otimes d\mathbf{z} \otimes \mathbf{b} \\ d\mathbf{y}' &= D_\mathbf{y}\mathbf{h} d\mathbf{y} + \mathbf{H} \otimes d\mathbf{a} \otimes \mathbf{b} + \mathbf{H} \otimes \mathbf{a} \otimes d\mathbf{b} + D_\mathbf{z}\mathbf{H} \otimes d\mathbf{z} \otimes \mathbf{a} \otimes \mathbf{b}, \end{aligned} \tag{4.6}$$

where $\mathbf{z} = (\mathbf{a}, \mathbf{b}, \mathbf{y})$. These equations can also be used to determine the evolution of the $k + \sigma$ dimensional hyperplanes that are the tangent spaces of the $k + \sigma$ dimensional tracked manifolds $\mathcal{M}_\varepsilon$. At any point in a neighborhood $\mathcal{B}$ of $\mathcal{S}_\varepsilon$, such a tangent plane is determined by $k + \sigma$ differential forms (dual to the $k + \sigma$ dimensional tangent spaces), and for the tracking here it is natural to separate the basis forms into two classes.

Let

$$B = \{\mathbf{e}_1^a, \ldots, \mathbf{e}_k^a, \mathbf{e}_1^b, \ldots, \mathbf{e}_m^b, \mathbf{e}_1^y, \ldots, \mathbf{e}_\ell^y\}$$

denote an ordered basis for $\mathbb{R}_a^k \times \mathbb{R}_b^m \times \mathbb{R}_y^\ell$. The first class of basis forms consists of those $k + \sigma$ forms that involve all of the k fast unstable one forms $d\mathbf{e}_1^a, \ldots, d\mathbf{e}_k^a$, and any subset of σ of the ℓ slow/center one forms $d\mathbf{e}_{r_1}^y, \ldots, d\mathbf{e}_{r_\sigma}^y$. We use the variable $\mathbf{Z}$ to denote any member of this first class. The second class consists of all other $k + \sigma$ forms, and we use $\mathbf{X}$ to denote its members.

From the equation of first variations (4.6), one obtains the following equations for $\mathbf{Z}$ and $\mathbf{X}$:

$$\begin{aligned} \mathbf{Z}' &= \left(\mathrm{Tr}\Lambda + \varphi + D^{(\sigma)}\mathbf{h} + \Theta_1\right)\mathbf{Z} + \Theta_2\mathbf{X}, \\ \mathbf{X}' &= (B + \Psi_2)\,\mathbf{X} + \Psi_1\mathbf{Z}, \end{aligned} \tag{4.7}$$

where $\varphi(\mathbf{z}, \varepsilon) = \Sigma_{i,I=1}^k a_I(\partial\Lambda_{iI}/\partial a_i)$, the matrix

$$\begin{pmatrix} \mathrm{Tr}\Lambda(\mathbf{z},\varepsilon) + D^{(\sigma)}\mathbf{h} & \mathbf{0} \\ \mathbf{0} & B(\mathbf{z},\varepsilon) \end{pmatrix} = \begin{pmatrix} \Lambda(\mathbf{z},\varepsilon) & \mathbf{0} & \mathbf{0} \\ \mathbf{0} & \Gamma(\mathbf{z},\varepsilon) & \mathbf{0} \\ \mathbf{0} & \mathbf{0} & D_\mathbf{y}\mathbf{h}(\mathbf{y},\varepsilon) \end{pmatrix}^{(k+\sigma)},$$

and the Ψ_i and Θ_i are matrcies of the appropriate orders. These equations are coupled but linear in $\mathbf{Z}$ and $\mathbf{X}$. After performing a scaling of $\mathbf{Z}$ and

$\mathbf{X}$ so that they remain bounded during the time intervals when orbits on $\mathcal{M}_\varepsilon$ are near $\mathcal{S}_\varepsilon$ (see Section 2.4 of [24]), one derives natural exponential upper bounds on the scaled $\mathbf{Z}(\tau)$ and $\mathbf{X}(\tau)$ in terms of α_0, λ_0, and $|\gamma_0|$ for the desired forward time intervals. In addition, one obtains a lower bound on the scaled $\mathbf{Z}(\tau)$ as follows. For each $q \in \mathcal{Q}_\varepsilon^\delta$, one may choose a point $x(q)$ in $\mathcal{N}_\varepsilon$ exponentially close to it, and this point may be chosen smoothly in q, see Lemma 3.10 in [24]. Let $\mathbf{P}(\tau)$ be a $k+\sigma$ form that satisfies the truncated system

$$\mathbf{P}' = D^{(\sigma)}\mathbf{h}\mathbf{P}, \tag{4.8}$$

with $\mathbf{P}(0) = \mathbf{Z}(T_{x(q)}\mathcal{M}_\varepsilon)$. This system governs the evolution of tangent planes in the fast unstable and center-slow directions for orbits restricted to $\mathcal{S}_\varepsilon$ Assumptions (T1) and (T2) guarantee that $\mathbf{P}(0)$ is bounded from below and then an estimate (see Section 3.2 of [24]) shows the desired result, namely that also the scaled $\mathbf{Z}$ form is bounded from below.

For a complete presentation of the proof of Theorem 4.2, we refer the reader to Sections 2.3, 2.4, and 3.2 of [24] and to [15].

4.5. Some alternatives to assumption (T1). There are important systems (2.1) in which the invariant manifolds $\mathcal{M}_\varepsilon$ that one wishes to track do not intersect $W^S_{\text{loc}}(\mathcal{S}_\varepsilon)$, *i.e.*, they do not satisfy assumption (T1). Nevertheless, Exchange Lemma type tracking estimates can be obtained for some of these.

The first example arises in the class of perturbed Hamiltonian systems analyzed in [2]. There, the tracked manifolds are instead C^1 $\mathcal{O}(\varepsilon^\eta)$ close (for some $\eta > 0$) to $W^S_{\text{loc}}(\mathcal{S}_\varepsilon)$ on the entry face of a neighborhood $\mathcal{B}$, and it is shown by the same type of estimates as used to prove Theorem 4.2 that the tracked manifolds are C^1 $\mathcal{O}(\varepsilon^{\eta-\zeta})$ close to the appropriate submanifold of $W^U_{\text{loc}}(\mathcal{S}_\varepsilon)$ on exit from $\mathcal{B}$, for any positive number ζ. In [2], the times $\tau_q^{\mathcal{B}}$ are typically of the order of $\mathcal{O}(\ln(1/\varepsilon))$. These C^1 tracking estimates are in turn then used to establish the existence of multiple-pulse orbits homoclinic to hyperbolic equilibria in resonance bands.

Another example arises in establishing the existence of multiple-pulse periodic orbits in special fast-slow systems and in general fast-slow systems, see [21, 22, 23]. The periodic orbits found in these works lie close to multiple-pulse and multi-bump homoclinic orbits, but not in the intersections of the stable and unstable manifolds of normally hyperbolic invariant manifolds. Hence, hypothesis (T1) is not satisfied. Instead, these periodic orbits are found by first considering an augmented system (in which the number of dependent variables is doubled, and the doubled variables satisfy trivial ODEs), and then by transforming the Poincaré map problem into a boundary value problem for the augmented system. Appropriately chosen sets of initial conditions, which are kept track of by the doubled dependent variables, are then identified in neighborhoods of slow manifolds, and a modified Exchange Lemma is developed that provides sharp estimates for tracking the manifolds of solutions through these sets. The conclusions of the modified Lemma are of the same type as that of Theorem 4.2, namely they give C^1 exponential closeness of the tracked invariant manifold to the appropriate submanifold of the local unstable manifold in forward time and a similar closeness result to a submanifold of the local stable manifold in backward time. See also Theorem 4.2 of [4] for an application to stationary, spatially-periodic solutions of a set of coupled reaction-diffusion equations.

REFERENCES

[1] A. Bose, Symmetric and antisymmetric pulses in parallel coupled nerve fibers, *SIAM J. Appl. Math.*, **55** (1995), 1650–1674.

[2] R. Camassa, G. Kovacic, and S.-K. Tin, A Melnikov method for homoclinic orbits with many pulses, *Arch. Rat. Mech. Anal.*, **143** (1998), 105–193.

[3] B. Deng, The Silnikov problem, exponential expansion, strong λ-Lemma, C^1 linearization, and homoclinic bifurcation, *J. Diff. Eq.*, **79** (1989), 189–231.

[4] A. Doelman, T.J. Kaper, and P. Zegeling, Pattern formation in the one-dimensional Gray-Scott model, *Nonlinearity*, **10** (1997), 523–563.

[5] N. Fenichel, Persistence and smoothness of invariant manifolds for flows, *Indiana Univ. Math. J.*, **21** (1971), 193–226.

[6] N. Fenichel, Asymptotic stability with rate conditions, *Indiana Univ. Math. J.*, **23** (1974), 1109–1137; Asymptotic stability with rate conditions, II, *Indiana Univ. Math. J.*, **26** (1977), 81–93.

[7] N. Fenichel, Geometrical singular perturbation theory for ordinary differential equations, *J. Diff. Eq.*, **31** (1979), 53–98.

[8] G. Haller, *Chaos near resonance*, Applied Mathematical Sciences Series, **138** (1999) (Springer-Verlag, New York).

[9] M. Hayes, T.J. Kaper, N. Kopell, and K. Ono, On the application of geometric singular perturbation theory to some classical two-point boundary value problems, *Int. J. Bif. Chaos*, **8** (1998), 189–209.

[10] M. Hirsch, C. Pugh, and M. Shub, *Invariant manifolds*, Lecture Notes in Mathematics, **583** (1977) (Springer-Verlag, New York).

[11] C.K.R.T. Jones, Geometric singular perturbation theory, in *Dynamical systems, Montecatini Terme*, Lecture Notes in Mathematics, R. Johnson, ed. (Springer-Verlag, New York), **1609** (1994), 44–118.

[12] C.K.R.T. Jones, T.J. Kaper, and N. Kopell, Tracking invariant manifolds up to exponentially small errors, *SIAM J. Math. Anal.*, **27** (1996), 558–577.

[13] C.K.R.T. Jones and N. Kopell, Tracking invariant manifolds with differential forms in singularly perturbed systems, *J. Diff. Eq.*, **108** (1994), 64–88.

[14] C.K.R.T. Jones, N. Kopell, and R. Langer, Construction of the FitzHugh-Nagumo pulse using differential forms, in *Patterns and dynamics in reactive media*, IMA volumes in Mathematics and its Applications, H. Swinney, R. Aris, and D. Aronson, eds. (Springer-Verlag, New York), **37** (1991), 101–116.

[15] C.K.R.T. Jones and S.K. Tin, Generalized Exchange Lemmas and orbits heteroclinic to invariant manifolds, submitted, *Mem. AMS*, 2000.

[16] T.J. Kaper, An introduction to geometric methods and dynamical systems theory for singular perturbation problems, in *Analyzing multiscale phenomena using singular perturbation methods*, Proceedings of Symposia in Applied Mathematics, J. Cronin and R.E. O'Malley, Jr., eds. (American Mathematical Society, Providence, RI), **56** (1999), 85–132.

[17] T.J. Kaper and G. Kovacic, Multi-bump orbits homoclinic to resonance bands, *Trans. AMS*, **348** (1996), 3835–3887.

[18] J. Palis and W. DeMelo, *Geometric theory of dynamical systems* (Springer-Verlag, New York), 1982.

[19] J.E. Rubin and C.K.R.T. Jones, Existence of standing pulse solutions to an inhomogeneous reaction-diffusion system, *J. Dyn. Diff. Eq.*, **10** (1998), 1–35.

[20] K. Sakamoto, Invariant manifolds in singular perturbation problems for ordinary differential equations, *Proc. Roy. Soc. Edin.*, **A 116** (1990), 45–78.

[21] C. Soto-Trevino and T.J. Kaper, Higher-order Melnikov theory for adiabatic systems, *J. Math. Phys.*, **37** (1996), 6220–6249.

[22] C. Soto-Trevino, Geometric methods for periodic orbits in singularly perturbed systems, Ph.D. thesis, Boston University, 1998.

[23] C. Soto-Trevino, A geometric method for periodic orbits in singularly perturbed systems, this volume, 2000.

[24] S.K. Tin, On the dynamics of tangent spaces near a normally hyperbolic invariant manifold, Ph.D. thesis, Division of Applied Mathematics, Brown University, 1994.

[25] S.K. Tin, C.K.R.T. Jones, and N. Kopell, Invariant manifolds and singularly perturbed boundary value problems, *SIAM J. Num. Anal.*, **31** (1994), 1558–1576.

[26] S. Wiggins, *Normally hyperbolic invariant manifolds in dynamical systems,* Applied Mathematical Sciences Series (Springer-Verlag, New York), **105** (1994).

GEOMETRIC ANALYSIS OF THE SINGULARLY PERTURBED PLANAR FOLD*

M. KRUPA† AND P. SZMOLYAN†

Abstract. The geometric approach to singular perturbation problems is based on powerful methods from dynamical systems theory. These techniques have been very successful in the case of normally hyperbolic critical manifolds. However, at points where normal hyperbolicity fails, e.g. fold points or points of self-intersection of the critical manifold, the well developed geometric theory does not apply. We present a method based on blow-up techniques which leads to a rigorous geometric analysis of these problems. The blow-up method leads to problems which can be analysed by standard methods from the theory of invariant manifolds and global bifurcations. The presentation is in the context of a planar singularly perturbed fold. The blow-up used in the analysis is closely related to the rescalings used in the classical analysis based on matched asymptotic expansions. The relationship between these classical results and our geometric analysis is discussed.

1. Introduction. In a rather broad sense singular perturbation problems may be defined as problems described by ordinary or partial differential equations with solutions varying on two or more very different temporal and/or spatial scales. In analogy with fluid dynamics the narrow regions of rapid temporal and/or spatial variation are called layers. Typically, there exists a small parameter ε in the governing differential equations such that the layer behaviour becomes more pronounced as ε tends to zero. By neglecting effects on some of these scales, i.e. by setting $\varepsilon = 0$ in the original problem, one obtains a - hopefully simpler - limit problem, more amenable to analysis and/or simulation. Typically, the solution of the limit problem does not satisfy all initial and/or boundary conditions of the original problem. Therefore, additional terms, varying on different scales and governed by different limit problems, must be introduced. This - often very sophisticated - procedure leads to the construction of a formal asymptotic expansion of the solution of the original problem (see [3, 15, 21]). In a second step, the validity of the asymptotic expansion must be proved, i.e. one has to prove that the asymptotic expansion approximates the solution of the original problem with an error tending to zero at a certain rate as ε goes to zero.

This general procedure is one of the basic principles of applied mathematics. Due to substantial technical difficulties the second step is often omitted. Even in these cases the formal asymptotic expansions can provide valuable insights. Yet we feel that obtaining a clear qualitative description of the dynamic phenomena occurring in this problem is very important and can serve as a guiding light in obtaining quantitative approximations or nu-

*Research supported by the Austrian Science Foundation under grant Y 42-MAT.

†Institut für Angewandte und Numerische Mathematik, TU Wien, Wiedner Hauptstraße 8-10/115/1, A-1040 Wien, Austria.

merical descriptions. In fact one of the goals of this article is to indicate how expansions of the solutions can be obtained based on the geometric method.

In the following we present a brief introduction to geometric singular perturbation theory which has proven to be very successful in this endeavor. We consider singularly perturbed ordinary differential equations in the standard form

$$\begin{aligned} \varepsilon\dot{x} &= f(x,y,\varepsilon), \\ \dot{y} &= g(x,y,\varepsilon), \end{aligned} \qquad x \in \mathbb{R}^n, \quad y \in \mathbb{R}^k, \quad 0 < \varepsilon \ll 1 \tag{1.1}$$

where f, g are C^k-functions with $k \geq 2$. Equations of this form arise frequently in applications, e.g. biology, chemistry, electrical engineering, economics and mechanics. By transforming from the slow time scale τ to the fast time scale $t := \tau/\varepsilon$ on obtains the equivalent system

$$\begin{aligned} x' &= f(x,y,\varepsilon), \\ y' &= \varepsilon g(x,y,\varepsilon). \end{aligned} \tag{1.2}$$

One tries to analyse the dynamics of (1.1) by suitably combining the dynamics of the *reduced problem*

$$\begin{aligned} 0 &= f(x,y,0), \\ \dot{y} &= g(x,y,0), \end{aligned} \tag{1.3}$$

and the dynamics of the *layer problem*

$$\begin{aligned} x' &= f(x,y,0), \\ y' &= 0, \end{aligned} \tag{1.4}$$

which are the limiting problems for $\varepsilon = 0$ on the slow and the fast time scales, respectively.

The foundation of the geometric approach to singular perturbation theory was given by N. Fenichel [9]. The basic reasoning is as follows. The reduced problem (1.3) is a dynamical system on the set $S := \{(x,y) \in \mathbb{R}^{n+k} : f(x,y,0) = 0\}$. In the following we refer to S as the *critical manifold*. A normally hyperbolic invariant manifold of fixed points $S_0 \subset S$ of the layer problem (1.4) persists as a locally invariant slow manifold S_ε of (1.1) for ε sufficiently small. The restriction of (1.1) to S_ε is a small smooth perturbation of the reduced problem (1.3).

Connections between critical manifolds $S_{i,0}$, $i = 1,2$ generated by heteroclinic orbits of the layer problem (1.4) correspond to layers of the full problem (1.1) realized as intersections of the unstable manifold of $S_{1,\varepsilon}$ and the stable manifold of $S_{2,\varepsilon}$ (see [16, 24]).

In many applications the dynamics of the lower-dimensional problems (1.3) and (1.4) can be analysed in sufficient detail to understand the existence and bifurcation of global singular structures, e.g. singular periodic orbits, singular heteroclinic and homoclinic orbits, composed of orbits of (1.3) and (1.4). Under appropriate hyperbolicity and transversality conditions this allows to analyse the persistence of these global structures and their bifurcations. For an excellent introduction to geometric singular perturbation theory and an overview of applications we refer to the survey by C. Jones [12]). We would also like to refer to related articles in this proceedings.

For the case of normally hyperbolic critical manifolds S the geometric theory is fairly complete. However, despite many efforts, points on the critical manifold S where normal hyperbolicity breaks down remained a major obstacle to the geometric theory. This was a definite shortcoming of the theory in view of the abundance of nonhyperbolic points in applications. Loss of normal hyperbolicity primarily occurs due to:

1. bifurcation points of the critical manifold S associated with zero eigenvalues of the Jacobian $\frac{\partial f}{\partial x}$, i.e. fold points or points of self-intersection of S;
2. points in S where the layer problem undergoes a Hopf bifurcation, due to a pair of eigenvalues of the Jacobian $\frac{\partial f}{\partial x}$ crossing the imaginary axis upon varying y.
3. branches of S along which the variable x becomes unbounded, causing an arbitrarily small angle between the fibers $y = \text{const.}$ of the layer problem and the tangent space of S.

A well known example involving 1.) are relaxation oscillations [10, 17], i.e. problems with solutions slowly moving towards a fold point, jumping from the fold point to another stable branch of S, following the slow dynamics again until another fold point is reached, jumping again etc., thus, possibly forming periodic solutions as in the case of van der Pol's equation.

The second scenario leads (in analytic systems) to *delayed Hopf bifurcation* described in A. Neishtadt's article in this volume (see also [4, 19]).

Phenomena due to an unbounded branch of S have been investigated in less detail. This situation is often due to inadequate scaling of the variables and can be treated by suitable rescaling to a certain extent, however some non-hyperbolicity is likely to remain. We refer to the article by A. Milik and the second author in this volume for an example where this occurs and to [25] for a detailed geometric analysis of such a problem.

A closely related - but even more delicate - phenomenon is that of *canards*. A canard solution is a solution of a singularly perturbed system which passes close to a bifurcation point of the critical manifold and follows, rather surprisingly, a repelling slow manifold for a considerable amount of time. In geometric terms a (maximal) canard solution corresponds to the intersection of an attracting and a repelling slow manifold near a nonhyperbolic point of S. The planar case is degenerate and canards can occur persistently only in one-parameter families of planar vector fields. However, in dimensions larger than two, canards can occur in generic situations.

The discovery and the first analysis of this phenomenon is based on nonstandard analysis and is due to a group of nonstandard analysts centered around M. Diener, F. Diener, E. Callot and E. Benoit. For more information on their approach and results we refer to [4, 5]. For a treatment of canards by means of matched asymptotic expansions we refer to [2, 8, 18]. See also the survey in [1].

For a long time it was not clear how to apply geometric theory in the case of loss of hyperbolicity, in particular to study canards. A breakthrough came with the work of F. Dumortier and R. Roussarie [7] on canard cycles. They give a detailed geometric explanation and proof of the canard phenomenon by using blow-up of singularities and foliations by center manifolds as main techniques. For more details on their approach we also refer to their article in this volume. The method is very powerful but also quite intricate. In order to be able to apply it to other possibly higher-dimensional singular perturbation problems and to obtain a better general understanding of singular perturbation problems we would like to

a) make the method as simple as possible,

b) link it to the existing geometric theory, and

c) relate it to classical methods of asymptotic expansions.

Recently, we have obtained many new results in this direction, some of which were announced at this workshop. Full details will be available in a series of forthcoming papers.

In this article we will analyse the dynamics and asymptotics of a planar singularly perturbed system in the neighborhood of a generic fold. Asymptotic analysis of this innocent looking problem has attracted a lot of attention in the past, starting with the work of Pontryagin [22]. A substantial part of the book by Mishchenko and Rozov [17] contains a rather involved description of the precise asymptotics (see also [10, 13, 20]).

We feel that our geometric analysis provides a new understanding of the problem. While some of our results are contained in [7] our detailed description seems to be new. Our approach, although similar in essence, draws much more on the existing geometric singular perturbation theory and thus can be naturally seen as its extension. This article can serve as an introduction to the blow-up method and can be used as a starting point for studying literature on more delicate problems or applications, e.g. [7, 14, 25, 26]. A complementary introduction to the theory is given in the article of Dumortier and Roussarie in this volume.

We now briefly explain the organization of the rest of the paper. In Section 2 we specify the problem which we will study and formulate our results. In Section 3 we define and explain the blow-up transformation which is used in the following. The nonhyperbolic point is "blown-up" to an invariant two sphere. To study the dynamics of the induced vector field in a neighborhood of the sphere we introduce three charts, which describe different parts of the sphere. A chart K_1 is used to study the attracting slow manifold as it approaches the fold point, in the second chart K_2 we analyse how the attracting slow manifold moves along the sphere. The third chart K_3 is used to study the departure from the sphere. This is done in Sections 4–6. In Section 7 these results are put together to prove the main results.

2. Assumptions and results. Consider the singularly perturbed ODE

$$\begin{aligned} x' &= f(x, y, \varepsilon), \\ y' &= \varepsilon g(x, y, \varepsilon), \end{aligned} \tag{2.1}$$

where $(x, y) \in \mathbb{R}^2$ and ε is a small real parameter. Suppose that (x_0, y_0) is such that

$$f(x_0, y_0, 0) = 0, \qquad \frac{\partial f}{\partial x}(x_0, y_0, 0) = 0. \tag{2.2}$$

Our goal is to obtain a characterization of the local dynamics and the asymptotics in the neighborhood of (x_0, y_0) for sufficiently small values of ε. We make the following non-degeneracy assumptions:

$$\begin{aligned} &\frac{\partial^2 f}{\partial x^2}(x_0, y_0, 0) \neq 0, \\ &\frac{\partial f}{\partial y}(x_0, y_0, 0) \neq 0, \\ &g(x_0, y_0, 0) \neq 0. \end{aligned} \tag{2.3}$$

We assume, without loss of generality, that

$$(x_0\, y_0) = (0, 0), \qquad \frac{\partial^2 f}{\partial x^2}(0, 0, 0) > 0, \qquad \frac{\partial f}{\partial y}(x_0, y_0, 0) < 0$$

hold. As before let

$$S = \{(x, y) : f(x, y, 0) = 0\}$$

be the *critical manifold*. The non-degeneracy assumptions imply that there exists a neighborhood U of the origin such that $(0, 0)$ is the only point in $U \cap S$, where $\frac{\partial f}{\partial x}$ vanishes and that the $S \cap U$ is approximately a parabola. Let S_a resp. S_r denote its left resp. right branch, so that $S = S_a \cup S_r$ (see Fig. 1). The assumption $\frac{\partial^2 f}{\partial x^2}(0, 0, 0) > 0$ implies that for $y > 0$ the branch S_a is attracting and the branch S_r is repelling for the layer problem, which also explains the notation. The origin is nonhyperbolic, weakly attracting from the left and weakly repelling to the right. The direction of the reduced flow is determined by the sign of $g(0,0,0)$. We treat the case $g(0,0,0) < 0$ and comment on the other case later. This implies $y' < 0$ for the full problem (2.1) as well as $\dot{y} < 0$ for the reduced problem, i.e. the reduced flow on S_a and S_r is directed towards the fold point. This is the situation relevant to relaxation oscillations. Actually, orbits on S_a and S_r reach the fold point in finite time. The only possibility to continue from there in the singular limit is along the (weakly) unstable fiber of the layer problem along the positive x-axis. Thus, the curve

$$S_a \cup \{(x, 0), x > 0\}$$

is expected to be a zeroth order approximation.

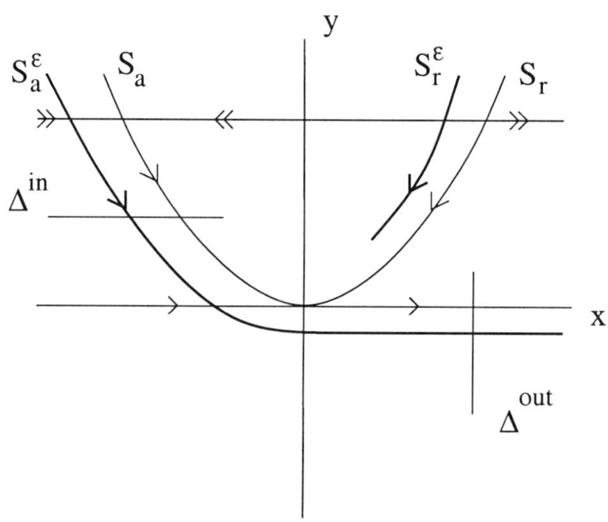

FIG. 1. *Critical manifold, slow manifolds and and sections for the fold point.*

It follows from the standard theory [9] that outside an arbitrarily small neighborhood V of $(0, 0)$ the manifolds S_a and S_r perturb smoothly to locally invariant manifolds $S_{a,\varepsilon}$ and $S_{r,\varepsilon}$ for sufficiently small $\varepsilon \neq 0$. We would like to point out that $S_{a,\varepsilon}$ and $S_{r,\varepsilon}$ are actually very simple. They consist of a single solutions.

Recall that the slow manifolds are obtained as sections $\varepsilon = const.$ of two-dimensional, locally invariant, center-like manifolds M_a resp. M_r of the extended system

$$\begin{aligned} x' &= f(x, y, \varepsilon), \\ y' &= \varepsilon g(x, y, \varepsilon), \qquad (2.4) \\ \varepsilon' &= 0, \end{aligned}$$

in the extended phase space $\mathbb{R}^3$.

For this extended system $S\times\{0\}$ is a manifold of equilibria. Outside of a neighbourhood of the fold point $(0,0,0)$ the linearization of system (2.4) at points $S_a\times\{0\}$ has a double zero eigenvalue and one uniformly hyperbolic (stable) eigenvalue. This allows to prove the existence of the attracting center-like manifold M_a, the manifold M_r is obtained in a similar way. At the fold point $(0,0,0)$ the linearization has a triple eigenvalue zero and the construction of the slow manifolds breaks down. Clearly, the slow manifolds can be extended in forward and backwards time by the flow, however, their behaviour is then not controlled by the singular limit problems introduced so far.

Note that S_r is strongly repelling and of less interest at least in the context of relaxation oscillations. We focus our attention on S_a and investigate how $S_{a,\varepsilon}$ as well as nearby solutions behave as they pass near the fold point. We expect that close to the fold point a transition from slow motion along $S_{a,\varepsilon}$ to a fast motion almost parallel to the unstable fibers occurs. A similar analysis could be carried out for $S_{r,\varepsilon}$.

REMARK 2.1. It is known that the slow manifolds M_a and M_r and hence their sections $S_{a,\varepsilon}$ and $S_{r,\varepsilon}$ are not unique and are only determined up to $O(e^{-C/\varepsilon})$, where, C is some positive constant. We make an arbitrary choice of M_a and M_r and indicate at the end that our results are independent of this choice.

We now view the previously introduced neighborhood U as a neighborhood of (0, 0, 0) in $\mathbb{R}^3$. We pick U sufficiently small, so that $g(x, y, \varepsilon) \neq 0$, $\forall(x, y, \varepsilon) \in U$. Before stating the main results we rewrite system (2.1) resp. (2.4) in a canonical form. By rescaling x, y, ε, and t we obtain

$$\begin{aligned} x' &= -y + x^2 + h(x, y, \varepsilon), \\ y' &= \varepsilon g(x, y, \varepsilon), \\ \varepsilon' &= 0, \end{aligned} \tag{2.5}$$

with

$$\begin{aligned} h(x, y, \varepsilon) &= axy + bx^3 + O(y^2, \varepsilon x, \varepsilon y), \\ g(x, y, \varepsilon) &= -1 + cx + O(x^2, y, \varepsilon), \end{aligned}$$

where the new function g is related to the original one by the rescaling. The terms with coefficients a, b, and c are singled out because they will show up in the sequel. This form of the equations will be used throughout the forthcoming analysis.

For small positive $\rho > 0$ and a suitable interval $J \subset \mathbb{R}$ let

$$\Delta^{in} = \{(x, \rho^2), \, x \in J\}$$

be a section transverse to S_a and let

$$\Delta^{out} = \{(\rho, y), \, y \in \mathbb{R}\}$$

be a section transverse to the fast fibers (see Fig. 1). Let $\pi : \Delta^{in} \to \Delta^{out}$ be the transition map for the flow of (2.1). The following two theorems are the main results of this paper.

THEOREM 2.1. *Under the assumptions made in this section there exists $\varepsilon_0 > 0$ such that the following assertions hold for $\varepsilon \in (0, \varepsilon_0]$:*

1. *The manifold $S_{a,\varepsilon}$ passes through Δ^{out} at a point $(\rho, h(\varepsilon))$ where $h(\varepsilon) = O(\varepsilon^{2/3})$.*
2. *The section Δ^{in} is mapped to an interval around $S_{a,\varepsilon} \cap \Delta^{out}$ of size $O(e^{-C/\varepsilon})$, where C is a positive constant.*

REMARK 2.2. This result also explains why the nonuniqueness of the slow manifold M_a resp. $S_{a,\varepsilon}$ does not affect our results. Two different choices of these manifolds are exponentially close at Δ^{in} and even more so at Δ^{out} due to the exponential contraction during the passage. In particular the asymptotic expansion given below does not depend on the choice of the slow manifold.

To relate our work to the existing analysis by means of matched asymptotic expansions we prove

THEOREM 2.2. *The function $h(\varepsilon)$ defined in Theorem 2.1 has the asymptotic expansion*

$$h(\varepsilon) = c_1\varepsilon^{2/3} + c_2\varepsilon \ln \varepsilon + c_3\varepsilon + O(\varepsilon^{4/3}\ln(\varepsilon)). \tag{2.6}$$

The constants c_i, $i = 1, 2, 3$ are computable functionals of the functions f and g.

REMARK 2.3. The expansion (2.6) is the beginning of the infinite asymptotic expansion given in the book by Mishchenko and Rozov [17, pp. 96, Eq. (14.4)]. We do not attempt to reproduce their impressive results in all details, but rather try to explain the particular form of the expansion, i.e. why the expansion is in powers of $\varepsilon^{1/3}$ and why certain terms are multiplied by powers of $\ln \varepsilon$.

3. Blow-up. In this section we define and describe the blow-up transformation. The basic observation is that the foldpoint (0, 0, 0) is a more degenerate equilibrium point of system (2.5) than the other points of the critical manifold S. The linearization of system (2.5) at the origin has a triple zero eigenvalue while the linearization at the other points of the critical manifold S has a double zero eigenvalue and one negative resp. positive eigenvalue for $x < 0$ resp $x > 0$.

The important insight in [7] is that blow-up techniques are the right tool to analyse nilpotent equilibria like the fold point, viewed as a degenerate equilibrium of the extended system (2.4). The blow-up method is essentially a clever coordinate transformation by which the degenerate equilibrium is "blown-up" to a two-sphere. In certain directions transverse to the sphere and even on the sphere one gains enough hyperbolicity to allow a complete analysis by standard techniques. In a sense the whole technique is a generalization of the well known blow-up methods for degenerate equilibria of planar vector fields [6]. In the simplest situations this corresponds to blowing-up the degenerate equilibrium to the circle $r = 0$ by rewriting the vector field in polar coordinates $(r, \vartheta) \in \mathbb{R} \times S^1$. The analysis is often simplified substantially by using a quasihomogeneous blow-up, i.e. by using different powers (weights) of r for different variables in the defining transformation.

The blow-up transformation for system (2.5) is:

$$x = \bar{r}\bar{x}, \qquad y = \bar{r}^2\bar{y}, \qquad \varepsilon = \bar{r}^3\bar{\varepsilon}\ . \tag{3.1}$$

with weights 1, 2, and 3. We define the manifolds $B = S^2 \times \mathbb{R}$ and $B_0 = S^2 \times [0, r_0]$ for $r_0 > 0$. We consider the blow-up transformation as a mapping

$$\Phi : B_0 \to \mathbb{R}^3, \tag{3.2}$$

with $(\bar{x}, \bar{y}, \bar{\varepsilon}) \in S^2$. Hence the blow up Φ is a *polar blow up*. We restrict attention to B_0 for sufficiently small $r_0 > 0$ such that system (2.4) is described by the canonical form (2.5) in the region $\Phi(B_0)$. We will be mainly interested in nonnegative values of $\bar{\varepsilon}$ and $\bar{r}$ but everything that follows makes sense for negative values as well.

Let X denote the vector field corresponding to (2.5). Since X vanishes at the point $(0,0,0)$ there exists a vector field $\bar{X}$ on B_0 such that

$$\Phi_* \bar{X} = X,$$

where Φ_* is induced by Φ. It remains to study the vector field $\bar{X}$ on the manifold B_0. Note that this suffices, since $\Phi(B_0)$ is a full neighborhood of the origin. In principle one could use spherical coordinates on S^2, however this would lead to rather lengthy computations. It is natural and almost mandatory to use different charts for the manifold B to simplify the analysis.

We will now introduce the charts used later on. Much of this could be done in a more abstract way, however, it is instructive to be as specific as possible. Loosely speaking, we will define a chart K_2, which describes a neighborhood of the upper half sphere defined by $\bar{\varepsilon} > 0$, and charts K_1 and K_3 which describe neighborhoods of the part of the equator of S^2 which are needed in the analysis. In problems where a neighborhood of the whole equator needs to be analysed two further charts must be defined analogously. The subscripts in K_1, K_2, and K_3 denote the order in which the charts are used later.

Let $B^+_{\bar{\varepsilon}}$ denote the submanifold of B defined by $\bar{\varepsilon} > 0$, the submanifolds $B^+_{\bar{x}}$ and $B^+_{\bar{y}}$ are defined analogously.

The chart $K_2 : B^+_{\bar{\varepsilon}} \to \mathbb{R}^3$ is defined by

$$x_2 = \bar{x}\bar{\varepsilon}^{-1/3}, \qquad y_2 = \bar{y}\bar{\varepsilon}^{-2/3}, \qquad r_2 = \bar{r}\bar{\varepsilon}^{1/3}\ , \tag{3.3}$$

where (x_2, y_2, r_2) are the coordinates in $\mathbb{R}^3$. Positive values of $\bar{r}$ correspond to positive values of r_2. The upper half of S^2 defined by $\bar{\varepsilon} > 0$ is mapped one to one onto the plane $r_2 = 0$. This mapping is very similar to a stereographic projection with slight differences caused by the different weights. The whole construction is reminiscent of the choice of affine coordinates for the projective plane, again with slight differences caused by the different weights, see Fig. 2 for a schematic view of these charts.

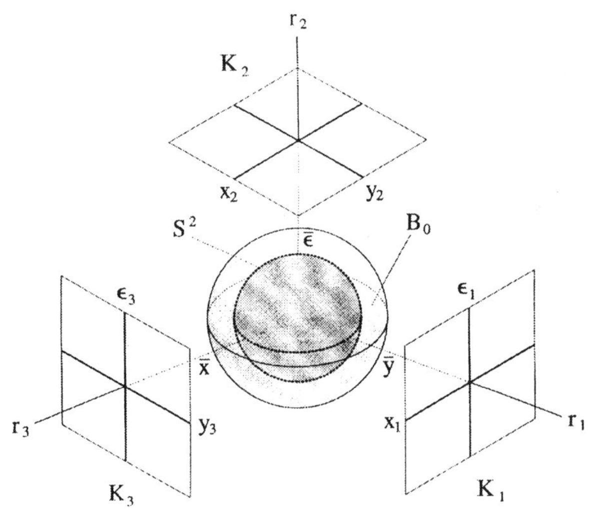

FIG. 2. *The manifold B_0 and the charts K_1, K_2, and K_3.*

In the chart K_2 the blow-up transformation (3.1) is

$$x = r_2 x_2, \qquad y = r_2^2 y_2, \qquad \varepsilon = r_2^3, \tag{3.4}$$

which is a *directional blow-up* in the direction of positive ε. In these variables the point $(0, 0, 0)$ is blown-up to the plane $r_2 = 0$.

The chart $K_1 : B_{\bar{y}}^{+} \to \mathbb{R}^3$ is defined by

$$x_1 = \bar{x}\bar{y}^{-1/2}, \qquad \varepsilon_1 = \bar{\varepsilon}\bar{y}^{-3/2}, \qquad r_1 = \bar{r}\bar{y}^{1/2}, \tag{3.5}$$

where $(x_1, \varepsilon_1, r_1)$ are the coordinates in $\mathbb{R}^3$. In the chart K_1 the blow-up transformation (3.1) is

$$x = r_1 x_1, \qquad y = r_1^2, \qquad \varepsilon = r_1^3 \varepsilon_1, \tag{3.6}$$

which is a *directional blow-up* in the direction of positive y. In these coordinates the point $(0,0,0)$ is blown-up to the plane $r_1 = 0$.

Similarly, the chart $K_3 : B_{\bar{x}}^+ \to \mathbb{R}^3$ is defined by

$$y_3 = \bar{y}\bar{x}^{-2}, \qquad \varepsilon_3 = \bar{\varepsilon}\bar{x}^{-3}, \qquad r_3 = \bar{r}\bar{x}, \tag{3.7}$$

where $(y_3, \varepsilon_3, r_3)$ are the coordinates in $\mathbb{R}^3$. In the chart K_3 the blow-up transformation (3.1) is

$$x = r_3, \qquad y = r_3^2 y_3, \qquad \varepsilon = r_3^3 \varepsilon_3 . \tag{3.8}$$

In our analysis we will need to change coordinates between these charts on their overlap domains. A simple computation gives

LEMMA 3.1. *Let κ_{12} denote the change of coordinates from K_1 to K_2. Then κ_{12} is given by*

$$x_2 = x_1 \varepsilon_1^{-1/3}, \qquad y_2 = \varepsilon_1^{-2/3}, \qquad r_2 = r_1 \varepsilon_1^{1/3}, \qquad \textit{for } \varepsilon_1 > 0, \tag{3.9}$$

and κ_{12}^{-1} is given by

$$x_1 = x_2 y_2^{-1/2}, \qquad \varepsilon_1 = y_2^{-3/2}, \qquad r_1 = r_2 y_2^{1/2}, \qquad \textit{for } y_2 > 0 . \tag{3.10}$$

Let κ_{23} denote the change of coordinates from K_2 to K_3. Then κ_{23} is given by

$$y_3 = y_2 x_2^{-2}, \qquad \varepsilon_3 = x_2^{-3}, \qquad r_3 = r_2 x_2, \qquad \textit{for } x_2 > 0, \tag{3.11}$$

and κ_{23}^{-1} is given by

$$x_2 = \varepsilon_3^{-1/3}, \qquad y_2 = y_3 \varepsilon_3^{-2/3}, \qquad r_2 = r_3 \varepsilon_3^{1/3}, \qquad \textit{for } \varepsilon_3 > 0 . \tag{3.12}$$

The above constructions make perfect sense if restricted to B_0.

We introduce the following notation: $\bar{P}$ denotes an object in the blow-up which corresponds to an object P in the original problem. If $\bar{P}$ is described in one of the charts then P_i denotes the object in chart K_i, $i = 1, 2, 3$. This notation is used only when necessary, mostly to distinguish between various invariant manifolds.

4. Dynamics in chart K_2. The transformation (3.4) is just a rescaling of (x, y), since $r_2 = \varepsilon^{1/3}$. By inserting (3.4) into system (2.5) we obtain the vector field $\bar{X}$ in the chart K_2

$$\begin{aligned} x_2' &= r_2(x_2^2 - y_2 + O(r_2)), \\ y_2' &= r_2(-1 + O(r_2)), \\ r_2' &= 0. \end{aligned} \tag{4.1}$$

Since $r_2' = 0$ this blown-up system is still a family of planar vector fields with parameter r_2. We now desingularize the equations by rescaling time $t_2 := r_2 t$, so that the factor r_2 disappears. This desingularization is necessary to obtain a nontrivial flow on the blown-up locus $r_2 = 0$. We obtain

$$\begin{aligned} x_2' &= x_2^2 - y_2 + O(r_2), \\ y_2' &= -1 + O(r_2), \\ r_2' &= 0. \end{aligned} \tag{4.2}$$

We first consider the case $r_2 = 0$ which gives

$$\begin{aligned} x_2' &= x_2^2 - y_2 \\ y_2' &= -1. \end{aligned} \tag{4.3}$$

This is a Riccati equation, whose solutions can be expressed in terms of special functions. The relevant results can be found in [17, pp. 68–72]. Here we restate the results needed in our analysis. For the sake of readability we omit the subscript 2 of the variables.

PROPOSITION 4.1 ([17]). *The Riccati equation (4.3) has the following properties:*

1. *Every orbit has a horizontal asymptote $y = y_r$, where y_r depends on the orbit, such that $x \to \infty$ as y approaches y_r from above.*
2. *There exists a unique orbit γ_2 which can be described as $(x, \tilde{y}(x))$ and is asymptotic to the left branch of the parabola $x^2 - y = 0$ for $x \to -\infty$. The orbit γ_2 has a horizontal asymptote $y = -\Omega_0 < 0$, such that $x \to \infty$ as y approaches $-\Omega_0$ from above.*
3. *The function $\tilde{y}(x)$ has the asymptotic expansions*

$$\begin{aligned} \tilde{y}(x) &= x^2 + \frac{1}{2x} + O(\frac{1}{x^4}), \qquad x \to -\infty, \\ \tilde{y}(x) &= -\Omega_0 + \frac{1}{x} + O(\frac{1}{x^3}), \qquad x \to \infty. \end{aligned}$$

4. *All orbits to the right of γ_2 are backward asymptotic to the right branch of the parabola $x^2 - y = 0$.*
5. *All orbits to the left of γ_2 have a horizontal asymptote $y = y_l > y_r$, where y_l depends on the orbit, such that $x \to -\infty$ as y approaches y_l from below.*

REMARK 4.1. The constant Ω_0 is the smallest positive zero of

$$J_{-1/3}(2z^{3/2}/3) + J_{1/3}(2z^{3/2}/3),$$

where $J_{-1/3}$ resp. $J_{1/3}$ are Bessel functions of the first kind.

The assertions of Proposition 4.1 are illustrated in Fig. 3. We will see that the special solution γ_2 is backward asymptotic to an equilibrium p_a

on the equator of S^2. The importance of the special solution γ_2 is that it "leads" the incoming attracting slow manifold across the upper half of the sphere S^2 to a point from where take-off in the direction of the fast flow occurs.

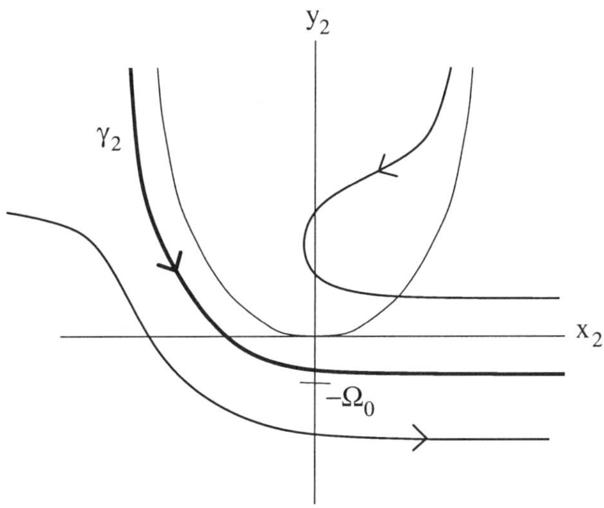

FIG. 3. *Solutions of the Riccati equation (4.3).*

We are interested in describing the transition map for (4.1) within a bounded domain D_2. Within such a domain we can deduce properties of the flow of (4.1) from Proposition 4.1 by using regular perturbation. A detailed study of the effect of the $O(r_2)$ perturbations outside D_2, i.e. close to infinity will be carried out in the charts K_1 and K_3. For $\delta > 0$ we define the following sections

$$\Sigma_2^{in} = \{(x_2, y_2, r_2) \, : \, y_2 = \delta^{-\frac{2}{3}}\},$$
$$\Sigma_2^{out} = \{(x_2, y_2, r_2) \, : \, x_2 = \delta^{-\frac{1}{3}}\}.$$

Let Π_2 be the transition map of the flow (4.2) from Σ_2^{in} to Σ_2^{out}. Let $q_0 = \gamma_2 \cap \Sigma_2^{in}$.

PROPOSITION 4.2. *The transition map Π_2 has the following properties:*

1. $\Pi_2(q_0) = (\delta^{-\frac{1}{3}}, -\Omega_0 + \delta^{\frac{1}{3}} + O(\delta), 0)$.

2. A neighborhood of q_0 is mapped diffeomorphically onto a neighborhood of $\Pi_2(q_0)$.

Proof. Follows directly from Proposition 4.1 and regular perturbation theory. □

5. Dynamics in chart K_1. In this section we study the blown-up system (2.5) in chart K_1. By inserting (3.6) into system (2.5) we derive differential equations for x_1, r_1, and ε_1, i.e. we obtain the vector field $\bar{X}$ in the chart K_1. A straightforward computation gives

$$\begin{aligned} x_1' &= r_1\Big(-1 + x_1^2 + \frac{1}{2}\varepsilon_1 x_1 + r_1\Big(ax_1 + bx_1^3 - \frac{c}{2}\varepsilon_1 x_1^2\Big) + O(r_1^2)\Big), \\ r_1' &= \frac{1}{2} r_1^2\varepsilon_1\Big(-1 + cr_1x_1 + O(r_1^2)\Big), \\ \varepsilon_1' &= \frac{3}{2} r_1\varepsilon_1^2\Big(1 - cr_1x_1 + O(r_1^2)\Big). \end{aligned} \tag{5.1}$$

REMARK 5.1. The equation for ε_1' is obtained from the relation $3r_1r_1'\varepsilon_1 + r_1^3\varepsilon_1' = 0$ This equation is a consequence of the fact that $\varepsilon = r_1^3\varepsilon_1$ is a constant of motion. As we will see it is useful to treat system (5.1) as a three-dimensional problem, however, we keep the constant of motion ε in our mind. This seemingly artificial construction is actually crucial for the whole approach.

We desingularize the vector field by dividing by r_1, This desingularization is necessary to obtain a nontrivial flow on the blown-up locus $r_1 = 0$. This gives

$$\begin{aligned} x_1' &= -1 + x_1^2 + \frac{1}{2}\varepsilon_1 x_1 + r_1\Big(ax_1 + bx_1^3 - \frac{c}{2}\varepsilon_1 x_1^2\Big) + O(r_1^2), \\ r_1' &= \frac{1}{2}r_1\varepsilon_1\Big(-1 + cr_1x_1 + O(r_1^2)\Big), \\ \varepsilon_1' &= \frac{3}{2}\varepsilon_1^2\Big(1 - cr_1x_1 + O(r_1^2)\Big). \end{aligned} \tag{5.2}$$

System (5.2) has two invariant subspaces, namely the plane $r_1 = 0$ and the plane $\varepsilon_1 = 0$. Their intersection is the invariant line $l_1 := \{(x_1, 0, 0) : x_1 \in \mathbb{R}\}$, see Fig. 4. The dynamics on l_1 is governed by

$$x_1' = -1 + x_1^2 .$$

There are two equilibria $p_a = (-1,\, 0,\, 0)$ and $p_r = (1,\, 0,\, 0)$. For the flow on the line l_1 both points are hyperbolic, the relevant eigenvalue is -2 for p_a and 2 for p_r, i.e. p_a is attracting and p_r is repelling.

The dynamics in the invariant plane $\varepsilon_1 = 0$ is governed by

$$\begin{aligned} x_1' &= -1 + x_1^2 + r_1(ax_1 + bx_1^3) + O(r_1^2), \\ r_1' &= 0, \end{aligned} \tag{5.3}$$

This system has a normally hyperbolic curve $S_{a,1}$ of equilibria emanating from p_a and a curve $S_{r,1}$ of equilibria emanating from p_r, see Fig. 4. For

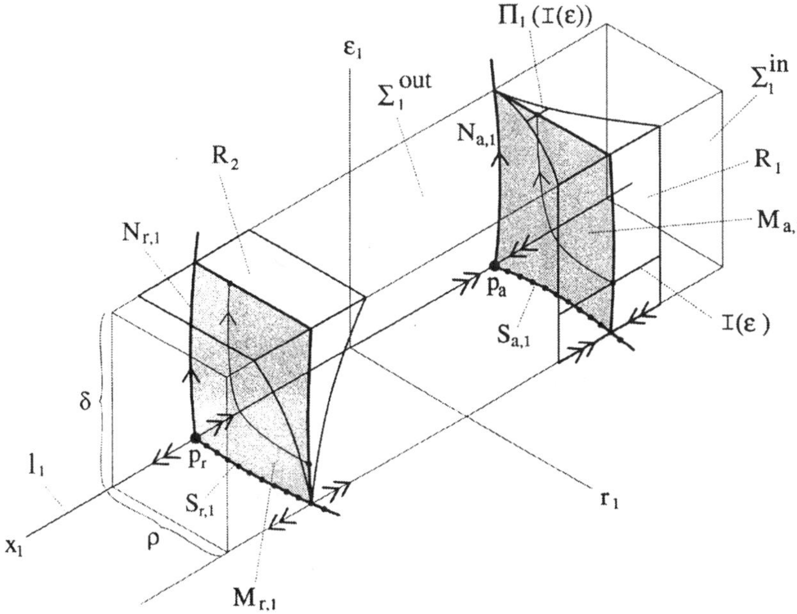

FIG. 4. *Geometry and dynamics in chart K_1.*

r_1 small, this follows from the implicit function theorem. Actually, $S_{a,1}$ and $S_{r,1}$ are precisely the branches of the critical manifold S described in Section 2, this explains also the notation. Along the curve $S_{a,1}$ the linearization of (5.3) has one zero eigenvalue, the other eigenvalue is negative and close to -2 for r_1 small. Along $S_{r,1}$ the situation is similar, however the non-zero eigenvalue is positive and close to 2 for r_1 small.

REMARK 5.2. Equation (5.3) is the directional (in the positive y-direction) blow-up of the fold point $(0,0)$ of the layer problem, i.e. system (2.1) with $\varepsilon = 0$. The line $l_1 = 0$ corresponds to the fold point. We have gained normal hyperbolicity of the lines of equilibria $S_{a,1}$ resp. $S_{r,1}$ at the points p_a resp. p_r due to the blow-up!

The dynamics in the invariant plane $r_1 = 0$ is governed by

$$\begin{aligned} x_1' &= -1 + x_1^2 + \frac{1}{2}\varepsilon_1 x_1, \\ \varepsilon_1' &= \frac{3}{2}\varepsilon_1^2 . \end{aligned} \tag{5.4}$$

We recover the fixed points p_a and p_r, however, there exists an additional zero eigenvalue due to the second equation. The corresponding eigenvector is $(-1,4)$ at both fixed points. Hence, there exist one-dimensional center manifolds $N_{a,1}$ at p_a and $N_{r,1}$ at p_r along which ε_1 increases for $\varepsilon_1 > 0$.

Note that the branch of the attracting center manifold $N_{a,1}$ at p_a in the half space $\varepsilon_2 > 0$ is unique, while the the repelling center manifold $N_{r,1}$ at p_r in the halfspace $\varepsilon_2 > 0$ is not unique.

We collect the information we have obtained so far in the following lemma.

LEMMA 5.1. *The linearization of system (5.2) at p_j, $j = a, r$ has the following real eigenvalues:*

1. *$\lambda_1 = -2$ at p_a and $\lambda_1 = 2$ at p_r with eigenvector $(1,0,0)$ corresponding to the flow on l_1,*
2. *$\lambda_2 = 0$ with an eigenvector tangent to $S_{j,1}$,*
3. *$\lambda_3 = 0$ corresponding with eigenvector $(-1,0,4)$ corresponding to the center direction in the invariant plane $r_1 = 0$.*

We restrict attention to the set

$$D_1 := \{(x_1, r_1, \varepsilon_1) : x_1 \in \mathbb{R}, 0 \le r_1 \le \rho, 0 \le \varepsilon_1 \le \delta\},$$

where $\rho > 0$ is the constant defining the sections Δ^{in} and Δ^{out} in Section 2 and $\delta > 0$ is the constant defining the sections Σ_2^{in} and Σ_2^{out} in Section 4.

Note that all objects defined later on extend smoothly to negative values of r_1 and ε_1, i.e. there are no problems due to the boundaries $r_1 = 0$ and $\varepsilon_1 = 0$. We have the following result:

PROPOSITION 5.1. *For ρ, δ sufficiently small the following assertions hold for system (5.2):*

1. *There exists an attracting two-dimensional C^k-center manifold $M_{a,1}$ at p_a which contains the line of fixed points $S_{a,1}$ and the center manifold $N_{a,1}$. In D_1 the manifold $M_{a,1}$ is given as a graph $x_1 = h_a(r_1, \varepsilon_1)$. The branch of $N_{a,1}$ in $r_1 = 0$, $\varepsilon_1 > 0$ is unique.*
2. *There exists a repelling two-dimensional C^k-center manifold $M_{r,1}$ at p_r which contains the line of fixed points $S_{r,1}$ and the center manifold $N_{r,1}$. In D_1 the manifold $M_{r,1}$ is given as a graph $x_1 = h_r(r_1, \varepsilon_1)$. The branch of $N_{r,1}$ in $r_1 = 0$, $\varepsilon_1 > 0$ is not unique.*
3. *There exists a stable invariant foliation $\mathcal{F}^s$ with base $M_{a,1}$ and one dimensional fibers. For any $c > -2$ there exists a choice of positive ρ and δ such that the contraction along $\mathcal{F}^s$ is stronger than e^{ct_1}.*
4. *There exists an unstable invariant foliation $\mathcal{F}^u$ with base $M_{r,1}$ and one dimensional fibers. For any $c < 2$ there exists a choice of positive ρ and δ such that the expansion along $\mathcal{F}^u$ is stronger than e^{ct_1}.*
5. *The unique branch of $N_{a,1}$ in $r_1 = 0$, $\varepsilon_1 > 0$ is the part of the special orbit γ_2 from Proposition 4.1 corresponding to $x_2 \in (-\infty, c_2]$, for some $c_2 < 0$, written in chart K_1.*

Proof. Assertions 1–4 follow from Lemma 5.1 and standard center manifold theory [11].

We know from Proposition 4.1 that the special orbit γ_2 written in the chart K_1 has the expansion

$$\left(x_2,\, x_2^2 + \frac{1}{2x_2} + O\left(\frac{1}{x_2^4}\right),\, 0\right)$$

as $x_2 \to -\infty$. The coordinate transformation (3.10) implies that $\kappa_{12}^{-1}(\gamma_2)$ has the expansion

$$\left(x_2\left(x_2^2 + \frac{1}{2x_2} + O\left(\frac{1}{x_2^4}\right)\right)^{-1/2},\, 0,\, \left(x_2^2 + \frac{1}{2x_2} + O\left(\frac{1}{x_2^4}\right)\right)^{-3/2}\right)$$

as $x_2 \to -\infty$. Expanding these terms in powers of x_2 shows that $\kappa_{12}^{-1}(\gamma_2)$ converges to p_a tangent to the center-direction $(-1, 0, 4)$ as $x_2 \to -\infty$. This and the uniqueness of the branch of $N_{a,1}$ in $r_1 = 0$, $\varepsilon_1 > 0$ imply Assertion 5. □

The assertions of Proposition 5.1 are illustrated in figure Fig. 4.

REMARK 5.3. Clearly, the center manifold $M_{a,1}$ in chart K_1 corresponds to a locally invariant manifold $\bar{M}_a$ of the blown-up vector field $\bar{X}$. The importance of Assertion 5 in the above proposition is that it allows us to track the manifold $\bar{M}_a$ as it moves across the sphere S^2 guided by the special orbit $\bar{\gamma}$ corresponding to the solution γ_2 of the Ricatti-equation.

We now define the following sections

$$\Sigma_1^{in} := \left\{(x_1, r_1, \varepsilon_1) \in D_1 \;:\; r_1 = \rho\right\},$$

$$\Sigma_1^{out} := \left\{(x_1, r_1, \varepsilon_1) \in D_1 \;:\; \varepsilon_1 = \delta\right\}.$$

REMARK 5.4. Note that Σ_1^{out} maps under the coordinate transformation (3.9) to Σ_2^{in}. An important part of our description of the flow near the fold is the description of the transition map from Σ_1^{in} to Σ_1^{out} near the center manifolds $M_{a,1}$ and $M_{r,1}$. Since the neighborhood of $M_{a,1}$ corresponds to the neighborhood of the attracting branch of the slow manifold of (2.1) we are more interested in understanding the dynamics near $M_{a,1}$. Yet the analysis of the two cases is very similar, so we handle them simultaneously.

Let R_1 be the rectangle in Σ_1^{in} defined by $|1 + x_1| \leq \beta$ and R_2 the rectangle in Σ_1^{out} $|1 - x_1| \leq \beta$ for sufficiently small $\beta > 0$. The constants ρ, δ, and β can be chosen such that $M_{a,1} \cap \Sigma_1^{in} \subset R_1$ and $M_{r,1} \cap \Sigma_1^{out} \subset R_2$. For $0 \leq \varepsilon_0 \leq \delta$ and $0 \leq r_0 \leq \rho$ let $I(\varepsilon_0)$ be the line $R_1 \cap \{\varepsilon = \varepsilon_0\}$ and $J(r_0)$ be the line $R_2 \cap \{r = r_0\}$.

In the neighborhood of p_j, $j = a, r$, the flow of (5.2) carries Σ_1^{in} to Σ_1^{out}. Let $\Pi_1 : \Sigma_1^{in} \to \Sigma_1^{out}$ be the transition map defined by the flow of (5.2). The map Π_1 is well defined on R_1, at least for small enough values of ρ, δ, and β. The map Π_1 is defined in a wedge-shaped set in Σ_1^{in} around $M_{r,1}$ that shrinks to $S_{r,1}$ for $\varepsilon_1 \to 0$. The reason for this difference is that $M_{a,1}$ is attracting and $M_{r,1}$ is repelling.

The evolution of ε_1 determines the transition time of solutions from Σ_1^{in} to Σ_1^{out}. The relevant equation is

$$\varepsilon_1' = \frac{3}{2}\varepsilon_1^2(1 - O(r_1)) \ . \tag{5.5}$$

We have the following estimate of transition times.

LEMMA 5.2. *The transition time T of a solution of system (5.2) from a point $p = (x_1, \rho, \varepsilon_1) \in \Sigma_1^{in}$ to the point $\Pi_1(p) \in \Sigma_1^{out}$ satisfies:*

$$T = \frac{2}{3}\left(\frac{1}{\varepsilon_1} - \frac{1}{\delta}\right)(1 + O(\rho)) \, . \tag{5.6}$$

Proof. The result follows immediately by integrating equation (5.5). □

PROPOSITION 5.2. *The transition map $\Pi_1 : \Sigma_1^{in} \to \Sigma_1^{out}$ defined by the flow of system (5.2) has the following properties:*

1. *$\Pi_1(R_1)$ is a wedge-like region in Σ_1^{out}. $\Pi_1^{-1}(R_2)$ is a wedge-like region in Σ_1^{in}.*
2. *More precisely, for fixed $c < 2$ there exists a constant K depending on the constants c, ρ, δ, and β such that*
   (i) for $\varepsilon_1 \in (0, \delta]$, the image $\Pi_1(I(\varepsilon_1))$ is an interval around $M_{a,1} \cap \Sigma_1^{out}$ whose width is bounded by $Ke^{-\frac{2c}{3}\left(\frac{1}{\varepsilon_1} - \frac{1}{\delta}\right)}$.
   (ii) For $r_1 \in (0, \rho]$, the preimage $\Pi_1^{-1}(J(r_1))$ is an interval around $M_{r,1} \cap \Sigma_1^{in}$ whose width is bounded by $Ke^{-\frac{2c}{3}\left(\frac{\rho^3}{r_1^3\delta} - \frac{1}{\delta}\right)}$.

Proof. The assertions follow from Proposition 5.1 and Lemma 5.2. The estimate for the width of $\Pi_1^{-1}(J(r_1))$ in the second assertion uses the identity $\varepsilon_1\rho^3 = \delta r_1^3$ for $p = (x_{1,in}, \rho, \varepsilon_1)$ and $\Pi_1(p) = (x_{1,out}, r_1, \delta)$ to express the transition time in terms of r_1. □

All our results concerning the dynamics in chart K_1 are illustrated in Fig. 4.

6. Dynamics in chart K_3. By inserting transformation (3.8) into system (2.5) we obtain the vector field $\bar{X}$ in chart K_3.

$$\begin{aligned} r_3' &= r_3(1 - y_3 + br_3 + ar_3y_3 + O(r_3^2)), \\ y_3' &= -\varepsilon_3 - 2y_3 + cr_3\varepsilon_3 + 2y_3^2 - 2by_3r_3 + 2ar_3y_3^2 + O(r_3^2(y_3 + \varepsilon_3)), \\ \varepsilon_3' &= -3\varepsilon_3(1 - y_3 + br_3 + ar_3y_3 + O(r_3^2)), \end{aligned} \tag{6.1}$$

where $'$ denotes differentiation with respect to $t_3 = r_3t$. The constants a, b, and c in these equations are the constants from system (2.5). The factor $F(r_3, y_3, \varepsilon_3) := 1 - y_3 + br_3 + ar_3y_3 + O(r_3^2)$ on the right hand side of the equation for r_3' resp. ε_3' is the same in both equations because of the constraint $\varepsilon = r_3^3\varepsilon_3$. The planes $\varepsilon_3 = 0$ and $r_3 = 0$ and the y_3-axis are invariant under the flow of system (6.1).

LEMMA 6.1. *The point $q = (0,0,0)$ is a hyperbolic equilibrium of system (6.1) with eigenvalues:*

1. $\lambda_1 = 1$ *with eigenvector* $(1,0,0)$ *corresponding to the flow in* $\varepsilon_3 = 0$.
2. $\lambda_2 = -2$ *with eigenvector* $(0,1,0)$ *corresponding to the flow on the* y_3*-axis,*
3. $\lambda_3 = -3$ *with eigenvector* $(0,1,1)$ *corresponding to the flow in* $r_3 = 0$,

Proof. Computation. □

LEMMA 6.2. *The special orbit γ_2 from Proposition 4.1 written in the chart K_3 lies in the plane $r_3 = 0$, converges to q as $x_2 \to \infty$, and is tangent at q to the vector* $(0,1,0)$.

Proof. The orbit γ_2 written in the chart K_2 has the expansion

$$\left(x_2,\, -\Omega_0 + \frac{1}{x_2} + O\Big(\frac{1}{x_2^3}\Big),\, 0 \right),$$

as $x_2 \to \infty$. The coordinate transformation (3.11) implies that $\kappa_{23}(\gamma_2)$ has the expansion

$$\left(0,\, \frac{-\Omega_0}{x_2^2} + O\Big(\frac{1}{x_2^3}\Big),\, O\Big(\frac{1}{x_2^3}\Big) \right),$$

as $x_2 \to \infty$. From this the assertions of the lemma follow immediately. □

This implies that parts of the manifold $\bar{M}_a$ corresponding to $\bar{r} > 0$ come close to the equilibrium q, see Fig. 5. Hence, we need a precise description of the dynamics of system (6.1) close to q. This is a somewhat delicate problem because of the *resonance* $\lambda_2 = \lambda_1 + \lambda_3$, which implies that there exists no smooth transformation of the nonlinear flow to the flow of the corresponding linearization, see [23]. For the description of the flow in a neighborhood of q we define sections Σ_3^{in} and Σ_3^{out} as follows

$$\Sigma_3^{in} = \Big\{ (r_3,\, y_3,\, \varepsilon_3) \;:\; r_3 \in [0,\rho],\, y_3 \in [-\beta,\beta],\, \varepsilon_3 = \delta \Big\},$$

$$\Sigma_3^{out} = \Big\{ (r_3,\, y_3,\, \varepsilon_3) \;:\; r_3 = \rho,\, y_3 \in [-\beta,\beta],\, \varepsilon_3 \in [0,\delta] \Big\},$$

where ρ and δ are the same constants as before, and $\beta > 0$ is sufficiently small.

REMARK 6.1. Note that Σ_2^{out} maps under the coordinate transformation (3.11) to Σ_3^{in}.

Let Π_3 be the transition map from Σ_3^{in} to Σ_3^{out}. Our goal is to obtain a formula for the map Π_3. Before stating the relevant result we need to discuss the structure of equation (6.1) in more detail. We first divide (6.1) by the multiplicative factor $F(r_3,\, y_3,\, \varepsilon_3)$ obtaining:

$$\begin{aligned} r_3' &= r_3, \\ y_3' &= -\varepsilon_3 - 2y_3 + \bar{c} r_3 \varepsilon_3 + G(r_3,\, y_3,\, \varepsilon_3), \qquad (6.2) \\ \varepsilon_3' &= -3\varepsilon_3. \end{aligned}$$

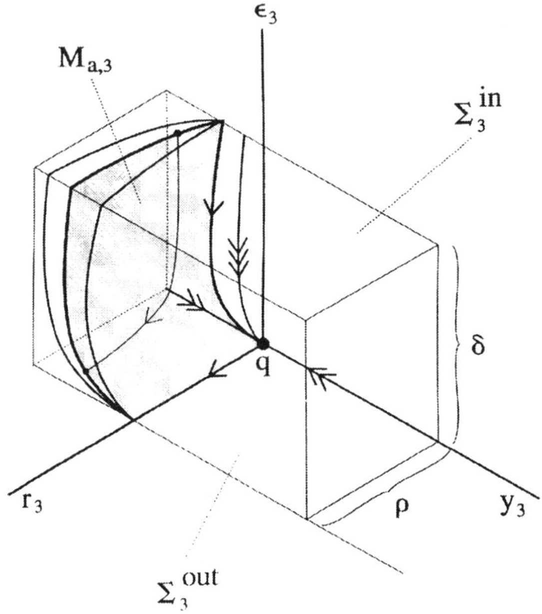

FIG. 5. *Geometry and dynamics of system (6.1) near the equilibrium q.*

Standard normal form theory [11] combined with techniques for removing the tail of the normal form [23] imply that there exists a near identity transformation given by

$$\begin{pmatrix} r_3 \\ \tilde{y}_3 \\ \varepsilon_3 \end{pmatrix} = \begin{pmatrix} r_3 \\ \Psi(r_3, y_3, \varepsilon_3) \\ \varepsilon_3 \end{pmatrix} \tag{6.3}$$

which brings (6.2) to the following form

$$r_3' = r_3, \tag{6.4a}$$

$$\begin{aligned}\tilde{y}_3' &= -\varepsilon_3 - 2\tilde{y}_3 + \bar{c}r_3\varepsilon_3 + r_3^2\tilde{y}_3^2\,G_1(r_3, \tilde{y}_3, \varepsilon_3)\\ &\quad + r_3^3\tilde{y}_3\varepsilon_3\,G_2(r_3, \tilde{y}_3, \varepsilon_3) + r_3^4\varepsilon_3^2\,G_3(r_3, \tilde{y}_3, \varepsilon_3),\end{aligned} \tag{6.4b}$$

$$\varepsilon_3' = -3\varepsilon_3. \tag{6.4c}$$

The function Ψ is C^k smooth, where k depends on the smoothness of (6.1). If (6.1) is C^∞ then k can be made arbitrarily large. We have the following result.

PROPOSITION 6.1. *The transition map* Π_3 *for system (6.1) has the form*

$$\Pi_3(r_3, y_3, \delta) = \begin{pmatrix} \rho \\ \Pi_{32}(r_3, y_3, \delta) \\ \left(\frac{r_3}{\rho}\right)^3 \delta \end{pmatrix},$$

with $\Pi_{32}(r_3, y_3, \delta)$ *given by*

$$\begin{aligned}\Pi_{32}(r_3, y_3, \delta) &= \Big(\Psi(r_3, y_3, \delta) - \delta + O(\rho)\Big)\left(\frac{r_3}{\rho}\right)^2\\ &\quad - (\bar{c} + O(\rho))\delta\rho\left(\frac{r_3}{\rho}\right)^3 \ln\left(\frac{r_3}{\rho}\right)\\ &\quad + (\delta + O(\rho))\left(\frac{r_3}{\rho}\right)^3 + O\left(r_3^2\left(\frac{r_3}{\rho}\right)^2 \ln\left(\frac{r_3}{\rho}\right)\right),\end{aligned} \tag{6.5}$$

where $\bar{c} = c + b$.

Proof. We rewrite the normal form equations (6.4) suppressing the subscript 3:

$$r' = r, \tag{6.6a}$$

$$\begin{aligned}\tilde{y}' &= -\varepsilon - 2\tilde{y} + \bar{c}r\varepsilon\\ &\quad + r^2\tilde{y}^2 G_1(r, \tilde{y}, \varepsilon) + r^3\tilde{y}\varepsilon G_2(r, \tilde{y}, \varepsilon) + r^4\varepsilon^2 G_3(r, \tilde{y}, \varepsilon),\end{aligned} \tag{6.6b}$$

$$\varepsilon' = -3\varepsilon. \tag{6.6c}$$

Consider vectors $(r_i, \tilde{y}_i, \varepsilon_i)$, $(r_o, \tilde{y}_o, \varepsilon_o)$ a solution $(r, \tilde{y}, \varepsilon)(t)$ and a number $T > 0$ such that

$$\begin{aligned} r(0) &= r_i, & r(T) &= r_o,\\ \tilde{y}(0) &= \tilde{y}_i, & \tilde{y}(T) &= \tilde{y}_o,\\ \varepsilon(0) &= \varepsilon_i, & \varepsilon(T) &= \varepsilon_o. \end{aligned}$$

We will now work out $(T, \tilde{y}_o, \varepsilon_o)$ as a function of $(r_i, \tilde{y}_i, \varepsilon_i, r_o)$. We first consider the vector field obtained from (6.4) by neglecting the higher order terms in (6.4c), namely

$$\begin{aligned} r' &= r, \\ \tilde{y}' &= -\varepsilon - 2\tilde{y} + \bar{c}r\varepsilon, \\ \varepsilon' &= -3\varepsilon. \end{aligned} \tag{6.7}$$

The solution of system (6.7) with initial conditions $(r_i, \tilde{y}_i, \varepsilon_i)$ is given by

$$r(t) = e^t r_i, \tag{6.8a}$$

$$\tilde{y}(t) = (\tilde{y}_i - \varepsilon_i + \bar{c}r_i\varepsilon_i t)e^{-2t} + \varepsilon_i e^{-3t}, \tag{6.8b}$$

$$\varepsilon(t) = e^{-3t}\varepsilon_i. \tag{6.8c}$$

The requirement $r(T) = r_o$ produces an expression for T, namely

$$T = \ln\left(\frac{r_o}{r_i}\right). \tag{6.9}$$

It follows that

$$\tilde{y}_o = \tilde{y}(T) = (\tilde{y}_i - \varepsilon_i)\left(\frac{r_i}{r_o}\right)^2 + \bar{c}\varepsilon_i\frac{r_i^3}{r_o^2}\ln\left(\frac{r_o}{r_i}\right) + \varepsilon_i(\frac{r_i}{r_o})^3,$$

$$\varepsilon_o = \left(\frac{r_i}{r_o}\right)^3 \varepsilon_i.$$

To obtain a solution of (6.6) we note that the r and ε coordinates are as given in (6.8a) and (6.8b), respectively. Substituting these expressions into (6.6c) we get the following equation

$$\tilde{y}' = -2y + \bar{c}r_i\varepsilon_i e^{-2t} - \varepsilon_i e^{-3t} + e^{-2t}r_i^2 H(\tilde{y}, r_i, \varepsilon_i, t), \tag{6.10}$$

where H is given by

$$\begin{aligned} H(\tilde{y}, r_i, \varepsilon_i, t) = {} & \tilde{y}_i^2 G_1(r_i e^t, \tilde{y}, \varepsilon_i e^{-3t}) \\ & + r_i\tilde{y}_i\varepsilon_i G_1(r_i e^t, \tilde{y}, \varepsilon_i e^{-3t}) \\ & + \varepsilon_i^2 r_i^2 G_2(r_i e^t, \tilde{y}, \varepsilon_i e^{-3t}). \end{aligned}$$

We introduce a new variable z by setting

$$\tilde{y} = e^{-2t}(z + \tilde{y}_i - \varepsilon_i + \bar{c}r_i\varepsilon_i t) + \varepsilon_i e^{-3t}.$$

Substituting into (6.10) we derive an equation for z

$$z' = r_i^2 \tilde{H}(z, r_i, \tilde{y}_i, \varepsilon_i, t), \tag{6.11}$$

where

$$\tilde{H}(z, r_i, \tilde{y}_i, \varepsilon_i, t) = H\Big(e^{-2t}(z + \tilde{y}_i - \varepsilon_i + \bar{c}r_i\varepsilon_i t) + \varepsilon_i e^{-3t}, r_i, \varepsilon_i, t\Big).$$

The transition time T is still given by (6.9). It is now easy to estimate using (6.11) that $z(T) = O(r_i^2 \ln(r_i/r_o))$. It follows that

$$\begin{aligned} \tilde{y}(T) = \quad & (\tilde{y}_i - \varepsilon_i)\left(\frac{r_i}{r_o}\right)^2 + \bar{c}\varepsilon_i \frac{r_i^3}{r_o^2} \ln\left(\frac{r_o}{r_i}\right) \\ & + \varepsilon_i \left(\frac{r_i}{r_o}\right)^3 + O\left(\frac{r_i^4}{r_o^2}\ln\left(\frac{r_o}{r_i}\right)\right). \end{aligned} \tag{6.12}$$

From (6.3) we conclude that $\tilde{y}_i = \Psi(r_i, y_i, \delta)$.

Finally wee need to apply the inverse of (6.3) which has the form

$$\begin{pmatrix} r \\ y \\ \varepsilon \end{pmatrix} = \begin{pmatrix} r \\ \tilde{\Psi}(r, \tilde{y}, \varepsilon) \\ \varepsilon \end{pmatrix}.$$

It follows from the form of (6.1) and (6.2) that $\tilde{\Psi}$ has the form

$$\tilde{\Psi}(r, \tilde{y}, \varepsilon) = \tilde{y} + O(\tilde{y}\epsilon, \epsilon^2, \tilde{y}r).$$

The result follows. □

We end this section by proving a lemma which will be useful in computing the coefficient c_1 in the expression for (2.6). Before stating the result we make the following definition:

$$\psi(y, \varepsilon) = \Psi(0, y, \varepsilon).$$

We have the following result.

LEMMA 6.3. *Suppose* $(0, y_i, \delta) = \kappa_{23}(\gamma_2 \cap \Sigma_2^{out})$. *Then*

$$(\psi(y_i, \delta) - \delta)\delta^{-\frac{2}{3}} = -\Omega_0.$$

Proof. Consider (6.2) with $r = 0$ (we suppress the subscript 3), namely

$$\begin{aligned} y' &= (-\varepsilon - 2y + 2y^2)/(1-y), \\ \varepsilon' &= -3\varepsilon. \end{aligned} \tag{6.13}$$

Equation (6.13) can also be obtained from the Ricatti equation (4.3) by applying the transformation κ_{23} and dividing out $\varepsilon^{1/3}(1-y)$. Note that the transformation

$$\begin{pmatrix} \tilde{y} \\ \varepsilon \end{pmatrix} = \begin{pmatrix} \psi(y, \varepsilon) \\ \varepsilon \end{pmatrix}$$

linearizes system (6.13), that is brings the equations to the form

$$\begin{aligned} \tilde{y}' &= -\varepsilon - 2\tilde{y}, \\ \varepsilon' &= -3\varepsilon. \end{aligned} \tag{6.14}$$

Any solution of (6.13) or (6.14) with non-zero initial condition in ε can be parametrized in terms of ε. Namely, if $(y(t), \varepsilon(t))$ is a solution of (6.13) with $(y(0), \varepsilon(0)) = (y_i, \varepsilon_i)$, $\varepsilon_i \neq 0$, then it is possible to solve for t as a function of ε and thus obtain a function $y(\varepsilon) = y(t(\varepsilon))$. Analogously one can obtain a function $\tilde{y}(\varepsilon)$, which in fact has the explicit form $\tilde{y}(\varepsilon) = (\tilde{y}_i - \varepsilon_i)\varepsilon_i^{-\frac{2}{3}}\varepsilon^{\frac{2}{3}} + \varepsilon$. Further, if $\tilde{y}_i = \psi(y_i, \varepsilon_i)$ then $\tilde{y}(\varepsilon) = \psi(y(\varepsilon), \varepsilon)$. Since

$$\psi(y, \varepsilon) = y + O(y^2, y\varepsilon)$$

it follows that

$$(\tilde{y}_i - \varepsilon_i)\varepsilon_i^{-\frac{2}{3}} = \lim_{\varepsilon \to 0} \tilde{y}(\varepsilon)\varepsilon^{-\frac{2}{3}} = \lim_{\varepsilon \to 0} y(\varepsilon)\varepsilon^{-\frac{2}{3}}.$$

If (y_i, ε_i) is the image by κ_{23} of a point on γ_2 then Proposition 4.1 implies that

$$\lim_{\varepsilon \to 0} y(\varepsilon)\varepsilon^{-2/3} = -\Omega_0.$$

The result follows. □

7. Proofs of the main results. In this section we prove the theorems stated in Section 2 by combining the results obtained in the individual charts. We begin by proving Theorem 2.2. The idea of the proof is to analyze the evolution of the center manifold $M_{a,1}$ under the flow of the blown-up vector-field $\bar{X}$. We denote the corresponding invariant manifold by $\bar{M}_a$.

Proof of Theorem 2.2. We define the map $\Pi : \Sigma_1^{in} \to \Sigma_3^{out}$ by

$$\Pi := \Pi_3 \circ \kappa_{23} \circ \Pi_2 \circ \kappa_{12} \circ \Pi_1,$$

where Π_j is the transition map from Σ_j^{in} to Σ_j^{out} in chart K_j, $j = 1, 2, 3$ and the maps κ_{12} and κ_{23} are the coordinate transformations from chart K_1 to K_3 and from chart K_2 to K_3, respectively. The map Π is the transition map from Σ_1^{in} to Σ_3^{out} for the flow induced by the blown-up vector-field $\bar{X}$ on B_0. We will analyze $\Pi(R_1 \cap M_{a,1})$ and then use the appropriate blow-down transformations to obtain the result.

We first consider the set $\Pi_1(R_1 \cap M_{a,1})$. It follows from Proposition 5.2 that $\Pi_1(R_1 \cap M_{a,1}) \subset \Sigma_1^{out}$ is a smooth curve transverse to the set $\{r_1 = 0\}$.

The transformation κ_{12} from K_1 to K_2 is given by (3.9). Recall that $\kappa_{12}(\Sigma_1^{out}) = \Sigma_2^{in}$. It follows that $\kappa_{12}(\Pi_1(R_1 \cap M_{a,1}))$ is a smooth (C^k, for some k) curve transverse to the plane $\{r_2 = 0\}$. We now apply the transition map Π_2 to this curve. It follows from Proposition 4.2 that the image of this curve under Π_2 has the form

$$\left\{(\delta^{-\frac{1}{3}}, h_2^{out}(r_2), r_2) \ : \ r_2 \in [0, \rho\delta^{1/3}]\right\}, \tag{7.1}$$

where $h_2^{out} : [0, \rho\delta^{1/3}] \to \mathbb{R}$ is a smooth function. Proposition 4.1 implies that

$$\left(\delta^{-\frac{1}{3}}, h_2^{out}(0), 0\right) = \gamma_2 \cap \Sigma_2^{out}.$$

The transformation κ_{23} described by equation (3.11) maps Σ_2^{out} to Σ_3^{in}. Under this transformation the curve (7.1) transforms to a smooth curve of the form

$$\left\{(r_3, h_3^{in}(r_3), \delta) \ : \ r_3 \in [0, \rho]\right\}$$

with

$$(0, h_3^{in}(0), \delta) = \kappa_{23}(\gamma_2 \cap \Sigma_2^{out}).$$

Proposition 6.1 now implies that $\Pi(R_1 \cap M_{a,1})$ has the form

$$\left\{(\rho, h_3^{out}(\varepsilon_3), \varepsilon_3) \ : \ \varepsilon_3 \in [0, \delta]\right\},$$

where

$$\begin{aligned} h_3^{out}(\varepsilon_3) &= (-\Omega_0 + O(\rho))\, \varepsilon_3^{\frac{2}{3}} - \frac{(\bar{c} + O(\rho))\rho}{3} \varepsilon_3 \ln(\varepsilon_3) \\ &\quad + (\tilde{c} + O(\rho))\, \varepsilon_3 + O(\varepsilon_3^{\frac{4}{3}} \ln(\varepsilon_3)), \end{aligned} \tag{7.2}$$

where $\tilde{c}$ is a coefficient which can be computed from the right hand side of (1.1). The assertion of Theorem 2.2 follows by applying transformation (3.2) restricted to Σ_1^{in} and Σ_3^{out}, since $\Phi(\Sigma_1^{in}) = \Delta^{in} \times [0, \rho^3\delta]$ and $\Phi(\Sigma_3^{out}) = \Delta^{out} \times [0, \rho^3\delta]$. □

REMARK 7.1. The sections Σ_1^{in} and Σ_3^{out} are defined by the constant ρ. In the original problem these are the sections Δ^{in} and Δ^{out}. Note that we have obtained the expression for c_1 and c_2 up to $O(\rho)$. Obviously a more precise result could be obtained if we worked out the near identity transformation (6.3) in more detail. We believe that asymptotic expansions like (2.6) can be conveniently computed and justified using the blow-up technique and intend to develop the method in this direction in future work.

Proof of Theorem 2.1. It follows from Proposition 5.2 that $\Pi_1(R_1)$ is a wedge-like region around $\Pi(R_1 \cap M_{a,1})$ of width $O(e^{-\frac{C}{\varepsilon_1}})$, where $C > 0$

is some constant. Since κ_{12}, Π_2 and κ_{23} are diffeomorphisms restricted to Σ_1^{out}, Σ_2^{in} and Σ_2^{out}, respectively, it follows that $\kappa_{23} \circ \Pi_2 \circ \kappa_{12} \circ \Pi_1(R_1)$ is also a wedge-like region of size $O(e^{-\frac{C}{\varepsilon_1}})$ around $\kappa_{23} \circ \Pi_2 \circ \kappa_{12} \circ \Pi_1(R_1 \cap M_{a,1})$. Finally we apply Proposition 6.1 to conclude that $\Pi(R_1)$ is a wedge-like region of width $O(e^{-\frac{C}{\varepsilon_1}})$ around $\Pi(R_1 \cap M_{a,1})$. The evolution of R_1 in the three charts is shown in Figure 6. The theorem follows by applying appropriate blow-down transformations. $\square$

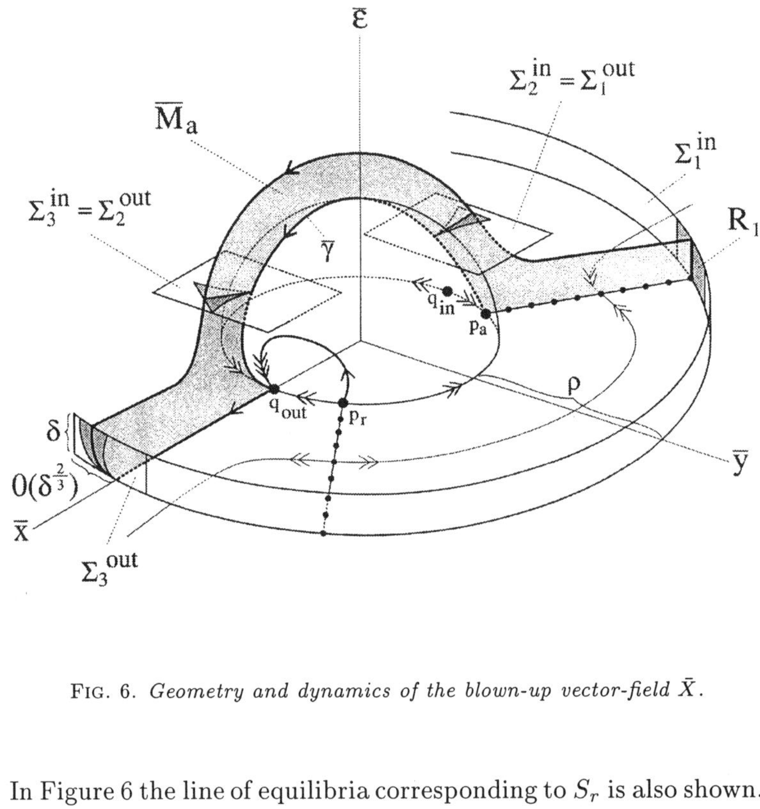

FIG. 6. *Geometry and dynamics of the blown-up vector-field $\bar{X}$.*

In Figure 6 the line of equilibria corresponding to S_r is also shown. The center-manifold $M_{r,1}$ at p_r can be extended to a locally invariant manifold $\bar{M}_r$, which can be studied analogously. The connecting orbit from p_r to q_{out} on the sphere is not unique. In Figure 6 we have drawn the unique connecting orbit tangent to the strong stable direction at q_{out}.

On the equator of S^2 there is an additional hyperbolic equilibrium q_{in} diametral opposite of q_{out}, which can be studied in a chart corresponding to a directional blow-up in the negative x direction. The equilibrium q_{in} is fully repelling for the flow on the sphere and has an attracting direction in the plane $\bar{\varepsilon} = 0$. This equilibrium did not enter our analysis, it is however important in the following situation. Recall that, in Section 2, we have made the assumption that the reduced flow is directed towards the fold, i.e. $g(0,0,0) < 0$. The other case $g(0,0,0) > 0$ can be treated in a similar way, however, in this case one has to analyze how solutions that reach the sphere close to this fourth equilibrium connect to the slow manifolds $\bar{M}_a$ and $\bar{M}_r$.

Acknowledgement. The authors would like to thank Freddy Dumortier, Alexandra Milik and Martin Wechselberger for helpful discussions. The authors also acknowledge support and hospitality of the Institute for Mathematics and Its Applications, where a part of this research was carried out.

REFERENCES

[1] V. I. ARNOLD (Ed.), *Dynamical Systems 5*, Encyclopedia of Mathematical Sciences, Springer (1989).

[2] B. BRAAKSMA, *Critical phenomena in dynamical systems of van der Pol type*, Thesis, University of Utrecht (1993).

[3] C. M. BENDER AND S. A. ORSZAG, *Advanced mathematical methods for scientists and engineers*, McGraw-Hill (1978).

[4] E. BENOIT (Ed.), *Dynamic Bifurcations*, Lecture Notes in Mathematics 1493, Springer, Berlin (1991).

[5] M. DIENER, *Regularizing microscopes and rivers*, SIAM J. Appl. Math. **25**, pp. 148–173 (1994).

[6] F. DUMORTIER, Techniques in the Theory of Local Bifurcations: Blow-Up, Normal Forms, Nilpotent Bifurcations, Singular Perturbations, in *Bifurcations and Periodic Orbits of Vector Fields* (ed. D. Szlomiuk), Kluwer **C408**, Dordrecht (1993).

[7] F. DUMORTIER AND R. ROUSSARIE, Canard cycles and center manifolds, Memoirs of the AMS **557** (1996).

[8] W. ECKHAUS, Relaxation oscillations including a standard chase on French ducks, in *Asymptotic Analysis II*, Springer Lecture Notes Math. **985**, pp. 449–494 (1983).

[9] N. FENICHEL, *Geometric singular perturbation theory*, J. Diff. Eq. **31**, pp. 53–98 (1979).

[10] J. GRASMAN, *Asymptotic methods for relaxation oscillations and applications*, Springer, New York (1987).

[11] J. GUCKENHEIMER AND P. HOLMES, *Nonlinear Oscillations, Dynamical Systems, and Bifurcations of Vector Fields*, Springer, New York (1983).

[12] C. K. R. T. JONES, Geometric singular perturbation theory, in *Dynamical Systems*, Springer Lecture Notes Math. **1609**, pp. 44–120 (1995).

[13] J. KEVORKIAN J. D. COLE, *Perturbation methods in applied mathematics*, Springer, New York (1981).

[14] M. KRUPA AND P. SZMOLYAN, *Extending geometric singular perturbation theory to nonhyperbolic points*, in preparation.

[15] P. A. LAGERSTROM, *Matched Asymptotic Expansions*, Springer (1988).

[16] X. B. LIN, *Heteroclinic bifurcation and singularly perturbed boundary value problems*, J. Diff. Eq. **84**, pp. 319–382 (1990).

[17] E. F. MISHCHENKO AND N. KH. ROZOV, *Differential equations with small parameters and relaxation oscillations*, Plenum, New York (1980).

[18] E. F. MISHCHENKO, YU. S. KOLESOV, A. YU. KOLESOV AND N. KH. ROZOV, *Asymptotic methods in singularly perturbed systems*, Consultants Bureau, New York and London (1994).

[19] A. I. NEISHTADT, *Persistence of stability loss for dynamical bifurcations I*, Differential Equations **23**, pp. 1385–1391 (1987).
stability loss for dynamical bifurcations II,

[20] K. NIPP, *Breakdown of stability in singularly perturbed autonomous systems I. Orbit equations*, SIAM J. Math. Anal. **17**, pp. 512–532 (1986).

[21] R. E. O'MALLEY, *Introduction to singular perturbations*, Academic Press, Inc., New York (1974).

[22] L. S. PONTRYAGIN, *Asymptotic properties of solutions of differential equations with small parameter multiplying leading derivatives*, Izv. AN SSSR, Ser. Matem. 21, **5**, pp. 605–626 (1957).

[23] S. STERNBERG, *On the nature of local homeomorphisms of Euclidean n-space II*, Am. J. Math. **80** pp. 623–631 (1958).

[24] P. SZMOLYAN, *Transversal heteroclinic and homoclinic orbits in singular perturbation problems*, J. Diff. Eq. **92**, pp. 252–281 (1991).

[25] P. SZMOLYAN, *Geometry of singular perturbations: A case study*, in preparation.

[26] M. WECHSELBERGER, *Singularly perturbed folds and canards in* $\mathbb{R}^3$, Thesis, TU-Wien (1998).

MULTIPLE TIME SCALES AND CANARDS IN A CHEMICAL OSCILLATOR

ALEXANDRA MILIK* AND PETER SZMOLYAN†

Abstract. We present a geometric singular perturbation analysis of a chemical oscillator. Although the studied three-dimensional model is rather simple, its dynamics are quite complex. In the original scaling the problem has a folded critical manifold which additionally becomes tangent to the fast fibers in a region relevant to the dynamics. Thus normal hyperbolicity of the critical manifold is lost in two regions. The dynamics depends crucially on effects due to the loss of normal hyperbolicity. In particular, canard solutions play an essential role. We outline how rescalings and blow-up techniques can be used to prove the existence of canards in this problem and to explain other qualitative aspects of the dynamics.

Key words. Geometric singular perturbation theory, chemical oscillator, canard solutions, blow-up.

AMS(MOS) subject classifications. Primary 34A36, 58F30, 34E15.

1. Introduction. Chemically reacting systems are prominent examples of dynamical systems evolving on often widely different time scales. They do this in a way which pushes the borders of the available mathematical techniques for the analysis of slow-fast systems. We demonstrate in this paper that even in rather simple equations a sophisticated singular perturbation analysis is needed to understand the dynamics. We do this in the context of a specific system the so called 3-dimensional autocatalator. However, the phenomena and also the techniques used in the analysis are of a more general nature.

The rate equations for the 3-dimensional autocatalator have the following dimensionless form [17]

$$
\begin{aligned}
\frac{da}{dt} &= \mu(\frac{5}{2} + c) - ab^2 - a \\
\varepsilon\frac{db}{dt} &= ab^2 + a - b \\
\frac{dc}{dt} &= b - c\,.
\end{aligned}
\tag{1.1}
$$

The variables a, b, and c represent concentrations of three different chemical species depending on time t. Hence, only nonnegative values of

*Institut für Angewandte und Numerische Mathematik, Technische Universität Wien, Vienna, AUSTRIA / IMA, University of Minnesota, U.S.; milik@ima.umn.edu; research supported by the Austrian Science Foundation under grant J 1449-MAT.

†Institut für Angewandte und Numerische Mathematik, Technische Universität Wien, Vienna, AUSTRIA; ps@umbriel.tuwien.ac.at; research supported by the Austrian Science Foundation under grant Y 42-MAT.

a, b, and c are considered in the following. The parameter ε is small. The parameter μ, which will be close to $2/7$ throughout this paper, is considered as a bifurcation parameter.

We decided to work with the constants $5/2$ and $2/7$ to remain close to the parameter values used in the detailed numerical study [17]. Depending on μ this model shows a rich variety of dynamical behaviour known collectively as mixed mode oscillations. Mixed-mode oscillations are solutions containing certain combinations of alternating large and small amplitude excursions. See Figure 1 which shows a typical time series with one large and two small peaks. This time series is obtained by a simulation of model (1.1). As μ varies in a tiny interval close to $2/7$ a Farey sequence of bifurcations of periodic orbits (of mixed mode type) leading from small Hopf-type periodic solutions to large relaxation type periodic solutions is observed. The periodic behaviour is intertwined with apparently chaotic behaviour [17]. In fact, our motivation for the study of this model was to find a geometric mechanism which explains the quite common phenomenon of mixed-mode oscillations [14].

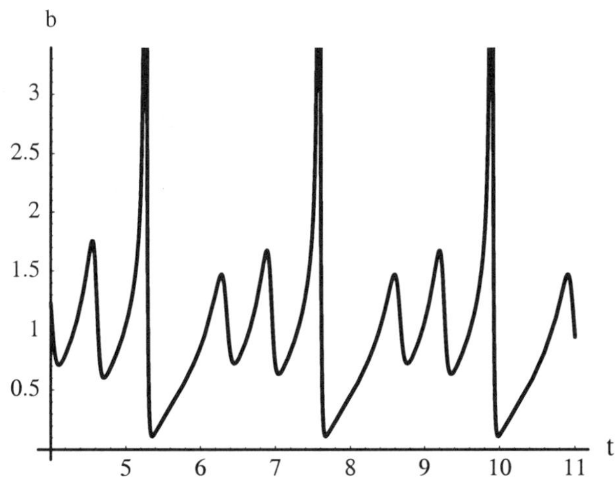

FIG. 1. *Time series of the variable b of the 3d-autocatalator. Parameters: $\mu = 0.299$, $\varepsilon = 0.013$.*

System (1.1) is already in a form which suggests the application of singular perturbation theory. We follow the geometric approach to singular perturbation theory, see N. Fenichel's original paper [9] or the recent survey by C. Jones [11]. By setting $\varepsilon = 0$ we obtain the equation $ab^2 + a - b = 0$ which defines the 2-dimensional critical manifold, which is a folded surface with one unbounded branch, see Figure 2. The lower branch of the critical manifold is attracting, the upper branch is repelling. Away from the fold with bounded values of b the critical manifold perturbs smoothly to two slow manifolds for small ε. The slow manifold close to the lower branch of the critical manifold is still strongly attracting, the other one is still strongly repelling.

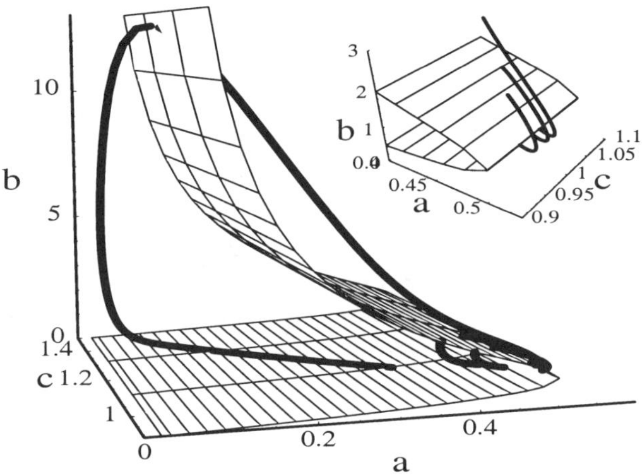

FIG. 2. *Critical manifold S with lower attracting branch S_a and upper repelling branch S_r and a typical solution of the 3-dimensional autocatalator. The parameters are $\varepsilon = 0.013$ and $\mu = 0.299$.*

However, recall that Fenichel's theory requires normal hyperbolicity of the critical manifold. Near the fold line and for $b \to \infty$ normal hyperbolicity of the critical manifold is lost. Thus the standard theory does not apply in these regions. Typically, a fold in a critical manifold leads to jump points, that is the orbit follows the attracting slow manifold until it reaches the fold and then jumps away almost vertically following the dynamics of the layer problem. This is the the well-known mechanism which leads to relaxation oscillations as, for instance, in van der Pol's equation, see [16]. Thus, one might expect to see solutions of system (1.1) following the attracting slow manifold until they come close to the fold, where an "explosive" growth of b would take place.

However, this behaviour is observed only at certain parts of the fold. To get an impression of often observed behaviour, as well as how the dynamics is organized by the slow manifold, we superimposed a numerically obtained solution onto a picture of the critical manifold in phase space (see Figure 2). We summarize the results which can be obtained by such simulations:

1. As one expects, all solutions starting between the attracting and the repelling slow manifold are quickly attracted to the attracting slow manifold. The flow on the attracting slow manifold is towards the fold.

2. There exist small loops where solutions first follow the attracting branch of the slow manifold, then the repelling branch for a while until they fall back to the attracting branch.
3. There exist large loops where solutions first follow the attracting branch, then the repelling branch for a while, then jump non-vertically away from the slow manifold, make a large excursion, and eventually fall back to the attracting branch of the slow manifold.
4. As a solution comes close to the fold the value of c seems to determine which behaviour occurs, i.e. there seems to exist a $c^*(\varepsilon) \approx 1$ such that for $c < c^*(\varepsilon)$ small loops occur, and for $c > c^*(\varepsilon)$ large loops occur. The transition between small and large loops is rather sharp.
5. After a small loop there is a net increase in the value of the variable c, while after a large loop there is a net decrease.
6. For solutions on the attracting slow manifold passing the fold at c distinctly larger than $c^*(\varepsilon)$ the typical transition to an almost vertical motion occurs. For solutions on the repelling slow manifold passing the fold (in backwards time !) at c distinctly smaller than $c^*(\varepsilon)$ the typical transition to an almost vertical motion occurs.
7. Observation 6. is actually only correct as long as $b = O(1)$, on the scale $b = O(1/\varepsilon)$ the fast dynamics is approximately along planes $a + b/\varepsilon = const.$.

Much of this can be seen in Figure 2. It can be explained by our main result:

THEOREM 1.1. *The forward in time continuation of the attracting slow manifold and the backward in time continuation of the repelling slow manifold intersect transversally in an orbit γ_ε. The orbit γ_ε intersects the plane $b = 1$ at a point $(a^*(\varepsilon), 1, c^*(\varepsilon))$. The intersection of the slow manifolds occurs in such a way that on the left side of γ_ε, i.e. on the side corresponding to smaller values of c, the continuation of the slow attracting manifold lies below the repelling slow manifold and vice-versa on the right side.*

See Figure 3 for a schematic view of the intersection of the slow manifolds.

At the point $(a^*(\varepsilon), 1, c^*(\varepsilon))$ the orbit γ_ε "turns around", i.e. there the variable a has its maximum along the orbit. We have introduced this point to have a precise description of the location of the orbit γ_ε close to the fold line. Clearly, this configuration of the slow manifolds explains the observations made above concerning the dynamics close to the fold. In the above description, and also in the above theorem, we have suppressed the smooth dependence of γ_ε and $c^*(\varepsilon)$ on the parameter μ.

In singularly perturbed systems an orbit like γ_ε is called a (maximal) canard. More generally, orbits that follow a repelling slow manifold for an $O(1)$ amount of time are called canards [4, 6]. Canards were discovered by E. Benoit, J.L. Callot, F. Diener, and M. Diener and first analyzed by

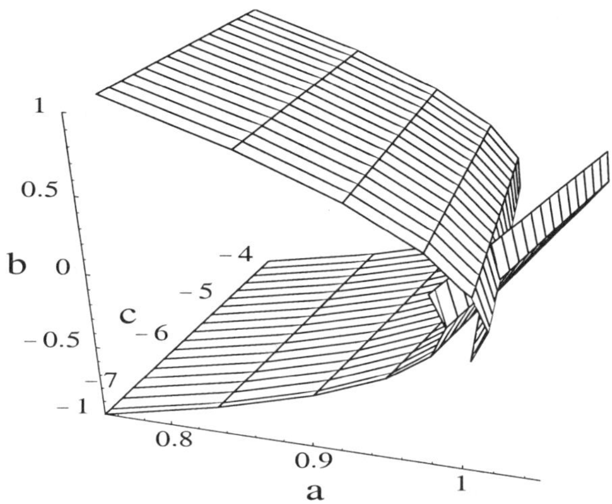

FIG. 3. *Schematic view of the intersection of the perturbed slow manifolds* S_a^ε *(lower surface) and* S_r^ε *(upper surface).*

means of nonstandard analysis [4]. Canards arise in one-parameter families of planar singularly perturbed systems if an equilibrium passes through the fold point as the parameter varies. For a treatment of this situation which also leads to the so called singular Hopf bifurcation by matched asymptotic expansions we refer to [2, 5, 8, 15]. Recently, F. Dumortier and R. Roussarie [7] gave a detailed geometric explanation and proof of the canard phenomenon in two dimensions by using the blow-up of singularities and foliations by center manifolds as the main techniques. For more details on their work we also refer to their article in this volume. We have worked out a complete blow-up analysis for the 2-dimensional autocatalator

$$
\begin{aligned}
\frac{da}{dt} &= \mu - ab^2 - a \\
\varepsilon\frac{db}{dt} &= ab^2 + a - b,
\end{aligned}
\tag{1.2}
$$

where the reactant corresponding to the variable c is fed into the reactor at a constant rate controlled by the parameter μ. Even in this much simpler problem the blow-up analysis of a planar canard (for μ close to 1) must be combined with another blow-up analysis, which is necessary to analyze the unbounded branch of the critical manifold, see [18].

Results on canards in higher dimensional systems by means of nonstandard analysis have been obtained by Benoit [3] and by means of matched asymptotic expansions in [15]. In three-dimensional problems with a folded critical manifold certain types of canards, which can be classified according the dynamics of the reduced flow on the critical manifold, occur generically. See also the survey in [1]. We will briefly comment on these types of canards in the next section.

In system (1.1) we have an even more degenerate case since there is a fixed point of the system close to the fold. In section 2 we will classify this case as a folded saddle-node. In [14] we argued analytically and numerically that mixed-mode oscillations can be explained convincingly by relating the dynamics of system (1.1) to the dynamics of a bimodal one-dimensional map. Reduction by one dimension is possible by introducing a suitable two-dimensional return map. The reduction to a one-dimensional map can be justified (somewhat heuristically) by the strong contraction of the flow near the attracting slow manifold. In the definition of the return map the canard associated with the folded saddle node plays the crucial role of separating the small loops from the large loops.

In this paper we discuss how certain parts of the construction in [14] can be put on a rigorous foundation. This requires extensions of the, by now well established, methods of geometric singular perturbation theory to parts of the critical manifold where normal hyperbolicity breaks down. Inspired by the pioneering work of F. Dumortier and R. Roussarie [7] on the geometric analysis of canard cycles in Van der Pol's equation such a theory for higher-dimensional problems is being developed. For an introduction we refer to the paper in this volume by M. Krupa and the second author. A systematic treatment of three-dimensional systems with a folded critical manifold is given in [19].

The essential tool is to blow-up degenerate equilibria corresponding to non-hyperbolic points of critical manifolds. Since the singular perturbation parameter ε plays an essential role in the blow-up a planar equation leads to a three-dimensional problem after the blow-up. For a three-dimensional equation the blown-up problem is four-dimensional. It can be even necessary to include parameters in the blow-up. Thus the complexity of the geometry, and hence of the analysis, increases with increasing dimensions. As we will see several different blow-ups have to be used for system (1.1).

In the 3-dimensional autocatalator the (slow) variable c behaves in a similar way to the parameter μ in (1.2); however, due to the coupling of all variables a more complicated situation arises. This paper can be seen as a report on work in progress, and also as an attempt to build a bridge from a more conventional asymptotic analysis to a geometric analysis based on blow-up methods.

We now outline the content of the rest of the paper. In Section 2 we study the reduced problem and the layer problem corresponding to system (1.1). We identify a degenerate singularity on the fold line as a folded saddle node. We show that there exists a unique solution of the reduced problem on the critical manifold which passes at the folded saddle-node from the attracting branch of the critical manifold to the repelling branch. We will prove that this solution is the singular limit of the canard.

All variables and even the singular perturbation parameter ε have to be rescaled in the regions where normal hyperbolicity is lost. Actually one needs to perform a blow-up which is more than just a rescaling. However, the essential global dynamics, i.e. connections between different slow manifolds, can be studied in a chart which corresponds to a rescaling of the variables. In Section 3, we introduce a rescaling suitable to study the existence of the canard. This leads to a singular perturbation of an integrable system, which can be roughly described as the normal form of a planar canard, where the control parameter is replaced by another slow variable. The unperturbed integrable system has one special solution which is asymptotic to both branches of the slow manifold. We show that this solution persists under perturbation. In the original coordinates this solution corresponds to the maximal canard γ_ε along which the continuations of the attracting and repelling slow manifolds intersect (see Figure 3).

In section 4 we show that the unboundedness of the critical manifold for $a \to 0+$ and the unboundedness of orbits of the layer problem require another rescaling which transforms system (1.1) into a different singularly perturbed system. Although this system is not in standard form it can be analysed by a straightforward application of geometric singular perturbation methods [9]. This leads to a mechanism that describes the large amplitude excursions. However, rigorous matching with solutions described in the original scaling also requires the blow-up method.

2. Reduced dynamics and slow manifolds. In the following we study system (1.1) for μ close to 2/7, which is the parameter range where mixed mode oscillations occur.

System (1.1) is a singularly perturbed problem on the slow time scale t with ε being the small perturbation parameter. The variable b is the fast variable while a and c are slow variables. Taking the limit $\varepsilon \to 0$ yields the reduced problem

$$\frac{da}{dt} = \mu(\frac{5}{2} + c) - ab^2 - a \tag{2.1}$$

$$0 = ab^2 + a - b \tag{2.2}$$

$$\frac{dc}{dt} = (b - c)\,. \tag{2.3}$$

This is a two-dimensional dynamical system on the critical manifold S defined by the equation $ab^2 + a - b = 0$. Solving for

$$a = \frac{b}{b^2 + 1} \tag{2.4}$$

shows that this manifold is a folded surface in (a, b, c)-space with the fold line defined by $a = 1/2$, $b = 1$. We denote the lower, i.e. $b < 1$, part of S by S_a and the upper part of S by S_r (see Figure 2).

Transforming system (1.1) to the fast timescale $\tau := t/\varepsilon$ yields the equivalent fast system

$$\begin{aligned} \frac{da}{d\tau} &= \varepsilon(\mu(\frac{5}{2}+c) - ab^2 - a) \\ \frac{db}{d\tau} &= ab^2 + a - b \\ \frac{dc}{d\tau} &= \varepsilon(b-c)\,. \end{aligned} \tag{2.5}$$

By taking the limit $\varepsilon \to 0$ in this system we obtain the layer problem

$$\begin{aligned} \frac{da}{d\tau} &= 0 \\ \frac{db}{d\tau} &= ab^2 + a - b \\ \frac{dc}{d\tau} &= 0\,. \end{aligned} \tag{2.6}$$

The layer problem is a one-dimensional dynamical system in the fast variables b with the slow variables a and c acting as parameters. The local stability of points in S as steady states of the layer problem (2.6) is determined by the derivative

$$\left.\frac{\partial(ab^2+a-b)}{\partial b}\right|_S = 2ab - 1|_S\,. \tag{2.7}$$

Thus, points in S_a are attracting and points in S_r are repelling. We say that S_a is an attracting critical manifold and S_r a repelling critical manifold. On the fold line the derivative (2.7) is zero. S_a is normally hyperbolic for $0 \le a \le a_1 < 1/2$, close to the fold line normal hyperbolicity is lost. Furthermore, we observe that the repelling critical manifold S_r becomes more and more aligned with the (vertical) unstable direction as $a \to 0+$. Hence S_r is normally hyperbolic only for $0 \le a_0 \le a \le a_1 < 1/2$. We conclude from Fenichel's theorems that normally hyperbolic parts of S_a and S_r perturb smoothly to locally invariant slow manifolds S_a^ε and S_r^ε of the same stability type as S_a and S_r. In the following we will think of these manifolds as being extended by the forward and backward flow, however, we cannot control their global dynamics by appealing to the singular limits used so far.

With a slight abuse of notation, we will sometimes think of the critical manifolds as three-dimensional manifolds in $(a, b, c, \varepsilon, \mu)$ space. Note, that the slow manifolds S_a^ε and S_r^ε are constructed as $\varepsilon = const.$ sections of four-dimensional, center-like manifolds M_a and M_r, see [9].

Next we investigate the dynamics of the reduced problem (2.1-2.3) on the critical manifold S. The description of S as a graph over the (b, c)-plane suggests projecting the reduced problem into the (b, c)-plane. By differentiating equation (2.4) with respect to time and substituting for da/dt we obtain

$$
\begin{aligned}
(1-b^2)\frac{db}{dt} &= (1+b^2)^2\left(\mu\left(\frac{5}{2}+c\right)-b\right) \\
\frac{dc}{dt} &= b-c\,.
\end{aligned}
\tag{2.8}
$$

System (2.8) is singular at the fold, i.e. at $b = 1$. The standard existence and uniqueness results for differential equations do not hold there. In particular different solutions of system (2.8) may approach the same point on the fold in finite forward or backward time. To obtain the phase portrait of this singular system we study the desingularized system

$$
\begin{aligned}
\frac{db}{dt} &= (1+b^2)^2\left(\mu\left(\frac{5}{2}+c\right)-b\right) \\
\frac{dc}{dt} &= (b-c)(1-b^2)\,,
\end{aligned}
\tag{2.9}
$$

which is obtained by multiplying the right hand side of (2.8) with the term $(1 - b^2)$. The new system has the same phase portrait as system (2.8) on S_a. On S_r, that is for $b > 1$, time has to be reversed in the phase portrait of system (2.9) to obtain the phase portrait of the reduced system (2.8). A computation proves:

LEMMA 2.1. *System (2.9) has two fixed points depending on μ:*

$$
p = \left(1, \frac{1}{\mu} - \frac{5}{2}\right), \quad q = \left(\frac{5\mu}{2(1-\mu)}, \frac{5\mu}{2(1-\mu)}\right).
$$

For $\mu < 2/7$ the point p is a saddle, q is a stable node. For $\mu = 2/7$ the two fixed points coalesce in the point $(1,1)$ in a transcritical bifurcation. For $\mu > 2/7$ the point p is a stable node, q is a saddle.

Note that the fixed point q corresponds to a genuine equilibrium of the full problem (1.1). The point p on the fold line is a singular point for the reduced problem (2.8), called a folded saddle (for $\mu < 2/7$) and a folded node (for $\mu > 2/7$). The direction of the flow of system (2.9) has to be reversed for $b > 1$ to obtain the flow of (2.8).

In this way the stable, respectively unstable, manifolds of the saddle become trajectories of (2.8) which pass from S_a to S_r, respectively from S_r to S_a, through p. At the folded node p all trajectories pass from S_a to S_r along the weaker direction of its stable manifold, the only exception is one trajectory which passes tangent to the strong stable direction from S_a to S_r. All solutions passing from S_a to S_r through the folded saddle or the folded node are candidates for canards.

However, we focus our attention on canards generated at the special value $\mu = 2/7$. For this value the genuine equilibrium q and the singular point p coalesce in the point $(1,1)$. Thus the situation is the 3-dimensional extension of the 2-dimensional scenario for canards, where an equilibrium passes through a fold point as a parameter varies. From this point of view the phenomena studied here in a concrete problem are clearly of a general nature.

Now we give a more detailed description of the dynamics of the reduced problem for $\mu = 2/7$. We start by analysing the flow of system (2.9).

LEMMA 2.2. *The center direction for the nonhyperbolic fixed point* $q = (1,1)$ *of system (2.9) with* $\mu = 2/7$ *is the vector* $(2,7)$. *There exists a center manifold on which the flow is weakly contracting towards* q *for* $b < 1$, *i.e. in* S_a, *and weakly repelling for* $b > 1$, *i.e. in* S_r. *There exists a strong stable manifold tangent to the vector* $(1,0)$.

Proof. The linearization at the equilibrium is the matrix

$$\begin{pmatrix} -4 & \frac{8}{7} \\ 0 & 0 \end{pmatrix},$$

which implies all assertions except the description of the center flow. The assertion about the center-flow follows from a center manifold computation. □

To obtain the phase portrait of the reduced problem (2.1-2.3) shown in Figure 4 one has to reverse the direction of the flow on S_r. We call the singular point $(1,1)$ a folded saddle-node. We remark that we use this terminology because of the phase portrait of problem (2.9) for fixed $\mu = 2/7$. As we have seen in Lemma 2.1 the unfolding of this degenerate singularity by μ is a transcritical bifurcation of equilibria for system (2.9). The folded saddle node is a singular point of the reduced problem. Solutions on the former center manifold converge to the folded saddle-node with an exponential rate as $t \to \infty$. This follows because of a cancellation between the quadratic zero of the flow of system (2.9) on the center manifold and the factor $(1-b)$ of the derivative db/dt in system (2.8).

The strong stable manifold becomes a solution γ which crosses the fold exactly at the folded saddle node. Moreover, γ is the only solution with this property. We call the orbit γ the singular canard. Trajectories on S_a on the left side of γ are attracted by the former center manifold. All solutions starting in S_a on the right side of γ approach points on the fold in finite time. The folded saddle node is the only point on the critical manifold S where a transition from S_a to S_r is possible. At points on the fold with $c > 1$ the reduced flow cannot be continued in forward time, that is a transition to a fast motion must occur. Similarly, at points on the fold with $c < 1$ the reduced flow cannot be continued backwards, that is a transition to a fast motion must occur in backwards time.

In the next section we analyse the behaviour of the slow manifolds for μ close to $2/7$ for small positive ε. In particular we will show that the

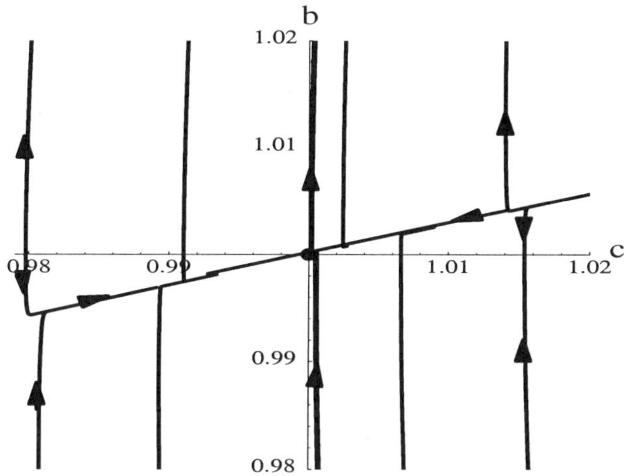

FIG. 4. *Computed phase portrait of the reduced problem in the* (c, b)*-plane with folded saddle-node* q *(black point) and singular canard* γ *(bold line).*

singular canard persists as a maximal canard γ_ε along which the extended slow manifolds S_a^ε and S_r^ε intersect.

3. Dynamics close to the folded saddle-node. The starting point of the analysis close to the folded saddle-node for small ε and μ close to 2/7 is equation (2.5)

$$
\begin{aligned}
\frac{da}{d\tau} &= \varepsilon(\mu(\frac{5}{2} + c) - ab^2 - a) \\
\frac{db}{d\tau} &= ab^2 + a - b \\
\frac{dc}{d\tau} &= \varepsilon(b - c) \qquad (3.1) \\
\frac{d\varepsilon}{d\tau} &= 0 \\
\frac{d\mu}{d\tau} &= 0\,,
\end{aligned}
$$

which are the equations on the fast scale extended by the trivial equations for the parameters ε and μ. All points on the fold line are equilibria of system (3.1) for $\varepsilon = 0$. The linearization of system (3.1) at a point $(1/2, 1, c, 0, \mu)$ on the fold line has $\lambda = 0$ as an eigenvalue of multiplicity five. At other points of the critical manifold S, which are also equilibria for $\varepsilon = 0$, the linearization has one nonzero (negative in S_r, positive in S_a) eigenvalue and an eigenvalue $\lambda = 0$ of multiplicity four. The degenerate equilibria on the fold must be treated with the blow-up method.

3.1. Blow-up and charts. In the following we focus on the folded saddle-node $(1/2, 1, 1, 0, 2/7)$. We first perform a coordinate transformation which shifts the folded saddle-node to the origin and blows up the shifted variables according to the transformation

$$\begin{aligned} a &= \frac{1}{2} + \bar{r}^2\bar{a} \\ b &= 1 + \bar{r}\bar{b} \\ c &= 1 + \bar{r}\bar{c} \\ \varepsilon &= \bar{r}^2\bar{\varepsilon} \\ \mu &= \frac{2}{7} + \bar{r}\bar{\mu}\,. \end{aligned} \tag{3.2}$$

Note that the weights, i.e. the powers of $\bar{r}$, used for the individual variables are the same as in the analysis of planar canards at a fold point. The variable c and the parameter μ both have weight one, they both play a role analogous to the parameter in the case of planar canards.

We define the manifold $B = S^4 \times \mathbb{R}$ and $B_0 = S^4 \times [0, r_0)$ for sufficiently small $r_0 > 0$ and regard the transformation (3.2) as the mapping

$$\phi : B_0 \to \mathbb{R}^5 \tag{3.3}$$

with $(\bar{a}, \bar{b}, \bar{c}, \bar{\varepsilon}, \bar{\mu}) \in S^4$ defined by $\bar{a}^2 + \bar{b}^2 + \bar{c}^2 + \bar{\varepsilon}^2 + \bar{\mu}^2 = 1$. We will be mostly interested in nonnegative values of $\bar{\varepsilon}$ corresponding to nonnegative ε. Note that all objects defined in the following extend smoothly to $\bar{r} < 0$, hence there are no problems associated with the boundary $\bar{r} = 0$ of B_0. Let X denote the vector field corresponding to (3.1). Since X vanishes at the point $(1/2, 1, 1, 0, 2/7)$ there exists a vector field $\bar{X}$ on B_0 such that

$$\phi_* \bar{X} = X\,,$$

where ϕ_* is induced by ϕ. It remains to study the vector field $\bar{X}$ on the manifold B_0. In fact to analyse system (3.1) in a neighborhood of the point $(1/2, 1, 1, 0, 2/7)$ it suffices to study $\bar{X}$ on B_0 for small $r_0 > 0$.

Instead of working directly on B we will use suitable charts. The most important chart K_1 is obtained by setting $\bar{\varepsilon} = 1$ in transformation (3.2). The corresponding coordinates $(a_1, b_1, c_1, \mu_1, r_1) \in \mathbb{R}^4 \times \mathbb{R}_0^+$ cover the part of B with $\bar{\varepsilon} > 0$. In these coordinates the blow-up transformation has the form

$$\begin{aligned} a &= \frac{1}{2} + r_1^2 a_1 \\ b &= 1 + r_1 b_1 \\ c &= 1 + r_1 c_1 \end{aligned} \tag{3.4}$$

$$\varepsilon = r_1^2$$
$$\mu = \frac{2}{7} + r_1\mu_1 .$$

As (a_1, b_1, c_1, μ_1) tends to infinity, the corresponding point in S^4 tends to a point $(\bar{a}, \bar{b}, \bar{c}, 0, \bar{\mu})$ on the "equator" of S^4. Hence, the dynamics at "infinity" in the chart K_1 is more conveniently studied in other charts. Actually, we will need just one other chart K_2 which is obtained analogously by setting $\bar{a} = -1$ in the blow-up transformation (3.2). The corresponding coordinates $(b_2, c_2, \varepsilon_2, \mu_2, r_2) \in \mathbb{R}^4 \times \mathbb{R}_0^+$ cover the part of B with $\bar{a} < 0$. In these coordinates the blow-up transformation has the form

$$\begin{aligned}
a &= \frac{1}{2} - r_2^2 \\
b &= 1 + r_2 b_2 \\
c &= 1 + r_2 c_2 \\
\varepsilon &= r_2^2 \varepsilon_2 \\
\mu &= \frac{2}{7} + r_2\mu_2 .
\end{aligned} \tag{3.5}$$

The change of coordinates between K_1 and K_2 in their overlap domain $a_1 < 0$ resp. $\varepsilon_2 > 0$ is described in the following table.

K_1 to K_2	K_2 to K_1
$b_2 = b_1(-a_1)^{-1/2}$	$a_1 = -\varepsilon_2^{-1}$
$c_2 = c_1(-a_1)^{-1/2}$	$b_1 = b_2\varepsilon_2^{-1/2}$
$\varepsilon_2 = -a_1^{-1}$	$c_1 = c_2\varepsilon_2^{-1/2}$
$\mu_2 = \mu_1(-a_1)^{-1/2}$	$\mu_1 = \mu_2\varepsilon_2^{-1/2}$
$r_2 = r_1(-a_1)^{1/2}$	$r_1 = r_2\varepsilon_2^{1/2}$

We introduce the following notation: $\bar{P}$ denotes an object in the blow-up which corresponds to object P in the original problem. If $\bar{P}$ is described in one of the charts then P_1 denotes the object in chart K_1 and similarly P_2 is the same object written in chart K_2. This convention is used only when necessary, mostly to distinguish between various invariant manifolds.

3.2. Study in K_1, family rescaling. The vector field $\bar{X}$ written in chart K_1 is obtained by inserting transformation (3.4) into system (3.1). A computation gives the equations

$$
\begin{aligned}
a_1' &= -b_1 + \frac{2}{7}c_1 + \frac{7}{2}\mu_1 + r_1\left(-2a_1 - \frac{1}{2}b_1^2 + \mu_1 c_1\right) + O(r_1^2) \\
b_1' &= 2a_1 + \frac{1}{2}b_1^2 + 2r_1 a_1 b_1 + O(r_1^2) \\
c_1' &= r_1(b_1 - c_1) \qquad (3.6) \\
\mu_1' &= 0 \\
r_1' &= 0\,,
\end{aligned}
$$

where superscript prime denotes differentiation with respect to the rescaled time $\tau_1 = r_1\tau$, that is we have divided the right-hand side of the equation by r_1. In the chart K_1, the system is still a two-parameter family of three-dimensional vector fields, the blow-up transformation in these coordinates is just a rescaling of variables and parameters.

It is interesting to observe that system (3.6) is again a singularly perturbed problem, however, with fast variables a_1, b_1 and the slow variable c_1. The singular perturbation parameter is r_1. A computation gives

LEMMA 3.1. *The singularly perturbed system (3.6) has a one-dimensional critical manifold*

$$
S_1 = \left\{\left(-\frac{1}{4}\left(\frac{2}{7}c_1 + \frac{7}{2}\mu_1\right)^2, \frac{2}{7}c_1 + \frac{7}{2}\mu_1, c_1\right) \;:\; c_1 \in \mathbb{R}\right\}.
$$

The branch of S_1 corresponding to $c_1 < -49\mu_1/4$ is attracting, the branch of S_1 corresponding to $c_1 > -49\mu_1/4$ is repelling. At $c_1 = -49\mu_1/4$ a (degenerate) Hopf bifurcation occurs. The reduced flow on S_1 is governed by the equation

$$
\dot{c}_1 = -\frac{5}{7}c_1 + \frac{7}{2}\mu_1\,,
$$

where the derivative is with respect to $t_1 := r_1\tau_1$.

The flow on S_1 has an attracting equilibrium point at $c_1 = 49\mu_1/10$. A computation shows that this equilibrium, written in the original variables, corresponds to the equilibrium q described in Lemma 2.1. We expect that a similar relation exists between the critical manifold S_1 and the center manifold described in Lemma 2.2. We have not checked this in detail.

For $\mu_1 \neq 0$, the reduced flow on the critical manifold S_1 passes through the Hopf point, for $\mu_1 = 0$ the equilibrium on the critical manifold and the Hopf point coalesce at $c_1 = 0$. This situation is reminiscent of the delayed Hopf bifurcation, see the paper by A. Neishtadt in this volume and references therein. Thus, a more detailed analysis of the local dynamics seems to be difficult. F. Dumortier has informed us that related problems are discussed in his paper with R. Roussarie in this volume. These connections will be investigated in future work. Fortunately, an understanding of these phenomena is not needed to prove the existence of the canard solution, since as we will see the canard solution stays away from S_1. However, it can very well be that these phenomena play a role in a complete understanding of mixed-mode oscillations.

The canard solution is obtained as a perturbation of a special solution of system (3.6) with $r_1 = 0$. This gives the equations

$$\begin{aligned} a_1' &= -b_1 + \frac{2}{7}c_1 + \frac{7}{2}\mu_1 \\ b_1' &= 2a_1 + \frac{1}{2}b_1^2 , \end{aligned} \tag{3.7}$$

where c_1 acts as a paramter in the same way as μ_1. For $2/7c_1 + 7/2\mu_1 = 0$ we obtain equation

$$\begin{aligned} a_1' &= -b_1 \\ b_1' &= 2a_1 + \frac{1}{2}b_1^2 , \end{aligned} \tag{3.8}$$

well known from the analysis of planar canards [2, 5, 7, 8]. The function

$$H(a_1, b_1) = \left(2 - 2a_1 - \frac{b_1^2}{2}\right)e^{a_1} \tag{3.9}$$

is a constant of motion. The level curves of H are shown in Figure 5. The origin is a center surrounded by periodic orbits. The bold line is the special orbit γ_1 corresponding to $H = 0$. A parametrization of γ_1 is given by $\{(-\tau_1^2 + 1, 2\tau_1) : \tau_1 \in \mathbb{R}\}$. Note that for $a_1 \ll 0$ the lower branch of γ_1 is strongly attracting, while the upper branch is strongly repelling. We will see that this behaviour is inherited from the stability properties of the critical manifolds S_a and S_r. In a suitable compactification of the plane the orbit γ_1 is a heteroclinic orbit connecting two equilibria. The nature of these "equilibria at infinity" will become clear in the following subsection.

LEMMA 3.2. *As $2/7c_1 + 7/2\mu_1$ passes through zero the orbit γ_1 of system (3.7) considered as a heteroclinic orbit connecting two "equilibria at infinity" breaks transversally.*

This crucial property of system (3.7) is already known from the analysis of planar canards in [7]. A proof of the Lemma based on a variant of the Melnikov method is given in Section 3.4. However, some care is needed to justify the Melnikov calculation, in particular we need a better understanding of system (3.6) at infinity. We will show that there exist four-dimensional invariant manifolds $M_{a,1}$ resp. $M_{r,1}$ of system (3.6), which are strongly attracting resp. repelling for $a_1 \ll 0$. We will show that these manifolds can be constructed as center-manifolds of equilibria of $\bar{X}$ on the "equator" of the sphere S^4. Recall that the sphere S^4 is the blown-up locus of the degenerate point on the fold line which we are studying. The necessary computations are done in chart K_2.

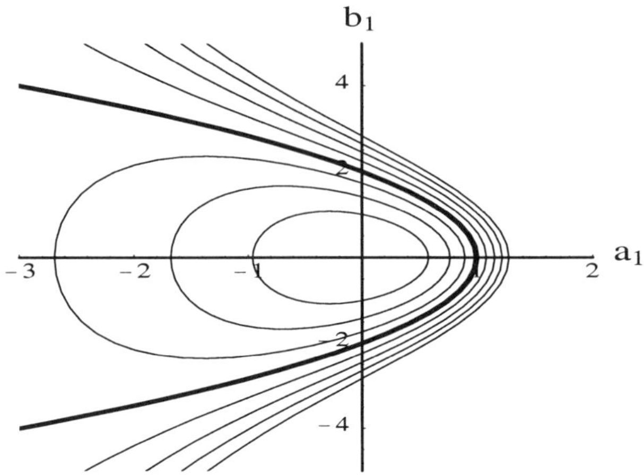

FIG. 5. *Level curves and the special orbit γ_1 (bold curve). Inside this orbit there are periodic orbits surrounding a center at the origin, outside there is a family of unbounded orbits of system (3.8).*

3.3. Study in K_2, phase-directional rescaling. The vector field $\bar{X}$ in the chart K_2 is obtained by inserting transformation (3.5) into system (3.1). A computation gives the equations

$$
\begin{aligned}
b_2' &= -2 + \frac{1}{2}b_2^2 + \frac{1}{2}\varepsilon_2 b_2\Big(- b_2 + \frac{2}{7}c_2 + \frac{7}{2}\mu_2 + O(r_2)\Big)\\
c_2' &= \frac{1}{2}\varepsilon_2 c_2\Big(- b_2 + \frac{2}{7}c_2 + \frac{7}{2}\mu_2 + O(r_2)\Big)\\
\varepsilon_2' &= \varepsilon_2^2\Big(- b_2 + \frac{2}{7}c_2 + \frac{7}{2}\mu_2 + O(r_2)\Big)\\
\mu_2' &= \frac{1}{2}\varepsilon_2 \mu_2\Big(- b_2 + \frac{2}{7}c_2 + \frac{7}{2}\mu_2 + O(r_2)\Big)\\
r_2' &= -\frac{1}{2}\varepsilon_2 r_2\Big(- b_2 + \frac{2}{7}c_2 + \frac{7}{2}\mu_2 + O(r_2)\Big),
\end{aligned}
\tag{3.10}
$$

where superscript prime denotes differentiation with respect to the rescaled time $\tau_2 = r_2\tau$, i.e. we have divided the right-hand side of the equation by r_2. In the chart K_2 we lose the property that the original problem is a two-parameter family of three-dimensional vector fields. The two parameters are now treated as dynamical variables. Note however, that the equations for ε_2 and μ_2 are obtained by differentiating the relations $\varepsilon = r_2\varepsilon_2$ and $\mu = 2/7 + r_2\mu_2$, and using $d\varepsilon/d\tau = 0$, $d\mu/d\tau = 0$. This implies that $r_2\varepsilon_2$ and $r_2\mu_2$ are two constants of motion.

System (3.10) has four invariant hyperplanes defined by $\varepsilon_2 = 0$, $r_2 = 0$, $c_2 = 0$, and $\mu_2 = 0$. In the plane $\varepsilon_2 = 0$ the equations become very simple

$$
\begin{aligned}
b_2' &= -2 + \frac{1}{2}b_2^2 + O(r_2) \\
c_2' &= 0 \\
\mu_2' &= 0 \\
r_2' &= 0\,.
\end{aligned}
\tag{3.11}
$$

Hence, there exist functions $h_a(r_2) = -2 + O(r_2)$ and $h_r(r_2) = 2 + O(r_2)$ both defined for $[0, \rho)$ for some $\rho > 0$ such that the equations $\varepsilon_2 = 0$ and $b_2 = h_a(r_2)$ resp. $b_2 = h_r(r_2)$ define three-dimensional manifolds $S_{a,2}$ resp. $S_{r,2}$ of equilibria of system (3.10). These manifolds are nothing but the branches S_a and S_r of the critical manifold (trivially extended in μ direction) of the layer problem (2.6). However, due to the blow-up we have gained normal hyperbolicity of these manifolds at $r_2 = 0$. The relevant eigenvalue is $\lambda = \pm 2$ at points $(\pm 2, c_2, \mu_2, 0)$.

In the plane $r_2 = 0$ the governing equations are

$$
\begin{aligned}
b_2' &= -2 + \frac{1}{2}b_2^2 + \frac{1}{2}\varepsilon_2 b_2\Big(- b_2 + \frac{2}{7}c_2 + \frac{7}{2}\mu_2\Big) \\
c_2' &= \frac{1}{2}\varepsilon_2 c_2\Big(- b_2 + \frac{2}{7}c_2 + \frac{7}{2}\mu_2\Big) \\
\varepsilon_2' &= \varepsilon_2^2\Big(- b_2 + \frac{2}{7}c_2 + \frac{7}{2}\mu_2\Big) \\
\mu_2' &= \frac{1}{2}\varepsilon_2 \mu_2\Big(- b_2 + \frac{2}{7}c_2 + \frac{7}{2}\mu_2\Big)\,.
\end{aligned}
\tag{3.12}
$$

All points $(\pm 2, c_2, 0, \mu_2)$ are equilibria with one hyperbolic eigenvalue $\lambda = \pm 2$ and a zero eigenvalue of multiplicity three.

We restrict our attention to a neighborhood of the equilibrium points $p_a = (-2, 0, 0, 0, 0)$ and $p_r = (2, 0, 0, 0, 0)$ since these are needed in the analysis of the canards. In the following theorem we denote the canonical basis of $\mathbb{R}^5$ by e_i, $i = 1, \dots, 5$.

THEOREM 3.1. *System (3.10) has the following properties:*

1. *There exists an attracting four-dimensional center-manifold $M_{a,2}$ at p_a with a stable foliation. The strong stable manifold at p_a is tangent to e_1. The center-manifold $M_{a,2}$ is tangent to span$\{-e_1 + e_3, e_2, e_4, e_5\}$ at p_a.*
2. *There exists a repelling four-dimensional center-manifold $M_{r,2}$ at p_r with an unstable foliation. The strong unstable manifold at p_r is tangent to e_1. The center-manifold $M_{r,2}$ is tangent to span$\{e_1 + e_3, e_2, e_4, e_5\}$ at p_r.*

3. *The intersection of $M_{a,2}$ with the invariant space defined by $r_2 = c_2 = \mu_2 = 0$, $\varepsilon_2 > 0$ is a single solution $\gamma_{a,2}$ of system (3.10) which converges to p_a tangent to $-e_1+e_3$ as $\tau_2 \to -\infty$. The solution $\gamma_{a,2}$ is the part of the special solution γ_1 of system (3.8) corresponding to $\tau_1 \in (-\infty, 0]$ rewritten in the chart K_2.*
4. *The intersection of $M_{r,2}$ with the invariant space defined by $r_2 = c_2 = \mu_2 = 0$, $\varepsilon_2 > 0$ is a single solution $\gamma_{r,2}$ of system (3.10) which converges to p_r tangent to $e_1 + e_3$ as $\tau_2 \to \infty$. The solution $\gamma_{r,2}$ is the part of the special solution γ_1 of system (3.8) corresponding to $\tau_1 \in [0, \infty)$ rewritten in the chart K_2.*

Proof. The first two assertions follow from standard center-manifold theory and the above observations on the spectrum of the linearization. In the invariant space $r_2 = c_2 = \mu_2 = 0$ system (3.10) simplifies to

$$b_2' = -2 + \frac{1}{2}b_2^2 - \frac{1}{2}\varepsilon_2 b_2^2$$

$$\varepsilon_2' = -\varepsilon_2^2 b_2 \,.$$

The center manifold of this planar system at $(-2, 0)$ is tangent to $(-1, 1)$. Its branch corresponding to $\varepsilon_2 > 0$ is unique. A computation based on the formulas for the change of coordinates between the charts K_1 and K_2 shows, that the orbit γ_1 rewritten in the chart K_2 converges to $(-2, 0)$ tangent to the center direction $(-1, 1)$ as $\tau_1 \to -\infty$. The uniqueness of the branch of the center manifold in $\varepsilon_2 > 0$ implies Assertion 3. Assertion 4 is proved the same way. □

Note that sections $\varepsilon = const.$ of the "blown-down" manifolds $\Phi(\bar{M}_a)$ and $\Phi(\bar{M}_r)$ are precisely the slow manifolds S_a^ε and S_r^ε of the original problem (1.1). The center-manifolds $M_{a,2}$ and $M_{r,2}$ are not unique, however all of them contain the special solution γ_1 written in chart K_2. Hence, our results hold for an arbitrary choice of these center-manifolds, respectively of the slow manifolds S_a^ε and S_r^ε.

3.4. Existence of canards. Now we are ready to prove Theorem 1.1. As a preparatory step we express the center-manifolds $M_{a,2}$ and $M_{r,2}$ from Theorem 3.1 in chart K_1 to obtain (locally) invariant manifolds $M_{a,1}$ and $M_{r,1}$ of system (3.6).

THEOREM 3.2. *The following assertions hold for system (3.6):*

1. *The forward in time continuation of $M_{a,1}$ and the backward in time continuation of $M_{r,1}$ intersect along the special orbit γ_1.*
2. *There exists a smooth distance function $d : U \to \mathbb{R}$, where U is a small neighborhood of $(0,0,0)$ in (c_1, μ_1, r_1)-space, such that in a tubular neighborhood of γ_1 the manifolds $M_{a,1}$ and $M_{r,1}$ intersect if and only if $d(c_1, \mu_1, r_1) = 0$.*

3. The manifolds $M_{a,1}$ and $M_{r,1}$ intersect transversally, the distance function satisfies

$$\frac{\partial d}{\partial c_1}(0,0,0) = \frac{2}{7}\sqrt{\pi}\,.$$

The equation $d(c_1, \mu_1, r_1) = 0$ can be solved for $c_1 = c_1^*(\mu_1, r_1)$ where c_1^* is a smooth function defined in a neighborhood of $(0,0)$.

Proof. The first assertion follows from Theorem 3.1 and the existence of the special orbit γ_1. The orbit γ_1 intersects the plane $b_1 = 0$ transversally, see Figure 5. We use this plane as a section and define

$$d(c_1, \mu_1, r_1) := a_1^-(c_1, \mu_1, r_1) - a_1^+(c_1, \mu_1, r_1)\,,$$

where a_1^- resp. a_1^+ are the a_1 coordinate of the points where $M_{a,1}$ resp. $M_{r,1}$ intersect the plane $b_1 = 0$, see Figure 6. This proves Assertion 2. Transversality of the intersection and the precise value of $\partial d/\partial c_1$ follow from the Melnikov calculation given below, which also proves Lemma 3.2. Solvability of the equation $d(c_1, \mu_1, r_1) = 0$ with respect to c_1 follows from the implicit function theorem.

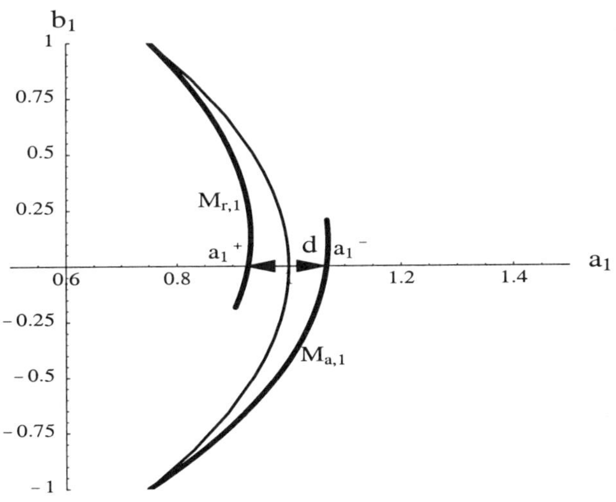

FIG. 6. *The special solution γ_1 and the distance function d for the manifolds $M_{a,1}$ and $M_{r,1}$ (bold curves).*

The linearization of system (3.7) around γ_1 is

$$\begin{pmatrix} a_1' \\ b_1' \end{pmatrix} = \begin{pmatrix} 0 & -1 \\ 2 & 2\tau_1 \end{pmatrix} \begin{pmatrix} a_1 \\ b_1 \end{pmatrix}. \tag{3.13}$$

The (up to a multiplicative constant) unique bounded solution of the corresponding adjoint equation is

$$\psi(\tau_1) = \begin{pmatrix} e^{-\tau_1^2} \\ \tau_1 e^{-\tau_1^2} \end{pmatrix}. \tag{3.14}$$

Note that $\psi(0)$ is a unit vector in the direction normal to γ_1, that is in the direction in which the distance is measured. Let f denoted the right hand side of system (3.7). It can be shown by adapting the usual derivation of the Melnikov Integral, see e.g. [10], that

$$\frac{\partial d}{\partial c_1}(0,0,0) = \int_{-\infty}^{\infty} \psi(\tau_1) \cdot \frac{\partial f}{\partial c_1}(\gamma_1(\tau_1))\, d\tau_1 \,. \tag{3.15}$$

A computation of the integral gives the value in Assertion 3. □

We remark that the derivative $\frac{\partial d}{\partial r_1}(0,0,0)$ can be computed by a similar Melinikov calculation. The justification of this calculation is more subtle, see [13].

The inequality $\frac{\partial d}{\partial c_1} > 0$ implies that the manifolds $M_{a,1}$ and $M_{r,1}$ intersect in the following way. In the section $b_1 = 0$ the inequality $a_1^- < a_1^+$ holds for $c_1 < c_1^*(\mu_1, r_1)$, i.e. the manifold $M_{a,1}$ lies to the "left" of $M_{r,1}$. For $c_1 > c_1^*(\mu_1, r_1)$ the manifold $M_{a,1}$ lies to the "right" of $M_{r,1}$, see Figure 3).

The following observation relates the break up of these manifolds to the small oscillations of mixed-mode oscillations. For simplicity we fix $\mu_1 = 0$. It is not hard to see that the part of $M_{a,1}$ corresponding to $r_1 = 0$, $c_1 < 0$ spirals towards the attracting part of the one-dimensional slow manifold S_1 described in Lemma 3.1. Similarly, the part of $M_{r,1}$ corresponding to $r_1 = 0$, $c_1 > 0$ spirals towards the repelling part of the one-dimensional slow manifold S_1 in backwards time. This behaviour will persist for $r_1 > 0$ at least for c_1 sufficiently far away from zero. For c_1 close to zero the situation is more complicated because of the Hopf bifurcations described in Lemma 3.1. We do not attempt a more detailed analysis at this point because of the complications described in the discussion following Lemma 3.1.

Clearly, all our results about the manifolds $M_{a,1}$ resp. $M_{r,1}$ and $M_{a,2}$ resp. $M_{r,2}$, are in fact results about the existence and intersection of invariant manifolds $\bar{M}_a$ resp. $\bar{M}_r$ of the blown up vector field $\bar{X}$ in B_0 for r_0 sufficiently small. This gives:

Proof. (of Theorem 1.1) All assertions follow from Theorem 3.1 and Theorem 3.2 by means of transformation (3.2). □

We conclude this section by summarizing the dynamics close to the fold. On the left side of the maximal canard the attracting slow manifold S_a^ε lies below the repelling slow manifold S_r^ε, while on the right side the attracting manifold lies above the repelling manifold. Here we refer to the side of the maximal canard corresponding to smaller values of the variable c as the left side and to the side corresponding to larger values of c as the right side. All solutions starting on the attracting manifold on the right side of the maximal canard will leave the vicinity of the fold. A solution starting on the left side of the maximal canard falls back to the attracting slow manifold (at a somewhat larger value of c) thus forming a small loop. Depending on which side of the canard the solution falls on, the solution either leaves the fold or goes through another small loop. Thus, left of the canard the attracting slow manifold scrolls up in a multilayered structure. Eventually, the solution lands on the right side of the canard and leaves the vicinity of the fold. The geometry of the repelling manifold is generated in a similar way in backwards time.

In the next section we analyse the dynamics after a solution leaves the vicinity of the fold.

4. Large relaxation oscillations. As mentioned in the introduction all orbits which leave the fold come back to the attracting manifold after a large excursion due to a global mechanism. It is clear that such a behaviour is not present in the singular limits we have studied so far. However, our approach of treating system (1.1) as singularly perturbed with slow variables a, c and fast variable b is only valid as long as $b = O(1)$. For $b = O(1/\varepsilon)$ new scales arise and a different asymptotic analysis is needed. This is best seen if the variables are rescaled according to

$$\begin{aligned} a &= A \\ b &= \frac{1}{\varepsilon}B \\ c &= C \\ t &= \varepsilon^2 T\,. \end{aligned} \tag{4.1}$$

This introduces the new time-scale T, which is the fastest time-scale in the problem. In the new variables system (1.1) has the form

$$\begin{aligned} \frac{dA}{dT} &= \varepsilon^2\mu\left(\frac{2}{7}+C\right) - AB^2 - \varepsilon^2 A \\ \frac{dB}{dT} &= AB^2 + \varepsilon^2 A - \varepsilon B \\ \frac{dC}{dT} &= \varepsilon B - \varepsilon^2 C\,. \end{aligned} \tag{4.2}$$

System (4.2) is also a singularly perturbed system, however not in standard form, that is the slow variables are not constant along the fibers of the layer problem. The layer problem is obtained by setting $\varepsilon = 0$

$$\begin{aligned}\frac{dA}{dT} &= -AB^2 \\ \frac{dB}{dT} &= AB^2 \\ \frac{dC}{dT} &= 0\,.\end{aligned} \tag{4.3}$$

For system (4.3) the planes $A = 0$ and $B = 0$ are manifolds of equilibria, that is they are critical manifolds. For $B \geq \delta_0$, $\delta_0 > 0$ the plane $A = 0$ is normally hyperbolic, the relevant eigenvalue is $\lambda = -B^2$. The critical manifold $B = 0$ is nowhere normally hyperbolic, the relevant linearization is zero. Along the line C-axis the problem is even more degenerate due to the intersection of the two critical manifolds.

The dynamics of the layer problem is rather simple. The two critical manifolds are connected by orbits of the layer problem. An arbitrary point $(A_0, 0, C_0)$ is connected to the point $(0, A_0, C_0)$ by a heteroclinic orbit of the layer problem in a straight line, since $A + B = const.$. The approach to $(0, A_0, C_0)$ as $T \to \infty$ is exponential, the approach to $(A_0, 0, C_0)$ as $T \to -\infty$ is in the center direction, hence only algebraic. These straight lines are the fibers of the stable manifold of the plane $A = 0$.

On the plane $A = 0$ a reduced problem can be defined by projecting the vector field (4.2) onto the plane $A = 0$ along the fibers of the stable manifold and dividing by ε, see [9]. This gives

$$\begin{aligned}\frac{dA}{d\tau} &= 0 \\ \frac{dB}{d\tau} &= -B \\ \frac{dC}{d\tau} &= B\,,\end{aligned} \tag{4.4}$$

where $\tau = \varepsilon T$. Note that τ is the "slow time-scale" for system (4.2) and the "fast time-scale" for system (1.1). The reduced problem (4.4) is also very simple. A solution starting at an arbitrary point $(0, B_0, C_0)$ converges for $\tau \to \infty$ exponentially to the point $(0, 0, B_0 + C_0)$ along a straight line. Fenichel's theorems imply that the critical manifold $A = 0$, the reduced dynamics and the stable foliation perturb smoothly for small ε as long as $B \geq \delta_0$, $\delta_0 > 0$.

This gives the following singular limit description of the mechanism generating the large loops. Assume that a solution of the original system (1.1) jumps away from the critical manifold S_r at a point (a_0, b_0, c_0). Written in the rescaled variables (4.1) this point is $(a_0, \varepsilon b_0, c_0)$. In the limit $\varepsilon \to 0$ this gives the initial condition $(a_0, 0, c_0)$ for the layer problem (4.3). The point $(a_0, 0, c_0)$ is connected with the point $(0, a_0, c_0)$ by a heteroclinic orbit of the layer problem (4.3). The point $(0, a_0, c_0)$ is connected to the point $(0, 0, a_0 + c_0)$ by an orbit of the reduced problem (4.4).

This end point has coordinates $(0, 0, a_0 + c_0)$ also in the original scaling and is thus a point lying on the critical manifold S_a of the original system (1.1). Thus, the the solution comes back to the attracting critical manifold S_a at a point which depends on the a and c coordinates of the point where the solution left the critical manifold S_r. On the critical manifold S_a the flow of the reduced system (2.2) carries the point $(0, 0, a_0 + c_0)$ back to the neighborhood of the fold line and the whole process starts again.

This singular limit analysis of the return mechanism cannot be made rigorous by standard arguments because of the loss of hyperbolicity near the critical manifold $B = 0$. Also the matching at take-off from the critical manifold S_r needs to be done more rigorously. As mentioned in the introduction we have worked out a rigorous analysis based on the blow-up method for system (1.2), see [18]. Work in progress indicates that the above arguments can be made rigorous by a variant of the blow-up method.

5. Conclusions and outlook. We presented a geometrical analysis of the dynamics of the autocatalator. One of our intentions was to present this prototypical example from theoretical chemistry to a mathematically oriented audience. We showed that this simple looking problem contains a surprisingly large number of scales. The other intention was to demonstrate that geometric singular perturbation theory combined with the blow-up method is well suited to analyse such problems.

In addition to making our analysis more complete, there are open questions we have not yet even touched. For instance, in a study of mixed mode oscillation in the BDK-model [12], which has a similar structure, the author proposes that the mixed-mode oscillations are related to a tangency of the stable and unstable manifolds of a periodic orbit. How does this relate to the present analysis? It seems possible to investigate this conjecture with the methods used in this paper. A related problem is to understand the origin of the apparently chaotic dynamics in problems which exhibit mixed-mode oscillations.

Acknowledgments. We would like to thank M. Krupa and M. Wechselberger for useful discussions. Part of this research was carried out while both authors were visiting the IMA. We would like to thank the IMA for support and for providing a stimulating research environment. A. M. also thanks the Geometry Center for technical support.

REFERENCES

[1] V. Arnol'd, *Dynamical Systems V*, Vol. **5** of Encyclopedia of Mathematical Sciences, Springer Verlag, Berlin, 1994.

[2] S. Baer and T. Erneux, *Singular hopf bifurcation to relaxation oscillations*, SIAM Journal of Applied Mathematics, **46** (1986), pp. 721–739.

[3] E. Benoit, *Systèmes lents-rapides dans $\mathbb{R}^3$ et leurs canards*, Société Mathématique de France Astérisque, **109–110** (1983), pp. 159–191.

[4] E. Benoit, J.-L. Callot, F. Diener, and M. Diener, *Chasse au canard*, Collect. Math., **31** (1981), pp. 37–119.

[5] B. Braaksma, *Critical phenomena in dynamical systems of van der Pol type*, Thesis, Rijksuniversiteit te Utrecht, Nederlands, 1993.

[6] M. Diener, *The canard unchained or how fast/slow dynamical systems bifurcate*, The Mathematical Intelligencer, **6** (1984), pp. 38–49.

[7] F. Dumortier and R. Roussarie, *Canard cycles and center manifolds*, Memoirs of the American Mathematical Society, **121** (1996).

[8] W. Eckhaus, *Relaxation oscillations including a standard chase on french ducks*, Vol. **985** of Springer Lecture Notes Math., Springer, 1983, pp. 449–494.

[9] N. Fenichel, *Geometric singular perturbation theory*, Journal of Differential Equations, **31** (1979), pp. 53–98.

[10] J. Guckenheimer and P. Holmes, *Nonlinear Oscillations, Dynamical Systems, and Bifurcations of Vector Fields*, Springer, New York, 1983.

[11] C. Jones, *Geometric singular perturbation theory*, Vol. **1609** of Springer Lecture Notes Math., Springer, 1995, pp. 44–120.

[12] M.T. Koper, *Bifurcations of mixed-mode oscillations in a three-variable autonomous van der pol-duffing model with a cross-shaped phase diagram*, Physica D, **80** (1995), pp. 72–94.

[13] M. Krupa and P. Szmolyan, *Extending geometric singular perturbation theory to nonhyperbolic points*, in preparation (1998).

[14] A. Milik, P. Szmolyan, H. Löffelmann, and E. Gröller, *The geometry of mixed-mode oscillations in the 3d-autocatalator*, International Journal of Bifurcation and Chaos, **8** (1998).

[15] E. Mishchenko, Y. Kolesov, A.Y. Kolesov, and N.K. Rozov, *Asymptotic methods in singularly perturbed systems*, Monographs in Contemporary Mathematics, Consultants Bureau, New York, A Division of Plenum Publishing Corporation 233 Spring Street, New York, N.Y. 10013, 1994.

[16] E. Mishchenko and N.K. Rozov, *Differential equations with small parameters and relaxation oscillations*, Plenum Press, New York, 1980.

[17] V. Petrov, S.K. Scott, and K. Showalter, *Mixed-mode oscillations in chemical systems*, Journal of Chemical Physics, **97** (1992), pp. 6191–6198.

[18] P. Szmolyan, *Geometry of singular perturbations: a case study*, in preparation (1998).

[19] M. Wechselberger, *Singularly perturbed folds and canards in $\mathbb{R}^3$*, Thesis, Technische Universität Wien, Austria, 1998.

A GEOMETRIC METHOD FOR PERIODIC ORBITS IN SINGULARLY-PERTURBED SYSTEMS

CRISTINA SOTO-TREVIÑO*

Abstract. In this work, we establish a modular geometric method to demonstrate the existence of periodic orbits in singularly perturbed systems of differential equations. These orbits have alternating fast and slow segments, reflecting the two time scales in the problems. The method involves converting the periodic orbit problem into a boundary value problem in an appropriately augmented system, and it employs several versions of the exchange lemmas due to Jones, Kopell, Kaper and Tin. It is applicable to models that arise in a wide variety of scientific disciplines, and applications are given to the FitzHugh-Nagumo, Hodgkin-Huxley, and Gray-Scott systems.

1. Introduction. In this paper, we study singularly perturbed systems of ordinary differential equations:

$$
\begin{aligned}
\dot{x} &= f(x,y;\varepsilon) \\
\dot{y} &= \varepsilon g(x,y;\varepsilon),
\end{aligned}
\tag{1.1}
$$

where $x \in \mathbb{R}^{k+\ell}$, $y \in \mathbb{R}^{p}$, k, ℓ, p are arbitrary positive integers, $0 < \varepsilon \ll 1$, and f and g are C^r, $r \geq 1$, in x, y and ε. These systems occur as models in a variety of scientific disciplines, including chemical reaction theory, fluid mechanics, Hamiltonian systems, neurophysiology, oceanography, optics, pattern formation theory, and population biology, to name but a few.

We are interested in those systems (1.1) for which there exist open sets $U \subset \mathbb{R}^{p}$ of y values such that the equation $f(x,y;0) = 0$ defines p-dimensional normally hyperbolic invariant manifolds, which we label in general by $\mathcal{S}^0$. Here, we recall that normal hyperbolicity means that the eigenvalues of the Jacobian of f at points on $\mathcal{S}^0$ all have nonzero real parts. For these systems, the Fenichel geometric singular perturbation theory (see [13, 30] and [49]) guarantees the existence of slow manifolds, generally labeled $\mathcal{S}^\varepsilon$ that are C^r $\mathcal{O}(\varepsilon)$ close to their unperturbed counterparts $\mathcal{S}^0$.

The focus in this work will be on systems of the form (1.1) that have periodic orbits which possess fast and slow segments, reflecting the different time (or length) scales in the equations. The fast segments are close to orbit segments of the reduced fast system associated to (1.1):

$$
\dot{x} = f(x,y;0), \qquad \dot{y} = 0,
\tag{1.2}
$$

while the slow segments are close to orbit segments of the reduced slow system(s)

$$
0 = f(x,y;0), \qquad y' = g(X(y),y;0)
\tag{1.3}
$$

*Department of Mathematics, Boston University, Boston, MA 02215. Current address: Volen Center, Brandeis University, Waltham, MA 02454.

associated to (1.1) and defined on the invariant manifold(s) $\mathcal{M}_0$, where (at least on each chart for the manifold) the manifold is given locally as the graph of the function $x = X(y)$. Here, the prime denotes differentiation with respect to the so-called slow time $\tau = \varepsilon t$.

By piecing together in alternation a finite sequence of orbit segments from the reduced fast and slow systems in such a way that they form a closed loop, one obtains what is called a singular periodic orbit. Perhaps the most well-known example of this is the singular periodic orbit constructed in the relaxation limit of the van der Pol equation, see [21] for example. Even though these singular orbits are not solutions of the full system (1.1), one expects, in some cases, that actual periodic orbits lie close to appropriate singular ones. It is one of the main goals of geometric singular perturbation theory to explore the conditions under which this is true.

Several methods have been used to demonstrate the existence of periodic orbits in systems of the form (1.1). Pioneering work was carried out by Tikhonov [55] and Levinson [40, 41], and in [39]. We refer the reader to [44] for presentations of the full theory and a historical overview. A nonexhaustive list of methods (with sample references) includes matched asymptotic expansions [46] and [36], analytic techniques [8, 14], topological approaches [6, 7, 10, 19, 23, 25, 26, 50, 54], the method of sub- and supersolutions [1], horseshoe constructions [20, 35, 37], exact calculations on a piecewise linear system [47], the construction of singular Poincaré maps [4, 44], and a geometric method [9, 45, 53]. In the topological methods, such as for example, topological shooting [23, 25, 26], isolating blocks [6, 7] and the Conley Index [19], as well as in the singular Poincaré map method [4, 44] when the slow manifolds are asymptotically stable, the existence of a singular periodic solutions guarantees the existence of true periodic orbit $\mathcal{O}(\varepsilon)$ close to it.

In this work, we present a modular geometric method that simultaneously yields the existence and local uniqueness of periodic orbits with alternating fast and slow segments in a broad class of systems of the form (1.1). This method was developed in [53]. The geometric structure we require of system (1.1) provides the backbone needed to construct useful singular periodic orbits. In particular, the backbone consists of a family of normally hyperbolic invariant manifolds, and a sequence of heteroclinic intersections between the stable and unstable manifolds of 'adjacent' manifolds in that family. A singular periodic orbit is then constructed by concatenating slow orbit segments lying on the invariant manifolds and fast segments that are the $\varepsilon \to 0$ limits of the heteroclinic orbits in these intersections. We show that, if the system (1.1) possesses a singular periodic orbit, and if the flow in the vicinity of the singular periodic orbit satisfies certain properties, then there exists a locally unique periodic orbit of (1.1) with $\varepsilon > 0$ sufficiently small that lies $\mathcal{O}(\varepsilon)$ close to the singular one.

The true periodic orbits in the system (1.1) with $0 < \varepsilon \ll 1$ make their fast excursions exponentially close to heteroclinic solutions that lie in the transverse intersection of the unstable manifold of a slow manifold with the stable manifold of another (or the same) slow manifold. The transverse intersections of the stable and unstable manifolds of the slow manifolds are in some cases already present in the $\varepsilon = 0$ limiting system. Examples include the FitzHugh-Nagumo equations [16, 33] and the Hodgkin-Huxley equations [6, 7, 24, 28]. In other cases, for example in perturbed Hamiltonian systems [34] and in the traveling wave problem for a coupled reaction-diffusion equation [9], they exist only for $\varepsilon > 0$, and tools like the Melnikov function are needed to find them. Furthermore, depending on the system, the existence of heteroclinic solutions with multiple fast excursions can be ascertained, for example, by methods like those in [5, 22, 34, 52].

The geometric method we develop in this work is modular. It transforms the Poincaré map problem into a boundary value problem. The transformation entails embedding system (1.1) into a higher-dimensional, or naturally augmented, system. In this augmented system, we find the solution of the boundary value problem as the transverse intersection of two appropriately constructed invariant manifolds. In some instances of (1.1), it is possible to exploit the symmetries of the system and avoid the need to augment the system, while still constructing the periodic orbit as the transverse intersection of two invariant manifolds [9, 51]. When symmetries are not present, then embedding the system in a higher-dimensional one is a key step in showing that the orbit that solves the boundary-value problem is periodic.

The idea of displaying the solution of a boundary-value problem in a singularly-perturbed system as the transverse intersection of invariant manifolds has been used in a number of works, see for example [27, 57]. Here, as well as in [57], the transverse intersection(s) of the relevant invariant manifolds is established by using technical tools to track each of them (and their respective tangent spaces) as orbits on them spend $\mathcal{O}(\frac{1}{\varepsilon})$ amounts of time in the neighborhood of the slow manifolds. These techniques include existing versions of the Exchange Lemma developed in [31–33, 56], as well as a modified version that we develop here to treat the specific problem of periodic orbits.

The general existence theorems given in this work are applicable to a wide variety of models. One application was already given in [9] to show the existence of stationary, large-amplitude, spatially periodic solutions in the 1-D Gray-Scott model. Here we treat two examples, demonstrating the existence and local uniqueness of periodic orbits in the FitzHugh-Nagumo [16] and Hodgkin-Huxley [28] models. The orbits in these systems correspond to large-amplitude, spatially-periodic traveling waves of the associated partial differential equations. First, we recover the results of [6] and [7] showing the existence of these solutions. Then, we also study the behavior of these orbits with respect to the wave speed, showing that, for each small but fixed value of ε, there exists an interval $(0, \theta^*(\varepsilon))$ of wave speeds such that there exists a periodic solution with wave speed θ for each $\theta \in (0, \theta^*(\varepsilon))$. Here the value $\theta^*(\varepsilon)$ is that associated to the homoclinic solution [38]. We note that stability analyses of these solutions are carried out in various contexts in [12, 17, 18, 42, 43, 47].

In many reaction-diffusion systems, small-amplitude periodic solutions are born in Turing or Ginzburg-Landau type bifurcations. This type of bifurcations was studied by Turing [58] (on finite intervals) and by Eckhaus [11] (on unbounded domains) and is also known as the Benjamin-Feir instability [3]. In some systems, such as for example in the 1-D Gray-Scott model the family of these small-amplitude periodic orbits can be continued (see [45]) into a regime where the periodic solutions are of large-amplitude, see also Section 4 of [9].

The paper is organized as follows. Section 2 contains the assumptions the system (1.1) must satisfy, and a statement of the main results for general systems (1.1). Section 3 to 6 contain the details of the construction of the periodic orbit. In particular, in Section 3, we give the augmented system and its Fenichel normal form. Then, we identify the domain of the Poincare map (and hence, for the augmented system, the set of points in which the solution of the boundary value problem will lie) in Section 4. The versions of the Exchange Lemma that are needed are given in Section 5, and finally, the tracking of the manifolds of solutions through the good initial conditions is carried out in Section 6. In Sections 7 and 8, the applications of our results to the existence and local uniqueness of periodic orbits in the FitzHugh-Nagumo and Hodgkin-Huxley models are given. Finally, Section 9 contains the proof of the modified Exchange Lemma.

2. Singular periodic orbits and statement of the main results. Before stating the main existence theorems, we make the following three assumptions:

ASSUMPTION 1. *The reduced fast system (1.2) has N manifolds of fixed points, where $N \geq 1$. We label them as critical manifolds, and we denote them by $\mathcal{S}_1^0, ..., \mathcal{S}_N^0$. The critical manifolds $\mathcal{S}_i^0$ are p-dimensional, compact, and normally hyperbolic. Specifically, for each $i = 1, ..., N$, the vector field linearized at any point on $\mathcal{S}_i^0$ has $k > 0$ eigenvalues with positive real parts bounded from zero and $\ell > 0$ eigenvalues with negative real parts bounded from zero.*

Here, we explicitly assume that there are functions $\phi_i(y)$ such that, for each $i = 1, \ldots, N$, the critical manifold $\mathcal{S}_i^0$ is given by the graph of function the $x = \phi_i(y)$, or that the manifold $\mathcal{S}_i^0$ can be covered by a finite number of charts such that on each component the manifold is given locally as the graph of a function over the slow variables.

With Assumption 1, invariant manifold theory implies that, in the system (1.1) with $\varepsilon = 0$, the manifolds $\mathcal{S}_i^0$ have C^r $k+p$-dimensional local unstable manifolds $W^u_{\text{loc}}(\mathcal{S}_i^0)$ and C^r $\ell+p$-dimensional local stable manifolds $W^s_{\text{loc}}(\mathcal{S}_i^0)$.

Now, in the case that $0 < \varepsilon \ll 1$, it follows from Fenichel theory [13] that, for each $i = 1, \ldots, N$, the normally hyperbolic invariant manifold $\mathcal{S}_i^0$, along with its local stable and unstable manifolds $W^s_{\text{loc}}(\mathcal{S}_i^0)$ and $W^u_{\text{loc}}(\mathcal{S}_i^0)$, will persist as locally invariant manifolds of the full system (1.1) with ε sufficiently small. See also Section 1 of [30]. In particular, there are p-dimensional, locally invariant slow manifolds $\mathcal{S}_1^\varepsilon, \ldots, \mathcal{S}_N^\varepsilon$, that lie $\mathcal{C}^r$ $\mathcal{O}(\varepsilon)$ close to the critical manifolds $\mathcal{S}_1^0, ..., \mathcal{S}_N^0$; and, there exist local stable and unstable manifolds $W^s_{loc}(\mathcal{S}_i^\varepsilon)$ and $W^u_{loc}(\mathcal{S}_i^\varepsilon)$ of the slow manifolds. The latter have dimension $\ell + p$ and $k + p$, respectively, and they are also $\mathcal{C}^r$ $\mathcal{O}(\varepsilon)$ close to their unperturbed counterparts. Moreover, since we assumed that the critical manifolds are given as the graphs of functions, the Fenichel persistence theory also implies that there exist functions $\phi_i^\varepsilon(y)$, for $i = 1, ..., N$ such that (at least locally) the slow manifolds are given by the graphs: $x = \phi_i^\varepsilon(y)$, for $i = 1, ..., N$.

Having briefly summarized the relevant geometric objects associated to the N slow manifolds both when $\varepsilon = 0$ and when $0 < \varepsilon \ll 1$, we make an assumption about the relative disposition of their perturbed stable and unstable manifolds. As stated in the Introduction, we are interested in the situations when the unstable and stable manifolds of pairwise adjacent slow manifolds intersect transversely. We make:

ASSUMPTION 2. *There exists a finite sequence of slow manifolds $\{\mathcal{S}_j^\varepsilon\}$, each element of which corresponds to one of the N manifolds $\mathcal{S}_1^\varepsilon, ..., \mathcal{S}_N^\varepsilon$, where j is taken from a (finite) index set J, such that: (i) the first and last element of this sequence correspond to the same slow manifold, (ii) any given $\mathcal{S}_j^\varepsilon$ occurs finitely often in the sequence, and (iii) one of the following holds:*

- **Case 1** (transversality at $\varepsilon = 0$): for every $j \in J$, $W^u(\mathcal{S}_j^0)$ transversely intersects $W^s(\mathcal{S}_{j+1}^0)$ in a p-dimensional heteroclinic manifold. Furthermore, there exist $(p-1)$-dimensional submanifolds $TO_j \subset \mathcal{S}_j^0$ and $TD_{j+1} \subset \mathcal{S}_{j+1}^0$ such that pairwise points that lie on TO_j and TD_{j+1} with the same y-coordinate are connected via a heteroclinic orbit in the transverse intersection of $W^u(\mathcal{S}_j^0)$ and $W^s(\mathcal{S}_{j+1}^0)$.
- **Case 2** (transversality at $\varepsilon > 0$): for every $j \in J$, and for $0 < \varepsilon \ll 1$ sufficiently small, $W^u(\mathcal{S}_j^\varepsilon)$ transversely intersects $W^s(\mathcal{S}_{j+1}^\varepsilon)$ in a p-dimensional heteroclinic manifold. Moreover, the $\varepsilon \to 0$ limiting stable and unstable fibers in this intersection have base points that constitute $(p-1)$-dimensional submanifolds $TO_j \subset \mathcal{S}_j^0$ and $TD_{j+1} \subset \mathcal{S}_{j+1}^0$, respectively.

REMARK 2.1. In both Case 1 and Case 2, the submanifolds TO_j and TD_{j+1} will be referred to as the take off and touch down sets. Each point in TO_j is, in the limit of $\varepsilon \to 0$, the base point of a fiber in $W^u(\mathcal{S}_j^0)$ that lies in the intersection with $W^s(\mathcal{S}_{j+1}^0)$, and hence the orbits in this fiber may be thought of as taking off from TO_j. See also [57], where the terminology of jump off and jump on sets is used.

REMARK 2.2. Case 2 can typically be verified via Melnikov theory. In particular, the sets TO_j and TD_{j+1} are given by the zeroes of the relevant Melnikov function, see for example [48]. Also, the homoclinic orbits may be of the multiple-pulse type, as long as they lie in the transverse intersections of the stable and unstable manifolds of a slow manifold. In the case of a multiple pulse homoclinic orbit, the associated fast orbit segment in the $\varepsilon = 0$ system consists of the same number of copies of the heteroclinic orbit of the reduced fast system.

Singular periodic orbits are now constructed by formally connecting orbit segments of the reduced fast (1.2) and slow (1.3) systems.

DEFINITION 1. *A singular periodic orbit of* (1.1) *with* $\varepsilon = 0$ *consists of orbit segments of* (1.3) *on* $\mathcal{S}_j^0$, *with* $\mathcal{O}(1)$ *length, that connect pairs of points on the touch down set* TD_j *and the take off set* TO_j, *interspersed with heteroclinic connections between the end point of the slow segment in* TO_j *and a point in* TD_{j+1}, *for every* j *in the index set* J. *Moreover, the succession of slow segments and heteroclinic connections is such that the resulting singular orbit is a closed, continuous curve.* See Figure 1 for an illustration.

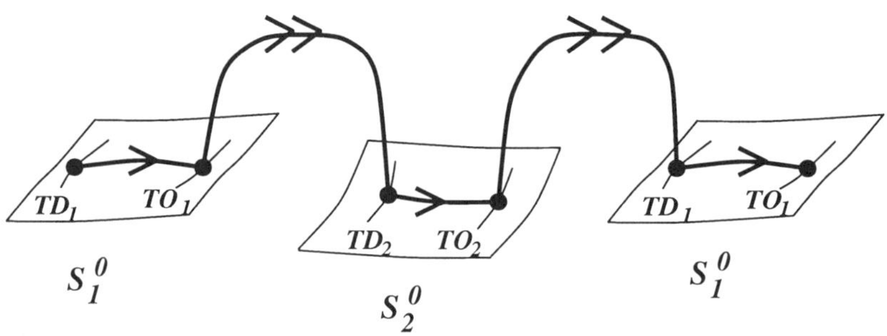

FIG. 1. *Schematic representation of a singular periodic orbit for* $p = 2$ *and* $N = 2$. *Dark segments with one (two) arrows correspond to slow (fast) segments.*

ASSUMPTION 3. *System (1.1) possesses a singular periodic orbit. Moreover, in neighborhoods of each of the slow segments of this singular periodic orbit, the slow flow* (1.3) *on* $\mathcal{S}_j^0$ *is transverse to the* $(p-1)$-*dimensional submanifolds* TD_j *and* TO_j, *for every* $j \in J$, see Figure 2 for an illustration.

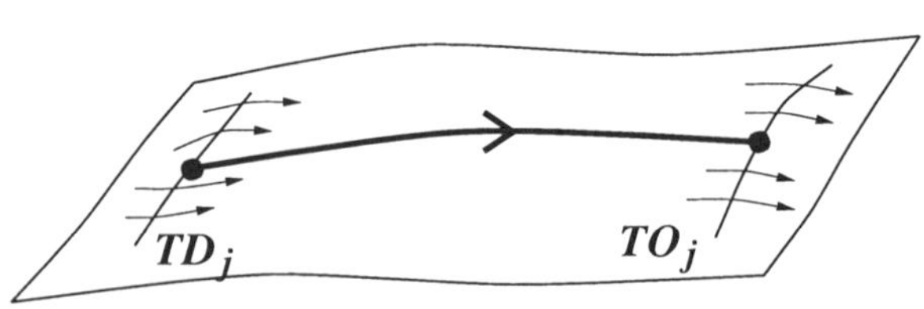

FIG. 2. *Schematic representation of Assumption 3 for* $p = 2$.

We are now ready to state the first main result for the case $p = 1$:

THEOREM 1. *If the system (1.1) satisfies assumptions 1–3 with* $p = 1$, *then, for* $\varepsilon > 0$ *sufficiently small, there exists a periodic orbit of (1.1) that is* $\mathcal{O}(\varepsilon)$ *close to the singular one. Moreover, the periodic orbit is locally unique.*

By locally unique, it is meant that the periodic orbit is unique in a neighborhood of the singular periodic orbit. The application of Theorem 1 to the FitzHugh-Nagumo equations is illustrated in Section 7.

For systems where $p > 1$, we shall require that the singular periodic orbit satisfies an extra condition. This condition is that the singular solutions around it are either attracted or repelled by it. Thus, the singular periodic orbit is locally unique. We make this precise by using the following definition of a singular periodic manifold and the singular map F associated to it. We remark that this extra requirement is not strictly necessary, and in Remarks 2.4 and 2.5 below we discuss less restrictive hypotheses.

A singular periodic manifold is simply the union of singular trajectories of (1.1) starting in a subset $\hat{T}D_1$ of TD_1 and coming back to it, where $\hat{T}D_1$ has the same dimension as TD_1. We will explicitly state below the requirement that the singular periodic manifold contains a singular periodic orbit. The singular map F associated to it is defined as follows. For each $i = 1, ..., N$, consider the reduced flow on $\mathcal{S}_i^0$. Let $f_s^i : TD_i \longrightarrow TO_i$ denote the (slow) map that associates to any point $p_i^{TD} \in TD_i$ the point at which the trajectory through p_i^{TD} under (1.3) intersects the submanifold TO_i, if such an intersection exists. For $i = 1, ..., N$, let $f_f^i : TO_i \longrightarrow TD_{i+1}$ denote the (fast) map that sends a point $p_i^{TO} \in TO_i$ to the point $p_{i+1}^{TD} \in TD_{i+1}$ where there is a heteroclinic orbit of the fast system (1.2) connecting p_i^{TO} to p_{i+1}^{TD}. Let L denote the penultimate member of the index set J of Assumption 2. For $j = 2, ..., L + 1$, let the map $F^j : TD_1 \longrightarrow TD_j$ denote the composition from $i = 1$ to $i = j$ of the slow map f_s^i followed by the fast map f_f^i:

$$F^j \equiv f_f^{j-1} \circ f_s^{j-1} \circ ... \circ f_f^2 \circ f_s^2 \circ f_f^1 \circ f_s^1 .$$

See Figure 3 for an illustration. Let F denote the associated singular map with $j = L + 1$, $F^{L+1} : TD_1 \longrightarrow TD_1$.

DEFINITION 2. *Assume there exists* $\hat{T}D_1$, *a submanifold of* $TD_1 \subset \mathcal{S}_1^0$ *of the same dimension as* TD_1 *such that for* $j = 2, ..., L + 1$, *the dimension of* $F^j(\hat{T}D_1)$ *is* $p - 1$, *and the singular map* F *has a fixed point* p_0. *A p-dimensional singular periodic manifold of* (1.1) *consists of the singular trajectories with slow segments of* $\mathcal{O}(1)$ *length between the manifolds* $F^j(\hat{T}D_1) \subset TD_j$ *and* $f_s^j(F^j(\hat{T}D_1)) \subset TO_j$, *and with fast segments between the manifolds* $f_s^j(F^j(\hat{T}D_1)) \subset TO_j$ *and* $F^{j+1}(\hat{T}D_1) \subset TD_{j+1}$, *for* $j = 1, ..., L$.

ASSUMPTION 4. *All eigenvalues of* $dF(p_0)$ *have norm bounded away from 1.*

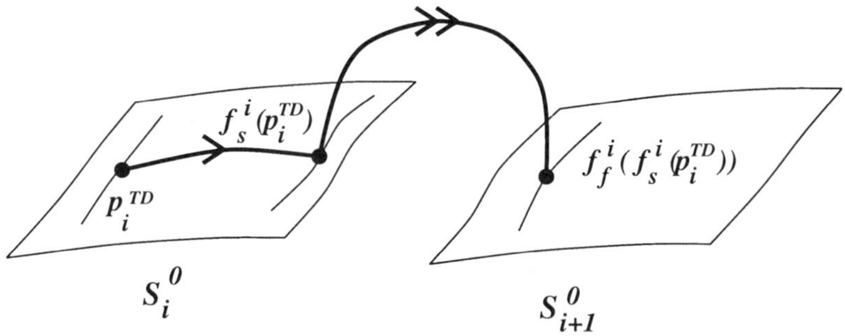

FIG. 3. *Schematic representation of the slow f_s^i and fast f_f^i maps.*

We are now ready to state the second main result:

THEOREM 2. *Let $p \geq 2$. Suppose* (1.1) *satisfies Assumptions* 1–3. *Suppose also that* (1.1) *has an p-dimensional singular periodic manifold whose singular periodic orbit satisfies Assumption* 4. *Then for $\varepsilon > 0$ sufficiently small* (1.1) *has a locally unique periodic orbit $\mathcal{O}(\varepsilon)$ close to the singular periodic orbit.*

We outline the general idea for the proof of Theorems 1 and 2 here. The main difficulty in finding periodic orbits is to determine the right initial conditions. This is typically done by constructing a Poincaré section for the full flow, and then by looking for a fixed point of the Poincaré map. We turn the Poincaré map problem into a boundary-value problem by embedding the system (1.1) in a higher dimensional system obtained by naturally appending equations to it that keep track of the initial conditions.

Then, we solve this boundary-value problem by looking for the transverse intersection of two invariant manifolds as follows. In the augmented system, the appropriate set of initial conditions is that which corresponds in the original system to the Poincaré section, near the slow manifold $\mathcal{S}_1^\varepsilon$. An invariant manifold is obtained from this set by flowing the initial conditions backward and forward in time. Let $\mathcal{M}_\varepsilon^+$ denote the component obtained by flowing the initial conditions forward in time, and let $\mathcal{M}_\varepsilon^-$ denote the component obtained by flowing the initial conditions backward in time. Finally, we show that $\mathcal{M}_\varepsilon^+$ transversely intersects $\mathcal{M}_\varepsilon^-$ *i.e.* that the invariant manifold intersects itself, in a one-dimensional curve that corresponds to the locally unique periodic orbit we seek.

Sections 3–5 contain the technical tools necessary for the construction and tracking of the manifolds $\mathcal{M}_\varepsilon^+$ and $\mathcal{M}_\varepsilon^-$. The proofs of Theorems 1 and 2 are given in Section 6.

REMARK 2.3. For systems of the form (1.1) with slow manifolds that have only attracting normal directions, such as the van der Pol equation in the relaxation limit, see for example [21], periodic orbits are obtained with the aid of a singular Poincaré map. In those systems, the boundaries ('knees') of the slow manifolds are formed by saddle-node equilibrium points of the fast system, where normal hyperbolicity is lost. The transitions between the fast and slow segments in the singular periodic orbit occur at these knees. Also, the domain of the singular Poincaré map is often chosen to be one of the knees. Pioneering work carried out in the 1930's–50's in this direction is described, for example, in [44]. The work in [4, 44] identify the conditions under which, if the singular Poincaré map has a fixed point and of its derivative at that fixed point has eigenvalues strictly less than one in absolute value, then for $\varepsilon > 0$ (1.1) has a unique stable periodic orbit near the singular one.

REMARK 2.4. For some systems (1.1) that possess singular periodic manifolds, Assumption 4 may be replaced with an alternative hypothesis that there exists a structurally stable, locally unique singular periodic orbit on the singular periodic manifold. The proof of Theorem 2 given in Sections 3–6 is readily modified when this alternative hypothesis is made. Examples in which this alternative hypothesis is useful include perturbed Hamiltonian systems, see for example [34]. In this class of problems, there are unique singular periodic orbits on the singular periodic manifolds, and these are found in the transverse intersections of invariant submanifolds on the singular periodic manifolds.

REMARK 2.5. The even more basic assumption that there is a singular periodic manifold, while sufficient with Assumption 4 or with the alternate hypothesis in Remark 2.4, may not be necessary in some examples for the existence of periodic orbits. In the fundamental work of [57], general conditions are given to establish the existence of fast-slow solutions to singularly-perturbed two-point boundary value problems. Given that we have converted the problem of demonstrating the existence of periodic orbits to such a boundary value problem, it may be possible to apply the general existence theorem of [57] in the context of systems (1.1), as well. The procedure for verifying that the hypotheses of that theory hold involve determining if, for each pair of adjacent slow manifolds $\mathcal{S}_j^\varepsilon$ in the sequence, there are invariant *submanifolds* of the slow manifolds such that the stable and unstable manifolds of these invariant submanifolds intersect each other transversely in one-dimensional curves in the augmented system (3.1) below. If it is possible to hook together these fast heteroclinics in alternation with appropriate slow orbit segments on the slow manifolds to form a singular periodic orbit, then by the theory in [57] the singular periodic orbit would then persist as a true periodic orbit of the full system (1.1) with $0 < \varepsilon \ll 1$.

3. The augmented system and its Fenichel normal form. In this section, we embed (1.1) into a higher dimensional system by appending some equations to (1.1). In [17] the idea of augmenting a system (specifically doubling its size) was used on a linear system (see eq. (2.7) in [17]) to compare initial and final values. Here, we show that a similar idea works for nonlinear systems. Finally, once we obtain the augmented system, we rewrite it in the neighborhood of the slow manifolds using the Fenichel normal form.

Let $\mathcal{S}_1^\varepsilon$ be the first manifold in the sequence $\{\mathcal{S}_j^\varepsilon\}$. Let $\mathbf{q_1^0}(t)$ be the segment of the singular periodic orbit on $\mathcal{S}_1^0$, and let $\mathbf{q_1^{0,s}} \in TD_1$, $\mathbf{q_1^{0,u}} \in TO_1$ be the endpoints of this segment. The existence of both points follows from Assumption 3. Chose a middle point $\mathbf{q_1^{0,m}}$ along $\mathbf{q_1^0}(t)$, so that the length of the two segments from $\mathbf{q_1^{0,m}}$ to each of the endpoints is $\mathcal{O}(1)$, see Figure 4.

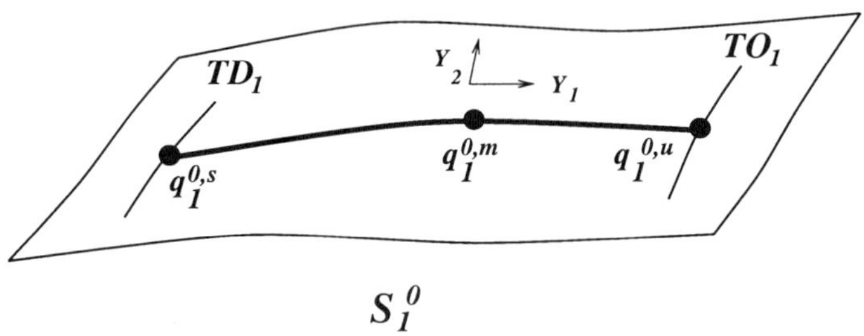

FIG. 4. *Schematic representation of $q_1^0(t)$, the initial segment of the singular periodic orbit, and the slow coordinates on $\mathcal{S}_1^0$, for $p = 2$.*

The following technical step, involving a change of the slow coordinates $\mathbf{y}$ and a rectification of the flow in the neighborhoods of the slow segments of the singular periodic orbit, will simplify our work later on. Let $\mathbf{q_1^0}(t)$ denote the set $\{\mathbf{Y} : Y_i = 0, i = 2, ..., p\}$. Here, the coordinate Y_1 indicates the direction along $\mathbf{q_1^0}(t)$, and the other slow variables $Y_2, ..., Y_p$ indicate the directions transverse to this segment. Without loss of generality, we can assume that $Y_1 = 0$ at $\mathbf{q_1^{0,m}}$.

The augmented system we study is:

$$
\begin{aligned}
\dot{x} &= f(x, Y; \varepsilon) \\
\dot{Y} &= \varepsilon g(x, Y; \varepsilon) \\
\dot{\hat{x}} &= 0 \\
\dot{\hat{\mathcal{Y}}} &= 0
\end{aligned}
\tag{3.1}
$$

where $x \in \mathbb{R}^{\mathbf{k}+\ell}$, $Y \in \mathbb{R}^{\mathbf{p}}$, $\hat{x} \in \mathbb{R}^{\mathbf{k}+\ell}$ and $\hat{\mathcal{Y}} = (\hat{Y}_2, ..., \hat{Y}_p) \in \mathbb{R}^{\mathbf{p-1}}$. The additional variables $\hat{x}$ have the same dimension as the fast variables x, and the variables $\hat{\mathcal{Y}}$ have one dimension less than the slow variables Y. In total, we have appended $k + \ell + p - 1$ parameters as slow variables. For any trajectory of (1.1), the $\hat{x}$ variables of (3.1) will correspond to the initial conditions of each of the x coordinates. Similarly, the $\hat{\mathcal{Y}}$ variables of (3.1) will correspond to the initial conditions of the last $p-1$ of the Y coordinates of this trajectory. The Y_1 coordinate of the initial condition will be taken as $Y_1 = 0$. Thus, instead of working in a $(k + \ell + p)$-dimensional space, we will be working in a $(2k + 2\ell + 2p - 1)$-dimensional space. As we will see in the next sections, working in this higher dimensional space offers us several advantages.

In the augmented system, the center manifolds $\mathcal{S}_1^\varepsilon, ..., \mathcal{S}_N^\varepsilon$ have dimensions $k+\ell+2p-1$, and their stable $W^s(\mathcal{S}_i^\varepsilon)$ and unstable $W^u(\mathcal{S}_i^\varepsilon)$ manifolds have dimensions $k + 2\ell + 2p - 1$ and $2k + \ell + 2p - 1$, respectively. When $W^u(\mathcal{S}_i^\varepsilon)$ transversely intersects $W^s(\mathcal{S}_{i+1}^\varepsilon)$, we obtain a $(k + \ell + 2p - 1)$-dimensional heteroclinic manifold. Similarly, the submanifolds TD_i and TO_i now have dimension $k + \ell + 2p - 2$.

The Fenichel normal form for (3.1) with $0 < \varepsilon \ll 1$ in the neighborhood of $\mathcal{S}_1^\varepsilon$ is:

$$\begin{aligned} \dot{a} &= \Lambda(a, b, y; \varepsilon)a \\ \dot{b} &= \Gamma(a, b, y; \varepsilon)b \\ \dot{y} &= \varepsilon m(y, \varepsilon) + \varepsilon h(a, b, y; \varepsilon)ab. \end{aligned} \tag{3.2}$$

Here, $a \in \mathbf{R}^\mathbf{k}$, $b \in \mathbf{R}^\ell$, $y \equiv (\hat{a}, \hat{b}, Y, \hat{\mathcal{Y}}) \in \mathbf{R}^{\mathbf{k}+\ell+\mathbf{2p}-\mathbf{1}}$, and Λ and Γ are square matrices of dimensions k and ℓ, respectively, m is a $k + \ell + 2p - 1$ component vector, and h is a rank three tensor. This normal form separates the exponentially growing, decaying and center directions; the sets $\{a = 0\}, \{b = 0\}$ correspond to $W_{loc}^s(\mathcal{S}_1^\varepsilon)$ and $W_{loc}^u(\mathcal{S}_1^\varepsilon)$, respectively. Fenichel normal forms can similarly be obtained in the neighborhoods of each of the $\mathcal{S}_j^\varepsilon$.

To finish this section, we state some important estimates that trajectories inside $\mathcal{B}_{\mathcal{S}_1}$ satisfy and that follow directly from (3.2):

$$|a(t)| \geq C_a |a(s)| \exp(\lambda_0^+(t - s)), \qquad t > s \tag{3.3}$$

$$|b(t)| \leq C_b |b(s)| \exp(-\lambda_0^-(t - s)), \qquad t > s \tag{3.4}$$

where $C_a > 0$ and $C_b > 0$, and $\lambda_0^+ > 0, \lambda_0^- > 0$ are appropriate bounds on the real part of the eigenvalues of the fixed points of $\mathcal{S}_1^0$. See [31, 32] for the derivation of these estimates.

4. Construction of the good set of initial conditions. Let $\hat{\mathcal{B}}_{\mathcal{S}_1}$ be a subset of the box $\mathcal{B}_{\mathcal{S}_1}$ in which the Y coordinates are rectified, and which is bounded by a tube of radius $\delta > 0$ around the singular segment $\mathbf{q_1^0}(t)$, where δ is small. Let $r_1^{0,u}$ denote a point in TO_1, and $Y_1(r_1^{0,u})$ denote the Y_1-coordinate of this point. The same notation will be used for a point $r_1^{0,s} \in TD_1$, see Figure 5.

By Assumption 3, the Y-coordinates of any trajectory that starts in $\hat{\mathcal{B}}_{\mathcal{S}_1}$ with $Y_1 = 0$ and that stays in $\mathcal{B}_{\mathcal{S}_1}$ will eventually coincide with the Y-coordinates of some point $r_1^{0,u^*} \in TO_1$. The time T_f in which the Y_1-coordinate of such trajectory goes from $Y_1 = 0$ to $Y_1(r_1^{0,u^*})$ is

$$T_f = \frac{1}{\varepsilon} \int_0^{Y_1(r_1^{0,u^*})} \frac{dY_1}{m_1(y, \varepsilon) + h_1(a, b, y; \varepsilon)ab}. \tag{4.1}$$

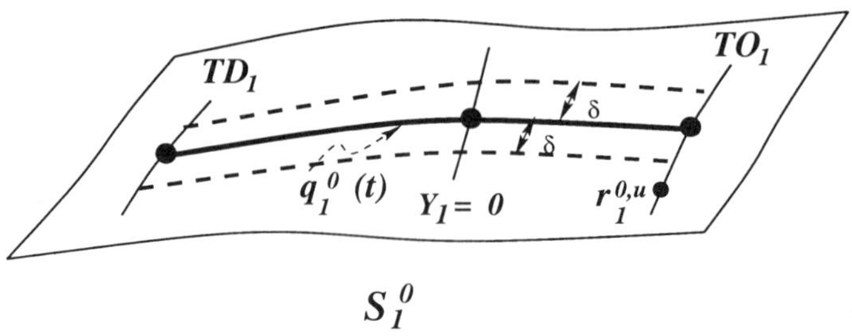

FIG. 5. *Schematic representation of a tube of radius δ around $q_1^0(t)$, for $p = 2$.*

Let $\underline{Y}_1(q_1^{0,u})$ and $\overline{Y}_1(q_1^{0,u})$ be the minimum and maximum, respectively, of the Y_1 coordinates of points $r_1^{0,u}$ in a neighborhood of radius less than or equal to 2δ around the point $q_1^{0,u}$, the endpoint of $q_1^0(t)$ in TO_1. A lower bound for T_f is

$$\underline{T}_f = \frac{1}{\varepsilon}\int_0^{\underline{Y}_1(q_1^{0,u})} \frac{dY_1}{\sup\{m_1(y,\varepsilon)+h_1(a,b,y;\varepsilon)ab\}} \equiv \frac{\underline{K}_f}{\varepsilon}, \tag{4.2}$$

where the sup is taken over all Y in $\hat{U}$, a sufficiently large closed subset of $\mathcal{S}_1^0$ containing all the slow trajectories that start at $Y_1 = 0$ and that are in a neighborhood of radius δ of the slow segment $\mathbf{q_1^0}(t)$, and over all $a, b \leq \Delta$. An upper bound is

$$\overline{T}_f = \frac{1}{\varepsilon}\int_0^{\overline{Y}_1(q_1^{0,u})} \frac{dY_1}{\inf\{m_1(y,\varepsilon)+h_1(a,b,y;\varepsilon)ab\}} \equiv \frac{\overline{K}_f}{\varepsilon}, \tag{4.3}$$

where the inf is also taken over all Y in $\hat{U}$, and over all $a, b \leq \Delta$. Similarly, one can compute the time T_b (as well as bounds $\underline{T}_b$ and $\overline{T}_b$ on it), which is the time it takes for the Y-coordinates of a trajectory starting in $\hat{\mathcal{B}}_{\mathcal{S}_1}$ and staying in $\mathcal{B}_{\mathcal{S}_1}$ to go from $Y_1 = 0$ to $Y_1 = Y_1(r_1^{0,s^*})$ in backward time, where $r_1^{0,s^*} \in TD_1$.

We now proceed to find the initial values of the fast variables. Let $a(0)$ be such that $C_a|a(0)|\exp(\lambda_0^+\underline{T}_f) = 2\Delta$. This choice implies $|a(0)| = \frac{2\Delta}{C_a}\exp(-\lambda_0^+\underline{T}_f)$, and we define $a_{max} \equiv |a(0)|$, so that a solution starting with this $a(0)$ satisfies $|a(\underline{T}_f)| \geq 2\Delta$ by (3.3). Therefore, the trajectory through $(a_{max}, b(0), Y_1 = 0, Y_r \in \hat{\mathcal{B}}_{\mathcal{S}_1}, \hat{a} = a_{max}, \hat{b} = b(0), \hat{\mathcal{Y}}_r = Y_r)$ for $r = 2, ..., p$ and $b(0) < \Delta$ leaves the box $\mathcal{B}_{\mathcal{S}_1}$ with its Y_1-coordinate less than $Y_1(r_1^{0,u})$ for all $r_1^{0,u} \in TO_1$.

In a similar fashion, but now using (3.4) in backward time, we can choose $b(0)$ so that $b(0) = \frac{2\Delta}{C_b}\exp(\lambda_0^-\underline{T}_b)$ and hence $b(\underline{T}_b) \geq 2\Delta$. We then define $b_{max} \equiv |b(0)|$. Therefore, a trajectory starting on $\hat{\mathcal{B}}_{\mathcal{S}_1}$ with b_{max} leaves the box $\mathcal{B}_{\mathcal{S}_1}$ with its Y_1-coordinate greater than $Y_1(r_1^{0,s})$ for all $r_1^{0,s} \in TD_1$.

Using a_{max}, b_{max} as defined above, we see that a solution through $(a_{max}, b_{max}, Y_1 = 0, Y_r \in \hat{\mathcal{B}}_{\mathcal{S}_1}, \hat{a} = a_{max}, \hat{b} = b_{max}, \hat{\mathcal{Y}}_r = Y_r)$, for $r = 2, ..., p$, leaves the box $\mathcal{B}_{\mathcal{S}_1}$ in forward time before its Y_1-coordinate equals the Y_1-coordinate of any point in TO_1, and in backward time before it equals the Y_1-coordinate of any point in TD_1. Now, by continuous dependence on initial conditions, we can take (a_{min}, b_{min}) small enough such that the solution through $(a_{min}, b_{min}, Y_1 = 0, Y_r \in \hat{\mathcal{B}}_{\mathcal{S}_1}, \hat{a} = a_{min}, \hat{b} = b_{min}, \hat{\mathcal{Y}}_r = Y_r)$, with $r = 2, ..., p$, leaves the box $\mathcal{B}_{\mathcal{S}_1}$ in forward time through its $|a| = \Delta$ side after its Y_1 coordinate equals $Y_1(r_1^{0,u})$ for some $r_1^{0,u} \in TO_1$ and after it $Y_1(r_1^{0,s})$ for some $r_1^{0,s} \in TD_1$ in backward time. (Recall that if a solution has initial conditions with $a = b = 0$, then the solution stays on $\mathcal{S}_1^{\varepsilon}$, and therefore it cannot leave the box through the $|a| = \Delta$ or $|b| = \Delta$ sides, and also that the set of points leaving the box is nonempty.)

We are now ready to define a good set of initial conditions. Let I_ε be defined as:

$$I_\varepsilon = \Big\{(a, b, y) : a \in I_a,\ b \in I_b,\\ Y_1 = 0,\ \hat{a} = a,\ \hat{b} = b,\ Y_r = \hat{\mathcal{Y}}_r \in I_{Y_r},\ r = 2, ..., p\Big\}, \tag{4.4}$$

where $I_a \equiv [a_{min}, a_{max}] \equiv \{a_i \in [a_{i_{min}}, a_{i_{max}}], i = i, ..., k\}$, $I_b \equiv [b_{min}, b_{max}] \equiv \{b_j \in [b_{j_{min}}, b_{j_{max}}], j = 1, ..., \ell\}$, $I_{Y_r} \equiv [Y_r(\mathbf{q}_m^0) + \delta, Y_r(\mathbf{q}_m^0) - \delta]$ and $Y_r(\mathbf{q}_1^{0,m})$ denotes the Y_r coordinate of the point $\mathbf{q}_1^{0,m}$. I_ε has dimension $k + \ell + p - 1$, and it is the product of $k + \ell + p - 1$ diagonal segments (see Figure 6), where the a and b coordinates are exponentially small. Also, the projection of this set onto the $\{b = 0\}$ space is $k + p - 1$ dimensional.

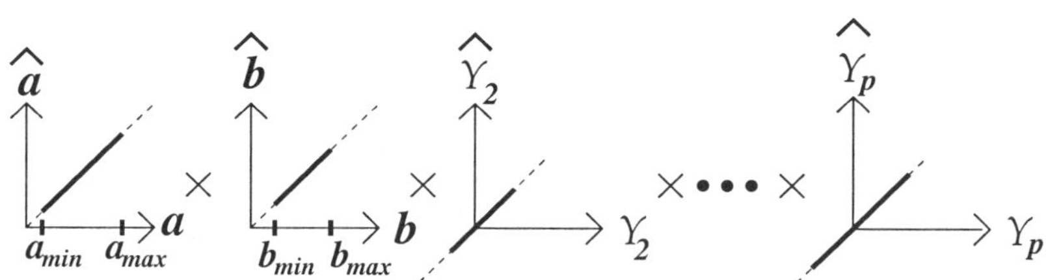

FIG. 6. *Representation of the good set of initial conditions I_ε. It corresponds to the set depicted taken in the $\{Y_1 = 0\}$ hyperplane.*

To finish this section, we point out that by construction, there are points in I_ε whose trajectories reach the boundary $|a| = \Delta$ of $\mathcal{B}_{\mathcal{S}_1}$ exactly when their Y_1-coordinates in forward time equal $Y_1(r_1^{0,u})$ for some $r_1^{0,u} \in TO_1$. This is because the map from I_ε to the exit face $|a| = \Delta$ of $\mathcal{B}_{\mathcal{S}_1}$ is continuous.

5. The Exchange Lemmas. Having constructed the good set of initial conditions I_ε in Section 4, we now state the main technical tools needed to track the forward trajectories of points in I_ε. The forward trajectories of points in I_ε will spend $\mathcal{O}(\frac{1}{\varepsilon})$ amounts of time in the neighborhoods of the slow manifolds $\mathcal{S}_j^\varepsilon$. The Exchange Lemmas stated in this section constitute the technical tools that we will use in Section 6 to track the invariant manifolds -and also their tangent spaces- formed by these trajectories to conclude that the tracked manifolds leave the neighborhoods $\mathcal{B}_{\mathcal{S}_j^\varepsilon}$ $\mathcal{C}^1$ exponentially close to the local unstable manifolds of the $\mathcal{S}_j^\varepsilon$.

The material presented in this section stands alone, independent of the other sections in this chapter. In Section 5.1, we state the existing versions of the Exchange Lemma. One is for transversality at $\varepsilon = 0$ and the other for $\varepsilon > 0$ transversality. In Section 5.2, we state a new, modified Exchange Lemma (MELESE) that we use to construct the periodic orbits. The proof of this lemma is presented in Section 9.

5.1. The Exchange Lemma. The Exchange Lemmas [30–33, 56] hold for singularly perturbed systems of the form

$$\begin{aligned} \dot{x} &= f(x, y; \varepsilon) \\ \dot{y} &= \varepsilon g(x, y; \varepsilon) \end{aligned} \tag{5.1}$$

where $x \in \mathbb{R}^{k+\ell}$, just as for (1.1), but here we take $y \in \mathbb{R}^{\mathbf{n}}$ so that we allow y to be a general vector of slow variables. We require that (5.1) possesses the structure stated in Assumption 1, with p replaced by n. Let $\mathcal{S}^\varepsilon$ be an n-dimensional slow manifold of (1.1), and $\mathcal{B}$ a neighborhood of $\mathcal{S}^\varepsilon$ where the Fenichel normal form (3.2) holds. Without loss of generality, we may take $\mathcal{B}$ to be the set $\mathcal{B} = \{(a, b, y) : |a| \leq \Delta, |b| \leq \Delta, y \in U\}$, where $\Delta > 0$ is small and U is an appropriate compact set.

Let $\mathcal{L}_\varepsilon$ be a $(k + \sigma)$-dimensional locally invariant manifold, with $1 \leq \sigma \leq n$ and $\varepsilon \geq 0$ for (5.1). Furthermore, for any subset X, let $\omega(X)$ denote the ω-limit set of X under the reduced fast flow of (5.1). Also, for any subset Y of the slow manifold $\mathcal{S}^0$ and for an interval $[\tau_1, \tau_2]$ of time, let $Y \cdot [\tau_1, \tau_2]$ denote the union of the images of Y under the reduced slow flow. Finally, for $\varepsilon \geq 0$, let q_ε be a point in the intersection of $\mathcal{L}_\varepsilon$ with the boundary $\partial\mathcal{B}$ of the box $\mathcal{B}$, if such an intersection exists.

The Exchange Lemmas give information about $\mathcal{L}_\varepsilon$ and its tangent space by following the evolution of the submanifold of $\mathcal{S}^\varepsilon$ given by the base points of the stable fibers of the transverse intersection $\mathcal{L}_\varepsilon \bigcap W^s_{loc}(\mathcal{S}^\varepsilon)$, see Figure 7.

We begin with the first version of the Exchange Lemma with Exponentially Small Error (ELESE) that we need, see [56]. It requires the following two tranversality hypotheses, and a hypothesis concerning the time of flight of a trajectory inside $\mathcal{B}$:

$(\mathbf{T1})_\mathbf{0}$ *There exists a point $q_0 \in \{|b| = \Delta\} \bigcap W^s_{loc}(\mathcal{S}^0)$ such that q_0 lies in the transverse intersection of $\mathcal{L}_0$ and $W^s_{loc}(\mathcal{S}^0)$.*

Let $N_0 \equiv \mathcal{L}_0 \cap W^s_{loc}(\mathcal{S}^0)$. Then N_0 has dimension σ. Moreover, we assume:

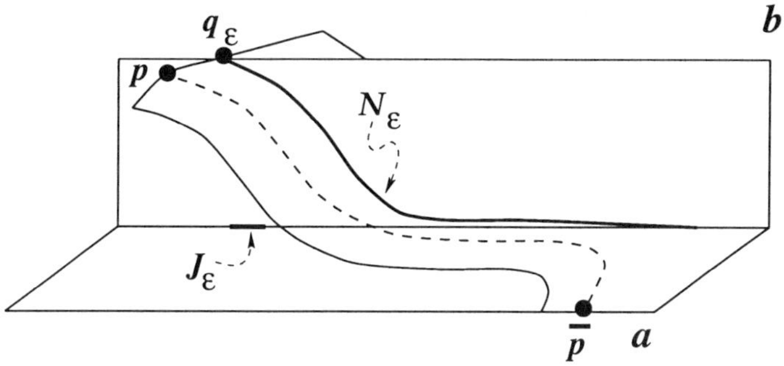

FIG. 7. *Illustration of the Exchange Lemma.*

(T2)$_0$ *$\omega(N_0)$ is a $\sigma-1$ dimensional submanifold of $\mathcal{S}^0$ containing the point $\overline{q}_0 \equiv \omega(q_0)$, and the flow of the reduced slow system* (1.3) *is transverse to $\omega(N_0)$ at $\overline{q}_0$.*

For a given point $p \in \mathcal{L}_\varepsilon$ on the "entrance" face of $\mathcal{B}$; *i.e.* $p \in \{|b| = \Delta\} \cap \mathcal{L}_\varepsilon$, let $\overline{p}$ denote the point at which the trajectory through p exits $\mathcal{B}$ on the face $\{|a| = \Delta\}$, and T_ε the time from p to $\overline{p}$. The "time" condition is:

(TE) $0 < \lim_{\varepsilon \to 0} \varepsilon T_\varepsilon = T_0 < \infty$.

LEMMA 1 (ELESE for transversality at $\varepsilon = 0$). *Let $\mathcal{L}_\varepsilon$ be a $k + \sigma$ dimensional invariant manifold for which $(T1)_0, (T2)_0$ and (TE) hold. Then, for $\varepsilon > 0$ sufficiently small and for $\gamma > 0$ sufficiently small and independent of ε, $\mathcal{L}_\varepsilon$ is $\mathcal{O}(e^{-\frac{c}{\varepsilon}})$ $\mathcal{C}^1$ close for some $c > 0$ at the exit point $\overline{p}$ to the $(k+\sigma)$ submanifold of $W^u_{loc}(\mathcal{S}^\varepsilon)$ given by $W^u_{loc}(\mathcal{S}^\varepsilon)|_{\omega(N_0)\cdot(T_0-\gamma, T_0+\gamma)}$.*

The first Exchange Lemmas are contained in [31–33], with $\sigma = 1$ and establish $\mathcal{O}(\varepsilon)$ $\mathcal{C}^1$ closeness. The closeness estimate was sharpened to $\mathcal{O}(e^{-\frac{c}{\varepsilon}})$ in [31], so that systems with transversality only when $\varepsilon > 0$ could also be treated (see also lemma 2 below). The version for general σ is given in [56]. We refer the reader to [30] for a more detailed history.

Lemma 1 allows us to track a locally invariant manifold $\mathcal{L}_\varepsilon$ and its tangent space while trajectories on $\mathcal{L}_\varepsilon$ spend $\mathcal{O}(\frac{1}{\varepsilon})$ amounts of time in the neighborhood $\mathcal{B}$ of the slow manifold $\mathcal{S}^\varepsilon$. The tracked manifold $\mathcal{L}_\varepsilon$ and its tangent space leave $\mathcal{B}$ exponentially close to a submanifold of $W^u_{loc}(\mathcal{S}^\varepsilon)$ and its tangent space, respectively, provided that $\mathcal{L}_\varepsilon$ transversely intersects $W^s_{loc}(\mathcal{S}^\varepsilon)$ on entry to $\mathcal{B}$.

It is also possible in a problem of the form (5.1) that $\mathcal{L}_\varepsilon$ and $W^s_{loc}(\mathcal{S}^\varepsilon)$ intersect transversely when $\varepsilon > 0$, but that this intersection is degenerate when $\varepsilon = 0$. This occurs, for example, in Hamiltonian systems and in systems of coupled reaction-diffusion equations. The second version of ELESE we state below in Lemma 2 treats this case, see [31, 56]. There are again two transversality conditions (see (T1) and (T2) below) required; and the "time" condition (TE) is the same as above.

(T1) *There exists a point $q_\varepsilon \in \{b = |\Delta|\} \bigcap W^s_{loc}(\mathcal{S}^\varepsilon)$, and $\mathcal{L}_\varepsilon$ intersects $W^s_{loc}(\mathcal{S}^\varepsilon)$ transversely at q_ε for $\varepsilon \neq 0$.*

Let $N_\varepsilon \equiv \mathcal{L}_\varepsilon \cap W^s_{loc}(\mathcal{S}^\varepsilon)$. Then N_ε has dimension σ. Let $J_\varepsilon \equiv \omega(N_\varepsilon)$ for $\varepsilon > 0$, and $J_0 = \lim_{\varepsilon\to 0}\omega(N_\varepsilon)$ and $\overline{q}_0 \equiv \omega(q_0)$.

(T2) *J_0 is a $\sigma - 1$ dimensional submanifold of $\mathcal{S}^0$ at $\overline{q}_0$, and the flow of the reduced slow system (1.3) is transverse to $\omega(N_0)$ at $\overline{q}_0$.*

The second version of ELESE is:

LEMMA 2 (ELESE for transversality at $\varepsilon > 0$). *Let $\mathcal{L}_\varepsilon$ be a $(k+\sigma)$ dimensional invariant manifold for which $(T1), (T2)$ and (TE) hold. Then for $\varepsilon > 0$ sufficiently small, $\mathcal{L}_\varepsilon$ is $\mathcal{O}(e^{-\frac{c}{\varepsilon}})$ $\mathcal{C}^1$ close for some $c > 0$ at the exit point $\overline{p}$ to the $(k+\sigma)$ submanifold of $W^u_{loc}(\mathcal{S}^\varepsilon)$ given by $W^u_{loc}(\mathcal{S}^\varepsilon)|_{J_\varepsilon\cdot T_\varepsilon}$, where the set $J_\varepsilon \cdot T_\varepsilon$ denotes the image under the flow of the set J_ε after a time T_ε.*

5.2. The modified Exchange Lemma.

In some applications, the tracked manifold $\mathcal{L}_\varepsilon$ does not intersect $W^s_{loc}(\mathcal{S}^\varepsilon)$ (see Figure 8). Instead, in some of these applications, it is possible to track $\mathcal{L}_\varepsilon$ by following its projection onto $W^u_{loc}(\mathcal{S}^\varepsilon)$. The modified ELESE (MELESE for short) given below makes this precise.

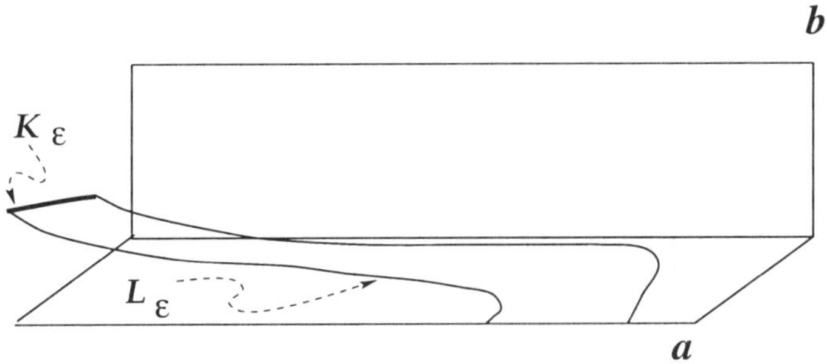

FIG. 8. *Illustration of the modified Exchange Lemma.*

We introduce more notation: let π denote the projection along the stable directions onto $W^u_{loc}(\mathcal{S}^\varepsilon)$. In Fenichel coordinates (3.2), if $r = (a, b, y) \in \mathcal{B}$, then $\pi r = (a, 0, y)$.

REMARK 5.1. When the stable and unstable fibers of the fixed points have been rectified, π corresponds to the projection along the stable Fenichel fibration, and $\pi r = \omega(r)$, the ω-limit of r under the fast reduced flow (1.2).

For the MELESE, we assume that the tracked manifold $\mathcal{L}_\varepsilon$ has dimension $k + \sigma$ and that it is obtained by the forward flow of a $(k + \sigma - 1)$ dimensional manifold K_ε. To be specific, we impose the following condition:

(M1) *There exists a $(k+\sigma-1)$-dimensional manifold K_ε for which πK_ε is a submanifold of $W^u_{\rm loc}(\mathcal{S}_\varepsilon)$ of dimension $k+\sigma-1$, such that the manifold $\mathcal{L}_\varepsilon$ is obtained by the union of trajectories starting on K_ε. Also, the projection of $\mathcal{L}_\varepsilon$ is a submanifold of $W^u_{\rm loc}(\mathcal{S}_\varepsilon)$ of dimension $k + \sigma$.*

For a point $p \in K_\varepsilon$, $T_p\mathcal{L}_\varepsilon$ denotes the tangent space of $\mathcal{L}_\varepsilon$ at the point p. Let $T_p\mathcal{L}_\varepsilon$ be spanned by the $k+\sigma$ tangent vectors $\{v_1, ..., v_{k+\sigma}\}$, where one of these tangent vectors corresponds to the vector field, and the others span the tangent space T_pK_ε. These tangent vectors generate a $k+\sigma$ dimensional parallelepiped. We are interested in the projection of this parallelepiped onto the different $k+\sigma$ dimensional subspaces of $\mathbb{R}^{\mathbf{k}+\ell+\mathbf{n}}$, which in Fenichel coordinates are generated by $k+\sigma$ of the unit vectors $\{e_{a_1}, ..., e_{a_k}, e_{b_1}, ..., e_{b_l}, e_{y_1}, ..., e_{y_n}\}$. These projected volumes can be obtained by computing the $k+\sigma$ subdeterminants of the $(k+\sigma)$ by $(k+\ell+n)$ matrix formed by the tangent vectors $\{v_1, ..., v_{k+\sigma}\}$.

Let $\mathcal{P}_{e_a, e_{y_1}, ..., e_{y_\sigma}}(v_1, ..., v_{k+\sigma})$ denote the projection of an $\mathcal{O}(1)$ parallelepiped generated by $\{v_1, ..., v_{k+\sigma}\}$ onto the $k+\sigma$ dimensional subspace that corresponds to the k unstable directions (indicated by the subscript $e_a \equiv (e_{a_1}, ..., e_{a_k})$) and σ of the n center directions $e_{y_1}, ..., e_{y_\sigma}$. Let $\mathcal{P}_{\mathcal{X}}(v_1, ..., v_{k+\sigma})$ correspond to the projection of the same parallelepiped onto any other $(k+\sigma)$-dimensional subspace. Let $\{\pi v_1, ..., \pi v_{k+\sigma}\}$ denote the projection of the tangent vectors along the stable directions. We require:

(M2) *For each combination of vectors $\{e_{y_1}, ..., e_{y_\sigma}\}$ taken from the collection $\{e_{y_1}, ..., e_{y_n}\}$, the projection $\mathcal{P}_{e_a, e_{y_1}, ..., e_{y_\sigma}}(v_1, ..., v_{k+\sigma})$ is exponentially close to $\mathcal{P}_{e_a, e_{y_1}, ..., e_{y_\sigma}}(\pi v_1, ..., \pi v_{k+\sigma})$. At least one of the projections $\mathcal{P}_{e_a, e_{y_1}, ..., e_{y_\sigma}}(v_1, ..., v_{k+\sigma})$ is $\mathcal{O}(1)$. Of the remaining projections $\mathcal{P}_{\mathcal{X}}(v_1, ..., v_{k+\sigma})$ onto the other $k+\sigma$ subspaces, there is at least one that is $\mathcal{O}(1)$. Similarly, for the subspace generated from the projected tangent vectors: at least one of the $\mathcal{P}_{\mathcal{X}}(\pi v_1, ..., \pi v_{k+\sigma})$ is $\mathcal{O}(1)$.*

This condition (see also Remark 5.3 below) means that the projection of the parallelepipeds generated from the tangent vectors and from the projected tangent vectors of $\mathcal{L}_\varepsilon$ onto the $k+\sigma$ dimensional subspaces of $W^u_{loc}(\mathcal{S}^\varepsilon)$ is practically the same. Furthermore, at least one of such projections must be significant. Finally, the "time" condition required for this modified version of ELESE is:

(ME) *There exists a point $p \in K_\varepsilon$ be such that its trajectory leaves $\mathcal{B}$ through the face $\{|a| = \Delta\}$ at a point $\overline{p}$, after an $\mathcal{O}(\frac{1}{\varepsilon})$ time T.*

The statement of the Modified ELESE (proven in Section 9) is:

LEMMA 3 (MELESE). *Let $\mathcal{L}_\varepsilon$ be a $(k+\sigma)$ dimensional invariant manifold, that satisfies* (M1), (M2), *and* (ME), *with the point p in* (ME) *being $q \in K_\varepsilon$. Then, for $\varepsilon > 0$ sufficiently small and $0 < \gamma$ independent of ε, $\mathcal{L}_\varepsilon$ is $\mathcal{O}(e^{-\frac{c}{\varepsilon}})$ $\mathcal{C}^1$ close at $\overline{q}$, for some $c > 0$, to the $k+\sigma$ dimensional submanifold of $W^u_{loc}(\mathcal{S}^\varepsilon)$ given by $W^u_{loc}(\mathcal{S}^\varepsilon)|_{\pi K_\varepsilon \cdot (T-\gamma, T+\gamma)}$.*

REMARK 5.2. A statement of MELESE in the special case $k+\ell = 2$ and $\sigma = 1$ was first given in [51].

REMARK 5.3. The conditions (M1) and (ME) are standard hypotheses in exchange lemmas. However, (M2) is different. The difference stems from the fact that the hypotheses here on the tracked manifold are verified inside the neighborhood $\mathcal{B}$, instead of on the entry face to $\mathcal{B}$. While this choice is natural for finding periodic orbits as is done in this work, it is inherently more technical. It is natural to ask how the hypothesis (M2) translates into a statement about the tracked manifold at the entry face to $\mathcal{B}$. The tracking work carried out in Section 6 shows that, at the entry face to $\mathcal{B}$, the segment of the manifold obtained by tracking the initial conditions I_ε backward in time is C^1 $\mathcal{O}(e^{-c/\varepsilon})$ close to $W^s_{\text{loc}}(\mathcal{S}_\varepsilon)$. This differentially close result is equivalent to condition (M2), and it is easier to verify in the context of examples.

6. Tracking of the manifolds $\mathcal{M}_\varepsilon^+$ and $\mathcal{M}_\varepsilon^-$. Let I_ε be the good set of initial conditions obtained in Section 4. If one projects I_ε onto the unaugmented or original variables (a, b, Y), one obtains the set $\Pi = \{(a, b, Y) : a \in I_a, b \in I_b, Y_1 = 0, Y_r \in I_{Y_r}, r = 2, ..., p\}$. This set is a section normal to the flow given by (1.1) in the neighborhood of $\mathcal{S}_1^\varepsilon$, or, in other words, a Poincaré section of the original system (1.1). Therefore, when analyzing the augmented system (3.1), any solution of (3.1) that starts in the full set I_ε and comes back to it is a periodic orbit. This is because any point in I_ε has distinct augmented $\hat{x}$ and $\hat{\mathcal{Y}}$ coordinates, and since the $\hat{x}$ and $\hat{\mathcal{Y}}$ coordinates of a trajectory do not change, the orbit can only come back to I_ε if it returns to its initial location. Therefore, if we solve the boundary value problem for (3.1) of finding a solution that starts on I_ε and returns to it, we will have found a fixed point of the Poincaré map of (1.1).

We now proceed to solve this boundary value problem using a geometric approach. Let $\mathcal{M}_\varepsilon^+$ be the manifold obtained by taking the union of all forward-time trajectories with initial conditions in I_ε and $\mathcal{M}_\varepsilon^-$ the one obtained by the backward-time trajectories with initial conditions in I_ε. By construction, both have dimension $k + \ell + p$. Our goal is to show that $\mathcal{M}_\varepsilon^+$ and $\mathcal{M}_\varepsilon^-$ intersect transversely in a one-dimensional manifold, and that this intersection corresponds to the periodic orbit we seek. To accomplish this goal, we first track the trajectories in $\mathcal{M}_\varepsilon^+$ while they stay in the neighborhood of $\mathcal{S}_1^\varepsilon$ until they leave this neighborhood. This is done in Section 6.1, with the modified Exchange Lemma as the key technical tool. In Section 6.2, we continue the tracking of the trajectories in $\mathcal{M}_\varepsilon^+$ using the Exchange Lemma, and show that there is a open set of trajectories on it that spend $\mathcal{O}(\frac{1}{\varepsilon})$ amounts of time in the neighborhood of each of the slow manifolds $\mathcal{S}_2^\varepsilon, ..., \mathcal{S}_L^\varepsilon$, where $\mathcal{S}_L^\varepsilon$ is the penultimate in the sequence $\{\mathcal{S}_j^\varepsilon\}$. Finally, in Section 6.3, we track in backward time the trajectories in $\mathcal{M}_\varepsilon^-$ until they leave the neighborhood of $\mathcal{S}_1^\varepsilon$, and show that $\mathcal{M}_\varepsilon^+$ and $\mathcal{M}_\varepsilon^-$ transversely intersect in the desired periodic orbit.

6.1. Tracking trajectories starting in the set I_ε inside $\mathcal{B}_{S_1}$. In this subsection, we first show that, in the neighborhood of $\mathcal{S}_1^\varepsilon$, the invariant manifold $\mathcal{M}_\varepsilon^+$ satisfies the hypotheses of MELESE (see Proposition 6.1 below). This will allow us to apply MELESE to conclude that $\mathcal{M}_\varepsilon^+$ leaves the neighborhood $\mathcal{B}_{S_1}$ C^1 exponentially close to a submanifold of $W^u_{loc}(\mathcal{S}_1^\varepsilon)$.

PROPOSITION 6.1. *The manifold* $\mathcal{M}_\varepsilon^+$ *satisfies the hypotheses of MELESE, Lemma 3, where the manifolds* K_ε *and* $\mathcal{L}_\varepsilon$, *are given by* I_ε *and* $\mathcal{M}_\varepsilon^+$, *respectively, and with* $n = k + \ell + 2p - 1$ *and* $\sigma = \ell + p$.

Proof. We will show that each of the conditions **(ME)**, **(M1)** and **(M2)** is satisfied. The augmented system (3.1) is of the form (5.1) with $n = k + \ell + 2p - 1$. As already stated, the good set of initial conditions I_ε has dimension $k+\ell+p-1$, and the locally invariant manifold $\mathcal{M}_\varepsilon^+$ obtained from the forward flow of I_ε is $(k + \sigma)-$dimensional, with $\sigma = \ell + p$. Also by the construction in Section 4, every trajectory that starts in I_ε leaves the neighborhood $\mathcal{B}_{\mathcal{S}_1}$ in an $\mathcal{O}(\frac{1}{\varepsilon})$ amount of time. Therefore the condition **(ME)** is satisfied.

We turn to verify **(M1)**. Let π_b denote the projection along the stable directions b onto the local unstable manifold $W^u_{loc}(\mathcal{S}_1^\varepsilon)$. The projection of I_ε onto $W^u_{loc}(\mathcal{S}_1^\varepsilon)$ is

$$\pi_b I_\varepsilon = \Big\{(a, b, y) : a \in I_a,\ b = 0, Y_1 = 0,\ \hat{a} = a,\ \hat{b} \in I_b, \quad Y_y = \mathcal{Y}_r \in I_{\mathcal{Y}_r},\ r = 2, ..., p\Big\} \tag{6.1}$$

where I_a, I_b and $I_{\mathcal{Y}_r}$ are as in Section 4. It is easy to see that $\pi_b I_\varepsilon$ is a submanifold of $W^u_{loc}(\mathcal{S}^\varepsilon)$ (*i.e.*,$\{b = 0\}$), and that its dimension is the same as that of I_ε: $k + \sigma - 1$, see Figure 9. Now, for all the points in the neighborhood $\mathcal{B}_{\mathcal{S}_1}$, the vector field has a component in the Y_1 direction. It follows that the projection of $\mathcal{M}_\varepsilon^+$, $\pi_b\mathcal{M}_\varepsilon^+$, which is a submanifold of $W^u_{loc}(\mathcal{S}^\varepsilon)$, has dimension $k + \sigma$. Therefore, I_ε and $\mathcal{M}_\varepsilon^+$ satisfy condition **(M1)**.

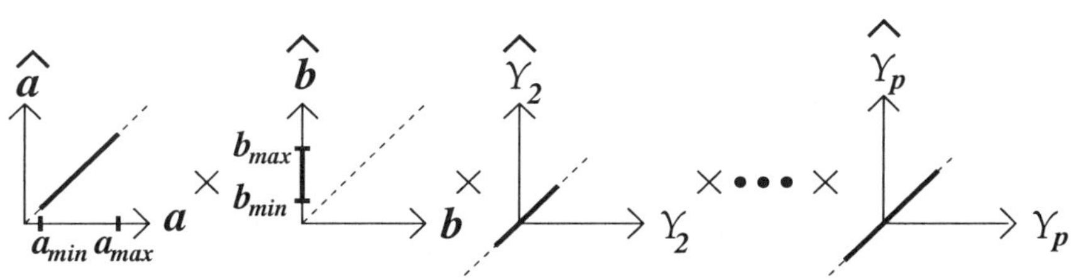

FIG. 9. *Representation of* $\pi_b I_\varepsilon$, *the projection of* I_ε *onto* $W^u_{loc}(\mathcal{S}_1^\varepsilon)$, $\pi_b I_\varepsilon$ *corresponds to the set depicted here, taken in the* $\{Y_1 = 0\}$ *hyperplane.*

Finally, and this is the main work of the proof, we verify condition **(M2)**. Let the standard vector basis of $\mathbb{R}^{2k+2\ell+2p-1}$ in Fenichel coordinates be

$$\Big\{e_{a_i}, e_{b_j}, e_{\hat{a}_i}, e_{\hat{b}_j}, e_{Y_s}, e_{\mathcal{Y}_m}; i = 1, ..., k, j = 1, ..., \ell, s = 1, ..., p, m = 2, ..., p\Big\}.$$

As above, e_r denotes the unit vector in the direction of the r-coordinate. Let e_{rs} denote a unit length vector along the diagonal of the plane spanned by the vectors e_r and e_s (see Figure 10). For a point p in the good set of initial conditions I_ε, let $\mathbf{v}$ denote the vector field at p, in Fenichel coordinates (3.2): $\mathbf{v} = (\Lambda a, \Gamma b, \varepsilon m(y,\varepsilon) + \varepsilon h(a,b,y;\varepsilon)ab)$.

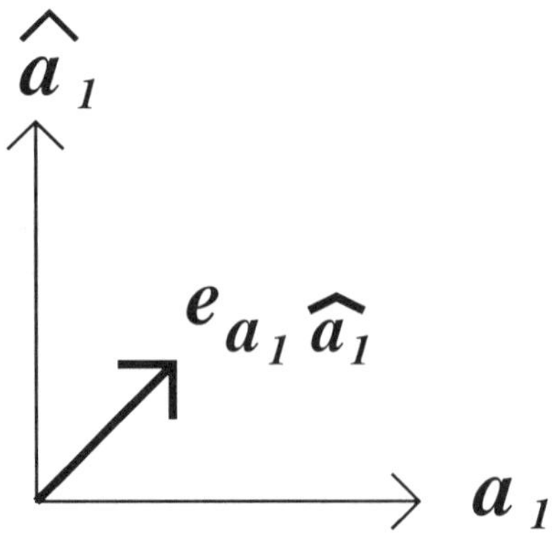

FIG. 10. *The vector $e_{a_1,\hat{a}_1}$ is the unit vector in the diagonal of $(a_1,\hat{a}_1)$ space.*

Consider now T_pI_ε, the tangent space of I_ε at p. By construction, T_pI_ε is spanned by:

$$\left\{e_{a_i\hat{a}_i}, e_{b_j\hat{b}_j}, e_{Y_r\hat{\mathcal{Y}}_r}; i = 1,\ldots,k, j = 1,\ldots,\ell, r = 2,\ldots,p\right\},$$

see Figure 6. Since $\mathcal{M}_\varepsilon^+$ is the union of trajectories starting in I_ε, the tangent space $T_p\mathcal{M}_\varepsilon^+$ is spanned by the above set of tangent vectors, plus the unit length vector $\hat{\mathbf{v}} \equiv \dfrac{\mathbf{v}}{|\mathbf{v}|}$:

$$\left\{e_{a_i\hat{a}_i}, e_{b_j\hat{b}_j}, e_{Y_r\hat{\mathcal{Y}}_r}; i = 1,\ldots,k, j = 1,\ldots,\ell, r = 2,\ldots,p, \hat{\mathbf{v}}\right\}, \tag{6.2}$$

where the a, b coordinates of p are exponentially small.

We will compute the projection of an $\mathcal{O}(1)$ volume of $T_p\mathcal{M}_\varepsilon^+$ onto each of the different $k+\ell+p$-dimensional subspaces of $\mathbb{R}^{2k+2\ell+2p-1}$. In addition, we compute the same projected volumes for the linear space generated by the following projected tangent vectors:

$$\begin{aligned}
\pi_b e_{a_i\hat{a}_i} &= e_{a_i\hat{a}_i}, & i &= 1,\ldots,k\\
\pi_b e_{b_i\hat{b}_i} &= \frac{1}{\sqrt{2}} e_{\hat{b}_i}, & i &= 1,\ldots,\ell\\
\pi_b e_{Y_i\hat{\mathcal{Y}}_i} &= e_{Y_i\hat{\mathcal{Y}}_i}, & i &= 2,\ldots,p\\
\pi_b \mathbf{v} &= \frac{1}{|\mathbf{v}|}\Big(\Lambda a, 0, \varepsilon m(y,\varepsilon) + \varepsilon h(a,b,y;\varepsilon)\Big).
\end{aligned} \tag{6.3}$$

Let $\Delta(T_p\mathcal{M}_\varepsilon^+)$ denote the rectangular matrix whose rows correspond to each of the tangent vectors in (6.2), and $\Delta(\pi_b T_p\mathcal{M}_\varepsilon^+)$ denote the rectangular matrix whose rows are each of the projected tangent vectors (6.3).

Consider next the $k + \sigma$ subdeterminants of $\Delta(T_p\mathcal{M}_\varepsilon^+)$ obtained by the first k columns and $\sigma = \ell + p$ of the last $n = k + \ell + 2p - 1$ columns. These are the $\mathcal{P}_{e_a, e_{y_1}, \dots, e_{y_\sigma}}(v_1, \dots, v_{k+\sigma})$ of condition **(M2)**. Similarly, from the first k columns and σ of the last n columns of $\Delta(\pi_b T_p\mathcal{M}_\varepsilon^+)$ we compute the $k+\sigma$ subdeterminants that correspond to the $\mathcal{P}_{e_a, e_{y_1}, \dots, e_{y_\sigma}}(\pi_b v_1, \dots, \pi_b v_{k+\sigma})$ of condition **(M2)**. Inspection of the matrices $\Delta(T_p\mathcal{M}_\varepsilon^+)$ and $\Delta(\pi_b T_p\mathcal{M}_\varepsilon^+)$, reveals $\mathcal{P}_{e_a, e_{y_1}, \dots, e_{y_\sigma}}(v_1, \dots, v_{k+\sigma}) = \mathcal{P}_{e_a, e_{y_1}, \dots, e_{y_\sigma}}(\pi_b v_1, \dots, \pi_b v_{k+\sigma})$, since their first k and last n columns are identical. (They differ only in the middle ℓ columns, which correspond to the b coordinates.) Furthermore, since $\mathbf{v}$ is $\mathcal{O}(\varepsilon)$, it is easy to see that at least one of the $\mathcal{P}_{e_a, e_{y_1}, \dots, e_{y_\sigma}}(v_1, \dots, v_{k+\sigma})$ is $\mathcal{O}(1)$. Also by direct inspection of $\Delta(T_p\mathcal{M}_\varepsilon^+)$ one readily sees that at least one of the remaining projections $\mathcal{P}_\mathcal{X}(v_1, \dots, v_{k+\sigma})$ is $\mathcal{O}(1)$, where $\mathcal{X}$ has the same meaning that in Section 5.2. Similarly for $\Delta(\pi_b T_p\mathcal{M}_\varepsilon^+)$: one of the remaining subdeterminants $\mathcal{P}_\mathcal{X}(\pi_b v_1, \dots, \pi_b v_{k+\sigma})$ is $\mathcal{O}(1)$. Thus, condition **(M2)** is satisfied, and the proposition is proven. □

Having established that the hypotheses of MELESE hold for $\mathcal{M}_\varepsilon^+$, we now apply its conclusion. Specifically, as in the above proof, let π_b denote the projection along the b directions onto the unstable manifold of $\mathcal{S}_1^\varepsilon$. Also, let $\pi_b I_\varepsilon \cdot [0, \infty)$ denote the union of all trajectories starting on $\pi_b I_\varepsilon$. By construction, $\pi_b I_\varepsilon \cdot [0, \infty)$ is a $(k + \ell + p)$-dimensional submanifold of $W_{loc}^u(\mathcal{S}_1^\varepsilon)$ that we will denote by $\mathcal{V}_{\varepsilon,1}^u$. (Note that $\mathcal{V}_{\varepsilon,1}^u$ is of the same dimension as $\mathcal{M}_\varepsilon^+$, but of lower dimension than $W_{loc}^u(\mathcal{S}_1^\varepsilon)$. Also, it need not be the same as $\pi_b\mathcal{M}_\varepsilon^+$, the projection of $\mathcal{M}_\varepsilon^+$ onto $W_{loc}^u(\mathcal{S}_1^\varepsilon)$). The MELESE, Lemma 3, implies that the tracked manifold $\mathcal{M}_\varepsilon^+$ leaves the box $\mathcal{B}_{\mathcal{S}_1}$ $\mathcal{C}^1$ exponentially close to $\mathcal{V}_{\varepsilon,1}^u$.

We will now exploit some of the main concepts of geometric singular perturbation theory to characterize $\mathcal{V}_{\varepsilon,1}^u$. We refer the reader to the original paper of Fenichel [13] and to the exposition in [32] for the definition of unstable and stable fibers and their base points. We shall assume that the vector field in the neighborhood of $\mathcal{S}_j^\varepsilon$ is in Fenichel normal form (3.2), *i.e.*, that the fibers are straightened out. Let π_a denote the projection along the unstable (or a directions). For any subset V of the local unstable manifold $W_{loc}^u(\mathcal{S}_j^\varepsilon)$, we take the union of all the unstable fibers of $\mathcal{S}_j^\varepsilon$ that intersect the set V. Then, $\pi_a V$ corresponds to the base points in $\mathcal{S}_j^\varepsilon$ of these unstable fibers. Consider now the restriction, $\mathcal{V}_{\varepsilon,1}^u|_{\{a=\Delta\}}$, of $\mathcal{V}_{\varepsilon,1}^u$ to the exit face $\{a = \Delta\}$ of $\mathcal{B}_{\mathcal{S}_1}$. We are interested in the $\varepsilon \to 0$ limit of the set of base points $\pi_a(\mathcal{V}_{\varepsilon,1}^u|_{\{a=\Delta\}})$, and we denote this limit by BV_1^u. In the following proposition, we show that BV_1^u transversely intersects the take off set TO_1. We will use this result in Proposition 6.3 in the next subsection to show that $\mathcal{M}_\varepsilon^+$ transversely intersects $W^s(\mathcal{S}_2^\varepsilon)$.

PROPOSITION 6.2. *The set of base points $\pi_a(\mathcal{V}_{\varepsilon,1}^u|_{\{|a|=\Delta\}})$ is an $(\ell+p)$-dimensional submanifold of $\mathcal{S}_1^\varepsilon$. Moreover, BV_1^u, its $\varepsilon \to 0$ limit, transversely intersects the take-off set TO_1. We label this intersection by TO_1^u, and its dimension is $\ell + p - 1$.*

Proof. Recall that trajectories inside $\mathcal{B}_{\mathcal{S}_1}$ are governed by (3.2). Since we have rectified the slow flow in a neighborhood of the singular periodic orbit before augmenting the system, the evolution of the slow components $y = (\hat{a}, \hat{b}, Y, \hat{\mathcal{Y}})$ of trajectories in $\mathcal{B}_{\mathcal{S}_1}$ is given by

$$\begin{aligned} \dot{\hat{a}} &= \varepsilon h_a(a,b,y,\varepsilon)ab \\ \dot{\hat{b}} &= \varepsilon h_b(a,b,y,\varepsilon)ab \\ \dot{Y} &= \varepsilon m(Y,\varepsilon) + \varepsilon h_y(a,b,y,\varepsilon)ab \\ \dot{\hat{\mathcal{Y}}} &= \varepsilon h_{\mathcal{Y}}(a,b,y,\varepsilon)ab. \end{aligned} \tag{6.4}$$

Now, rather than tracking all of the trajectories starting from the $(k+\ell+p-1)$-dimensional set $\pi_b I_\varepsilon$ at once, we focus on a particular k-dimensional subset first. Let $\mathcal{I}_\varepsilon$ be a k-dimensional subset of $\pi_b I_\varepsilon$ in which the a variable is any element of I_a, $\hat{a} = a$, and the variables $b, \hat{b}, Y, \hat{\mathcal{Y}}$ are fixed, with $b = 0, Y_1 = 0$. The union of all the trajectories starting in $\mathcal{I}_\varepsilon$, denoted by $\mathcal{I}_\varepsilon \cdot [0,\infty)$, is a $(k+1)$-dimensional submanifold of $W^u(\mathcal{S}_1^\varepsilon)$. Therefore, $b = 0$ for trajectories in $\mathcal{I}_\varepsilon \cdot [0,\infty)$, and by (6.4) we have that the evolution of the slow variables $(\hat{a}, \hat{b}, Y, \hat{\mathcal{Y}})$ of these trajectories does not depend on the fast (a,b) variables. Furthermore, the evolution of the slow variables Y along one of these trajectories is exactly that of the evolution of the Y coordinates of a singular solution that has the same initial Y values.

Consider next the intersection of $\mathcal{I}_\varepsilon \cdot [0,\infty)$ with the exit face $\{|a| = \Delta\}$; this intersection is a k-dimensional manifold. Moreover, by construction of $\mathcal{I}_\varepsilon$, the trajectories with initial coordinates $a, \hat{a} = a_{\max}$ reach the boundary $\{|a| = \Delta\}$ before their Y_1 coordinate reaches $Y_1(r_1^{0,u})$ for any $r_1^{0,u} \in TO_1$. Those starting with $a, \hat{a} = a_{min}$ reach $\{a = \Delta\}$ after their Y_1 coordinate reaches $Y_1(r_1^{0,u})$. Therefore, the k-dimensional manifold $\{\mathcal{I}_\varepsilon \cdot [0,\infty)\}|_{\{|a|=\Delta\}}$ intersects the hyperplane $\{Y_1 = Y_1(r_1^{0,u})\}$ for some $r_1^{0,u} \in TO_1$, where $r_1^{0,u}$ depends on the fixed $\hat{b}, Y, \hat{\mathcal{Y}}$ coordinates. Furthermore, the $\varepsilon \to 0$ limit of the base points $\pi_a(\{\mathcal{I}_\varepsilon \cdot [0,\infty)\}|_{\{|a|=\Delta\}})$ is a one-dimensional submanifold of $\mathcal{S}_1^\varepsilon$, and by Assumption 3, it transversely intersects the hyperplane $\{Y_1 = Y_1(r_1^{0,u})\}$ restricted to $\mathcal{S}_1^0$. Thus, these base points transversely intersect TO_1.

Finally, consider all the trajectories through $\pi_b I_\varepsilon$ instead of just the ones through $\mathcal{I}_\varepsilon$. Then the restriction $\mathcal{V}_{\varepsilon,1}^u|_{\{|a|=\Delta\}}$ is an $(k+\ell+p-1)$-dimensional manifold, and therefore BV_1^u forms an $(\ell+p)$-dimensional submanifold of $\mathcal{S}_1^0$. Assumption 3 guarantees that these base points transversely intersect TO_1. Moreover, this intersection, denoted by TO_1^u, is an $(\ell+p-1)$ submanifold of $\mathcal{S}_1^0$, since TO_1 has dimension $k+\ell+p-2$. Hence, the proof is complete also in the general case. □

REMARK 6.1. One can also track $\mathcal{M}_\varepsilon^+$ and $\mathcal{M}_\varepsilon^-$ in the box $\mathcal{B}_{\mathcal{S}_1}$ by using the appropriate version of ELESE (*i.e.* Lemma 1 or Lemma 2, depending on the $\varepsilon = 0$ geometry of the involved manifolds). In this alternative approach, the set I_ε must be extended, so that the tranversality conditions **(T1)** or $(\mathbf{T1})_0$ are satisfied. I_ε is extended to

$$\hat{I}_\varepsilon = \left\{(a,b,y) : a \in \hat{I}_a, b \in \hat{I}_b, Y_1 = 0, \hat{a} = a, \hat{b} = b, Y_r = \mathcal{Y}_r \in I_{Y_r}, r = 2, ..., p\right\}$$

where $\hat{I}_a \equiv [-a_{min}, a_{max}]$, $\hat{I}_b = [-b_{min}, b_{max}]$, and the I_{Y_r} are as in I_ε, see Section 4, Figure 6. Thus, $I_\varepsilon \subset \hat{I}_\varepsilon$. Since $0 \in \hat{I}_a$, $\hat{I}_\varepsilon$ transversely intersects $\{a = 0\} = W^s_{loc}(\mathcal{S}^\varepsilon_1)$, so does $\hat{I}_\varepsilon \cdot [0, \infty)$. One then easily checks that $(\mathbf{T1})_0$, $(\mathbf{T2})_0$ and **(TE)** (or that **(T1)**, **(T2)** and **(TE)**) are satisfied for the locally invariant manifold $\hat{\mathcal{M}}^+_\varepsilon \equiv \hat{I}_\varepsilon \cdot [0, \infty)$. It follows that the conclusion of ELESE holds for $\mathcal{M}^+_\varepsilon$, since $\mathcal{M}^+_\varepsilon \subset \hat{\mathcal{M}}^+_\varepsilon$.

6.2. Tracking $\mathcal{M}^+_\varepsilon$ outside $\mathcal{B}_{\mathcal{S}_1}$. In this section, we first show in Proposition 6.3 that $\mathcal{M}^+_\varepsilon$ and $W^s(\mathcal{S}^\varepsilon_2)$ intersect transversely. This directly implies that there exist open sets of trajectories on $\mathcal{M}^+_\varepsilon$ that lie in a neighborhood of $\mathcal{S}^\varepsilon_2$. In Proposition 6.4 we show that $\mathcal{M}^+_\varepsilon$ leaves the neighborhood $\mathcal{B}_{\mathcal{S}_2}$ C^1 exponentially close to a certain submanifold, to be described below, of $W^u_{loc}(\mathcal{S}^\varepsilon_2)$. We then use arguments similar to those in Propositions 6.3 and 6.4 to track trajectories $\mathcal{M}^+_\varepsilon$ in their passage near the remaining submanifolds of the sequence: $\mathcal{S}^\varepsilon_3$,..., $\mathcal{S}^\varepsilon_L$.

PROPOSITION 6.3. *$\mathcal{M}^+_\varepsilon$ transversely intersects $W^s(\mathcal{S}^\varepsilon_2)$ in an $(\ell + p)$-dimensional manifold.*

Proof. First, we recall that it follows from Proposition 6.1 that $\mathcal{M}^+_\varepsilon$ is $\mathcal{C}^1$ exponentially close to $\mathcal{V}^u_{\varepsilon,1}$, a submanifold of $W^u_{loc}(\mathcal{S}^\varepsilon_1)$. Using Gronwall-type estimates, the $\mathcal{C}^1$ exponential closeness of $\mathcal{M}^+_\varepsilon$ and the global manifold $W^u(\mathcal{S}^\varepsilon_1)$ is maintained outside the box $\mathcal{B}_{\mathcal{S}_1}$ for a finite (fast) time. Second, by Assumption 2, $W^u(\mathcal{S}^\varepsilon_1)$ transversely intersects $W^s(\mathcal{S}^\varepsilon_2)$ in a $k + \ell + 2p - 1$ dimensional manifold. We now put these two items together. Specifically, since BV^u_1 transversely intersects TO_1, the union of unstable fibers with base points in BV^u_1 transversely intersects $W^s(\mathcal{S}^0_2)$. Moreover, this transversality persists for sufficiently small $\varepsilon > 0$, and hence $\mathcal{V}^u_{\varepsilon,1}$ transversely intersects $W^s(\mathcal{S}^\varepsilon_2)$. Now, from the $\mathcal{C}^1$ exponential closeness of $\mathcal{M}^+_\varepsilon$ and $\mathcal{V}^u_{\varepsilon,1}$, it follows that $\mathcal{M}^+_\varepsilon$ must also transversely intersect $W^s(\mathcal{S}^\varepsilon_2)$. The dimension of the transverse intersection of $\mathcal{M}^+_\varepsilon$ and $W^s(\mathcal{S}^\varepsilon_2)$ is $\ell + p$. □

REMARK 6.2. The angle of intersection of $\mathcal{M}^+_\varepsilon$ and $W^s_{loc}(\mathcal{S}^\varepsilon_2)$ depends on the angle of intersection of $\mathcal{V}^u_{\varepsilon,1}$ and $W^s_{loc}(\mathcal{S}^\varepsilon_2)$. If system (1.1) satisfies Assumption 2 with Case 1, then this angle is $\mathcal{O}(1)$. If, instead, Case 2 holds, then the angle is $\mathcal{O}(\varepsilon^\beta)$, for some $\beta > 0$.

Next, we track $\mathcal{M}^+_\varepsilon$ in its passage near $\mathcal{S}^\varepsilon_2$. By Proposition 6.3, $\mathcal{M}^+_\varepsilon$ and $W^s(\mathcal{S}^\varepsilon_2)$ intersect transversely. If the angle of intersection is $\mathcal{O}(1)$, we will use Lemma 1 with $\sigma = \ell + p$ to show that $\mathcal{M}^+_\varepsilon$ leaves the box $\mathcal{B}_{\mathcal{S}_2}$ C^1 exponentially close to an appropriate submanifold of $W^u_{loc}(\mathcal{S}^\varepsilon_2)$. We denote this submanifold, to be defined below, by $\mathcal{V}^u_{\varepsilon,2}$. If, on the other hand, $\mathcal{M}^+_\varepsilon$ and $W^s(\mathcal{S}^\varepsilon_2)$ intersect at an angle of $\mathcal{O}(\varepsilon^\beta)$ with $\beta > 0$, we use Lemma 2 to obtain the same closeness result for $\mathcal{M}^+_\varepsilon$ and $\mathcal{V}^u_{\varepsilon,2}$. It also follows from Assumption 3 and either Lemma 1 or Lemma 2 that there is a subset of trajectories in $\mathcal{M}_\varepsilon^+$ (this subset is also of dimension $k+p+\ell$) that leaves the box $\mathcal{B}_{\mathcal{S}_2}$ such that the Y_1 coordinates of the trajectories are in the neighborhood of the Y_1 coordinates of TO_2. This is stated more precisely in Proposition 6.4 below, where we also give a more detailed characterization of $\mathcal{V}^u_{\varepsilon,2}$. We denote by BV_2^u the base points of the unstable fibers in $\mathcal{V}^u_{\varepsilon,2}$, in the $\varepsilon \to 0$ limit.

We first borrow some of the notation introduced in Section 2, specifically that of the fast and slow maps, but now in the context of the augmented system (3.1). The (fast) map $f_f^i : TO_i \longrightarrow TD_{i+1}$ is the map that associates to a point $p_i^{TO} \in TO_i$ the point $p_{i+1}^{TD} \in TD_{i+1}$ such that there is a heteroclinic connection between p_i^{TO} and p_{i+1}^{TD}. Similarly, the slow map $f_s^i : TD_i \longrightarrow TO_i$ associates to any point $p_i^{TD} \in TD_i$ the point $p_i^{TO} \in TO_i$ at which the slow trajectory through p_i^{TD} intersects the take off set TO_i. We use Fenichel coordinates (a, b, y) in the neighborhood of each of the slow manifolds $\mathcal{S}_j^\varepsilon$ and we denote the entrance and exit faces of the box $\mathcal{B}_{\mathcal{S}_j}$ by $\{|b| = \Delta\}_j$ and $\{|a| = \Delta\}_j$, respectively.

Let T be the (slow) time it takes for the slow segment of the singular periodic orbit to go from the touch down set TD_2 to the take off set TO_2. We recall, from Section 5, that $R \cdot \tau$ denotes the time τ (slow) flow for a subset R of $\mathcal{S}_2^\varepsilon$, and that $R \cdot (\tau_1, \tau_2) = \cup_{\tau \in (\tau_1,\tau_2)} H \cdot \tau$. Let TD_2^s be the $(\ell + p - 1)$-dimensional submanifold of TD_2 given by $TD_2^s \equiv f_f^1(TO_1^u)$. Let $\mathcal{V}^u_{\varepsilon,2}$ be the submanifold of $W^u_{loc}(\mathcal{S}_2^\varepsilon)$ given by $W^u_{loc}(\mathcal{S}_2^\varepsilon)|_{TD_2^s \cdot (T-\gamma, T+\gamma)}$, where $\gamma > 0$ will be specified in the following proposition.

PROPOSITION 6.4. *$\mathcal{M}_\varepsilon^+$ leaves $\mathcal{B}_{\mathcal{S}_2}$ $\mathcal{C}^1$ exponentially close to the to the $(k+\ell+p)$-dimensional submanifold $\mathcal{V}^u_{\varepsilon,2}$ of $W^u_{loc}(\mathcal{S}_2^\varepsilon)$. The set of base points BV_2^u transversely intersects the take off set TO_2. This transverse intersection is denoted by TO_2^u, and it is an $(\ell + p - 1)$-dimensional submanifold of TO_2.*

Proof. Depending on the angle of intersection of $\mathcal{M}_\varepsilon^+$ and $W^s_{loc}(\mathcal{S}_2^\varepsilon)$, we will use either Lemma 1 or Lemma 2 to track the trajectories on the manifold $\mathcal{V}^u_{\varepsilon,1}$ inside the neighborhood $\mathcal{B}_{\mathcal{S}_2}$. In particular, depending on this angle, we will show that either the hypotheses of Lemma 1 or those of Lemma 2 are satisfied. Then, from the conclusion of these lemmas, we know $\mathcal{V}^u_{\varepsilon,1}$ exits $\mathcal{B}_{\mathcal{S}_2}$ $\mathcal{C}^1$ exponentially close to $\mathcal{V}^u_{\varepsilon,2}$. Hence, due to the closeness of $\mathcal{M}_\varepsilon^+$ and $\mathcal{V}^u_{\varepsilon,1}$ on entry to $\mathcal{B}_{\mathcal{S}_2}$ that was established in the above propositions, we also directly know that $\mathcal{M}_\varepsilon^+$ is $\mathcal{C}^1$ exponentially close to $\mathcal{V}^u_{\varepsilon,2}$ upon exiting $\mathcal{B}_{\mathcal{S}_2}$.

Let Z_ε denote the transverse intersection of $\mathcal{V}^u_{\varepsilon,1}$ and $W^s_{loc}(\mathcal{S}_2^\varepsilon)$. By Proposition 6.3, TD_2^s is the ω-limit set of the fast heteroclinic connections in Z_0.

Suppose the angle of intersection of $\mathcal{V}^u_{\varepsilon,1}$ and $W^s_{loc}(\mathcal{S}_2^\varepsilon)$ is $\mathcal{O}(1)$. We will verify that conditions $(\mathrm{T1})_0$, $(\mathrm{T2})_0$, and (TE) of Lemma 1 hold, with Z_ε assuming the role of $\mathcal{L}_\varepsilon$, $\sigma = \ell + p$, and $n = k + \ell + 2p - 1$. Condition $(\mathbf{T1})_\mathbf{0}$ is satisfied by hypothesis. Next, by the discussion above, we already know that $TD_2^s = \omega(Z_0)$, and it is of dimension $\ell + p - 1$. Furthermore, by Assumption 3, the slow flow is transverse to $\omega(Z_0)$. Therefore, condition $\mathbf{(T2)_0}$ is satisfied, with Z_0 assuming the role of J_0. Finally, condition **(TE)** holds, since points $\mathcal{V}^u_{\varepsilon,1}$ in the entry face $\{|b| = \Delta\}_2$ that are exponentially close to $W^s_{loc}(\mathcal{S}^\varepsilon_2)$ take $\mathcal{O}(\frac{1}{\varepsilon})$ amounts of (fast) time to reach the exit face $\{|a| = \Delta\}_2$. Since all three conditions hold, we apply the conclusion of Lemma 1 to $\mathcal{V}^u_{\varepsilon,1}$: it leaves $\mathcal{B}_{\mathcal{S}_2}$ $\mathcal{C}^1$ exponentially close to $W^u_{loc}(\mathcal{S}^\varepsilon_2)|_{TD^s_2\cdot(T-\gamma_1,T+\gamma_1)}$, where $\gamma_1 > 0$. Therefore, by the argument above, $\mathcal{M}^+_\varepsilon$ does, as well.

If instead the angle of intersection of $\mathcal{V}^u_{\varepsilon,1}$ and $W^s_{loc}(\mathcal{S}^\varepsilon_2)$ is $\mathcal{O}(\varepsilon^\beta)$ for some $\beta > 0$, we verify that the conditions of Lemma 2 are satisfied, with Z_ε again assuming the role of $\mathcal{L}_\varepsilon$ and with the same σ and n as above: **(T1)** holds by hypothesis. **(T2)** holds for the same reasons as above: $\omega(Z_0) = TD^s_2$ has dimension $\ell + p - 1$, and Assumption 3 guarantees the second part of the condition. Finally, Condition **(TE)** also holds. Therefore, it follows from Lemma 2 that $\mathcal{V}^u_{\varepsilon,1}$ leaves the box $\mathcal{B}_{\mathcal{S}_2}$ $\mathcal{C}^1$ exponentially close to $W^u_{loc}(\mathcal{S}^\varepsilon_2)|_{\omega(Z_\varepsilon)\cdot T}$. Since $\lim_{\varepsilon\to 0}\omega(Z_\varepsilon) = TD^s_2$, we can always find $\gamma_2 > 0$ such that $\omega(Z_\varepsilon)\cdot T \subset TD^s_2 \cdot (T-\gamma_2, T+\gamma_2)$. This means that $\mathcal{V}^u_{\varepsilon,1}$ is $\mathcal{C}^1$ exponentially close to the submanifold $W^s_{loc}(\mathcal{S}^\varepsilon_2)|_{TD^s_2\cdot(T-\gamma_2,T+\gamma_2)}$, and hence also $\mathcal{M}^+_\varepsilon$ does.

From the previous two paragraphs, we conclude that whatever the intersection angle of $\mathcal{M}^+_\varepsilon$ and $W^s_{loc}(\mathcal{S}^\varepsilon_2)$ is, $\mathcal{M}^+_\varepsilon$ is $\mathcal{C}^1$ exponentially close to the submanifold $W^s_{loc}(\mathcal{S}^\varepsilon_2)|_{TD^s_2\cdot(T-\gamma,T+\gamma)}$ upon exiting $\mathcal{B}_{\mathcal{S}^\varepsilon_2}$, where $\gamma = \min\{\gamma_1, \gamma_2\}$. This submanifold is denoted by $\mathcal{V}^u_{\varepsilon,2}$, and it has dimension $k + \ell + p$ by construction. Moreover, the base points of the unstable fibers in $\mathcal{V}^u_{\varepsilon,2}$, given by $\pi_a(\mathcal{V}^u_{\varepsilon,2})$, constitute precisely the set $TD^s_2 \cdot (T-\gamma, T+\gamma)$. In the $\varepsilon \to 0$ limit, we have that $BV^u_2 = TD^s_2 \cdot (T-\gamma, T+\gamma)$.

Now, consider the forward image of TD^s_2 under the slow flow on $\mathcal{S}^0_2$. Using Assumption 3 and the fact that the $\hat{a}, \hat{b}$ and $\hat{\mathcal{Y}}$ coordinates do not change (b, $a = 0$ in (6.4)), we have $TD^s_2\cdot(T-\gamma, T+\gamma)$ transversely intersects TO_2. This intersection manifold has dimension $(\ell+p-1)$, and it is denoted by TO^u_2. Thus, the proposition is proven. □

REMARK 6.3. Each point p^{TO}_2 in TO^u_2 corresponds to the intersection of a trajectory starting on $p^{TD}_2 \in TD^s_2$ with the set TO_2, or, using the slow map notation, $p^{TO}_2 = f^2_s(p^{TD}_2)$. Therefore, each point in TO^u_2 is associated with a point in TD^s_2 in such a way that only their $Y_1, ..., Y_p$ coordinates differ.

Now, by the same arguments as in Proposition 6.3, we know that $\mathcal{M}^+_\varepsilon$ will also transversely intersect $W^s(\mathcal{S}^\varepsilon_3)$. Moreover, by arguments similar to those used in Proposition 6.4, we know that $\mathcal{M}^+_\varepsilon$ leaves $\mathcal{B}_{\mathcal{S}_3}$ $\mathcal{C}^1$ exponentially close to $\mathcal{V}^u_{\varepsilon,3}$, a $(k+\ell+p)$-dimensional submanifold of $W^u_{loc}(\mathcal{S}^\varepsilon_3)$ defined in a fashion analogous to the definition of $\mathcal{V}^u_{\varepsilon,2}$. Also, let BV^u_3 be the base points of the $\varepsilon \to 0$ limit of the unstable fibers in $\mathcal{V}^u_{\varepsilon,3}$. By the same arguments as in Proposition 6.4, we have that BV^u_3 transversely intersects the take-off manifold TO_3. We denote this intersection manifold by TO_3^u.

Proceeding inductively, we track the trajectories in $\mathcal{M}_\varepsilon^+$ during their fast jumps between the neighborhoods of the slow manifolds (using the transversal intersections of $W^u(\mathcal{S}_j^\varepsilon)$ and $W^s(\mathcal{S}_{j+1}^\varepsilon)$ and arguments similar to those in Proposition 6.3) and during their passage in the vicinity of the slow manifolds themselves, using the same arguments as in Proposition 6.4, until we finally get to $\mathcal{S}_L^\varepsilon$, the penultimate in our sequence $\{\mathcal{S}_j\}$. Again, we know that $\mathcal{M}_\varepsilon^+$ leaves the box $\mathcal{B}_{\mathcal{S}_L}$ $\mathcal{C}^1$ exponentially close to $\mathcal{V}_{\varepsilon,L}^u$, a $(k+\ell+p)$-dimensional submanifold of $W_{loc}^u(\mathcal{S}_L^\varepsilon)$ defined in the same fashion as $\mathcal{V}_{\varepsilon,2}^u$. Moreover, let BV_L^u denote the base points of the unstable fibers in $\mathcal{V}_{\varepsilon,L}^u$, in the $\varepsilon \to 0$ limit. By the same arguments as in Proposition 6.4, we have that BV_L^u transversely intersects the take off set TO_L. We denote this intersection by TO_L^u; it is a submanifold of TO_L of dimension $(\ell+p-1)$.

REMARK 6.4. By the arguments at the end of Proposition 6.4, the set TO_L^u is the image of the set TO_1^u under the composition of the fast and slow maps:

$$TO_L^u = f_s^L \circ f_f^{L-1} \circ \ldots \circ f_f^2 \circ f_s^2 \circ f_f^1(TO_1^u).$$

6.3. $\mathcal{M}_\varepsilon^+$ transversely intersects $\mathcal{M}_\varepsilon^-$. In this section, we first track the trajectories on $\mathcal{M}_\varepsilon^-$ in backward time until they leave the box $\mathcal{B}_{\mathcal{S}_1}$. Then, we show that $\mathcal{M}_\varepsilon^+$ and $\mathcal{M}_\varepsilon^-$ intersect transversely. To track $\mathcal{M}_\varepsilon^-$, we consider the backward-time trajectories of all points starting on I_ε. Recall π_a denotes the projection along the unstable directions onto $W_{loc}^s(\mathcal{S}_1^\varepsilon)$. By arguments similar to those in Proposition 6.1, we know that I_ε and $\pi_a I_\varepsilon$ satisfy the hypotheses of MELESE (in backward time). Let $\mathcal{V}_{\varepsilon,1}^s$ denote $\pi_a I_\varepsilon \cdot [0,-\infty)$. Thus, we conclude from MELESE that (in backward time) $\mathcal{M}_\varepsilon^-$ leaves $\mathcal{B}_{\mathcal{S}_1}$ $\mathcal{C}^1$ exponentially close to $\mathcal{V}_{\varepsilon,1}^s$, a submanifold of $W_{loc}^s(\mathcal{S}_1^\varepsilon)$. Furthermore, $\mathcal{V}_{\varepsilon,1}^s$ has dimension $k+\ell+p$. Let BV_1^s denote, in the $\varepsilon \to 0$ limit, the set of base points $\pi_b(\mathcal{V}_{\varepsilon,1}^s|_{\{|b|=\Delta\}_1})$. By the same arguments as in Proposition 6.2, we know that BV_1^s is a manifold of dimension $k+p$ that transversely intersects TD_1. We label this intersection manifold by TD_1^s, and note that it has dimension $k+p-1$.

Now, from the previous section, we know that upon exiting $\mathcal{B}_{\mathcal{S}_L}$, $\mathcal{M}_\varepsilon^+$ is $\mathcal{C}^1$ exponentially close to $\mathcal{V}_{\varepsilon,L}^u$, a submanifold of $W_{loc}^u(\mathcal{S}_L^\varepsilon)$. Using Gronwall-type estimates, one sees that the exponential closeness of $\mathcal{M}_\varepsilon^+$ and $W^u(\mathcal{S}_L^\varepsilon)$ is maintained outside $\mathcal{B}_{\mathcal{S}_L}$ for a finite time until trajectories of $\mathcal{M}_\varepsilon^+$ intersect the entry face to $\mathcal{B}_{\mathcal{S}_1}$. In the following two propositions, we will show that $\mathcal{V}_{\varepsilon,1}^s$ and $\mathcal{V}_{\varepsilon,L}^u$ intersect transversely. We will show this by examining these two manifolds at the entry face of $\mathcal{B}_{\mathcal{S}_1}$. Since $\mathcal{M}_\varepsilon^-$ and $\mathcal{M}_\varepsilon^+$ are $\mathcal{C}^1$ exponentially close, respectively, to $\mathcal{V}_{\varepsilon,1}^s$ and $\mathcal{V}_{\varepsilon,L}^u$, this will imply that $\mathcal{M}_\varepsilon^-$ transversely intersects $\mathcal{M}_\varepsilon^+$ for ε sufficiently small.

We divide the proof that $\mathcal{V}_{\varepsilon,1}^s$ transversely intersects $\mathcal{V}_{\varepsilon,L}^u$ in two parts. Proposition 6.5 states that this intersection occurs, provided an intersection condition regarding BV_1^s and BV_L^u holds. In Proposition 6.6, we verify that this intersection condition is indeed satisfied by $\mathcal{V}^s_{\varepsilon,1}$ and $\mathcal{V}^u_{\varepsilon,L}$.

PROPOSITION 6.5. *If the manifolds $f_f^L(TO_L^u)$ and TD_1^s intersect transversely in the touch down set TD_1, then $\mathcal{V}^s_{\varepsilon,1}$ and $\mathcal{V}^u_{\varepsilon,L}$ intersect transversely.*

Proof. By Proposition 6.2, BV_1^s and BV_L^u have dimensions $k+p$ and $\ell+p$, respectively. Also, we know that the manifold TO_L^u is the transverse intersection of BV_L^u and the take off set TO_L. Similarly, TD_1^s is the transverse intersection of BV_1^s and TD_1.

Now, suppose that Case 1 of Assumption 2 holds. Then the only points of BV_L^u that are connected to points in BV_1^s via a heteroclinic connection of the fast system are those in TO_L^u. These heteroclinic connections lie in the transverse intersection of $W^u(\mathcal{S}_L^0)$ and $W^s(\mathcal{S}_1^0)$. Thus, we have that each point in TO_L^u is connected to a point in TD_1 via a heteroclinic connection of the fast system. The union of these connections forms an $(\ell+p)$-dimensional submanifold of the transverse intersection of $W^u(\mathcal{S}_L^0)$ and $W^s(\mathcal{S}_1^0)$ that we will denote by $H^u|_{TO_L^u}$. In turn, each point in TD_1^s is connected to a point in TO_L^u via a heteroclinic connection. The union of these fast orbits is denoted by $H^s|_{TD_1^s}$, and it is a $(k+p)$-dimensional submanifold of the transverse intersection of $W^u(\mathcal{S}_L^0)$ and $W^s(\mathcal{S}_1^0)$.

We now show that $H^u|_{TO_L^u}$ and $H^s|_{TD_1^s}$ intersect transversely. By assumption, $f_f^L(TO_L^u)$ and TD_1^s intersect transversely in a 0-dimensional manifold, since their respective dimensions are $\ell+p-1$ and $k+p-1$, and TD_1 is of dimension $k+\ell+2p-2$. Note that by definition of the map f_f^L, the existence of this transverse intersection implies that the only points in TD_1^s that are connected to a point in TO_L^u via a fast heteroclinic connection are precisely those in $f_f^L(TO_L^u)\cap TD_1^s$. This implies that $H^s|_{TO_1^u}$ transversely intersects $H^u|_{TD_L^s}$ in a one-dimensional submanifold of $W^u(\mathcal{S}_L^0)\cap W^s(\mathcal{S}_1^0)$. Furthermore, since only points in BV_L^u which are also in TO_L^u can be connected to a point in TD_1^s, we have that $W^u|_{BV_L^u}$ must transversely intersect $W^s|_{BV_1^s}$ also in a one-dimensional manifold. This transversality persists for $\varepsilon>0$, and therefore $\mathcal{V}^u_{\varepsilon,L}$ transversely intersects $\mathcal{V}^s_{\varepsilon,1}$ in a one-dimensional manifold. This proves the proposition when Case 1 holds.

Now, suppose Case 2 of Assumption 2 holds. Then the only heteroclinic connections between points in BV_L^u and BV_1^s that will persist for $\varepsilon>0$ as part of the transverse intersection of $W^u(\mathcal{S}_j^\varepsilon)$ and $W^s(\mathcal{S}_1^\varepsilon)$ are those whose base points are in TO_L and TD_1. Let $H^u|_{TO_L^u}$ be the union of all the heteroclinic connections between $\mathcal{S}_L^0$ and $\mathcal{S}_1^0$ with base points in TO_L^u. This is a manifold of dimension $\ell+p$, and for $\varepsilon>0$ it is a submanifold of the transverse intersection of $W^u(\mathcal{S}_L^\varepsilon)$ and $W^s(\mathcal{S}_1^\varepsilon)$. Similarly, $H^s|_{TD_1^s}$ denotes the union of all heteroclinic connections between $\mathcal{S}_1^0$ and $\mathcal{S}_L^0$ with base points in TD_1^s. The heteroclinic manifold $H^s|_{TD_1^s}$ has dimension $k+p$ and for $\varepsilon>0$ it is a submanifold of the transverse intersection of $W^u(\mathcal{S}_L^\varepsilon)$ and $W^s(\mathcal{S}_1^\varepsilon)$.

Since $f_f^L(TO_L^u)$ and TD_1^s intersect transversely, then so do $H^u|_{TO_L^u}$ and $H^s|_{TD_1^s}$. The latter intersection manifold is a one-dimensional set that consists of a finite number of heteroclinic connections between points in TO_L^u and TD_1^s. In fact, for $\varepsilon > 0$, these are the only persisting heteroclinic connections between points in BV_L^u and BV_1^s. Therefore, $\mathcal{V}_{\varepsilon,L}^u$ must transversely intersect $\mathcal{V}_{\varepsilon,1}^s$, also in Case 2, and the proposition is proven. □

PROPOSITION 6.6. *In the manifold TD_1, the $(\ell+p-1)$-dimensional manifold $f_f^L(TO_L^u)$ transversely intersects the $(k+p-1)$-dimensional manifold TD_1^s. Their transverse intersection is a zero-dimensional manifold.*

REMARK 6.5. For $p \geq 2$, the proof of this proposition exploits the local uniqueness of the singular periodic orbit to show that the sets $f_f^L(TO_L^u)$ and TD_1^s intersect and are not identical. They intersect transversely, because the singular map F, as defined in Section 2, is such that all eigenvalues of $dF(p_0)$ are bounded away from the unit circle.

Proof. We prove the proposition for $p = 1$ in part i), and the case for general p in part ii).

i) Assume $p = 1$. As a preliminary step, we need to describe TO_1^u and TD_1^s as graphs in $(\hat{a}, \hat{b})$ space. We first concentrate on TO_1^u and show that it can be parameterized by $\hat{b}$. Fix $\hat{b} \in I_b$ and consider $\mathcal{I}_\varepsilon$, the k-dimensional subset of $\pi_b I_\varepsilon$ obtained by fixing $\hat{b}$ in $\pi_b I_\varepsilon$. We know that $\{\mathcal{I}_\varepsilon \cdot [0,\infty)\}|_{\{|a|=\Delta\}_1}$ is a k-dimensional submanifold of the exit face $\{|a| = \Delta\}_1$. Let $BV_1^u|_{\hat{b}}$ denote, in the $\varepsilon \to 0$ limit, the set of base points $\pi_a(\{\mathcal{I}_\varepsilon \cdot [0,\infty)\}|_{\{|a|=\Delta\}_1})$. We know from Proposition 6.2 that $BV_1^u|_{\hat{b}}$ is a one-dimensional submanifold of $\mathcal{S}_1^0$ and that it transversely intersects the take-off set TO_1 at a single point. We denote this point by $TO_1^u|_{\hat{b}}$. Since $\dot{\hat{b}} = 0$ in (6.4), the $\hat{b}$ coordinate of $TO_1^u|_{\hat{b}}$ is the same as the fixed $\hat{b} \in I_b$. We now let $\hat{b}$ vary in I_b, to obtain $BV_1^u = \bigcup_{\hat{b}\in I_b} (BV_1^u|_{\hat{b}})$, and the ℓ dimensional submanifold TO_1^u given by $TO_1^u = \bigcup_{\hat{b}\in I_b} (TO_1^u|_{\hat{b}})$. Thus, TO_1^u is parameterized $\hat{b}$.

We now describe TO_1^u as a graph α in $(\hat{a}, \hat{b})$ space over the $\hat{b}$ variables. Recall that $\mathcal{S}_1^\varepsilon$ is $(Y_1, \hat{a}, \hat{b})$ space, and that $TO_1 = \{(Y_1, \hat{a}, \hat{b}) | Y_1 \text{ is fixed}\}$. This means that the take-off set TO_1 can be thought of as $(\hat{a}, \hat{b})$ space. Since $TO_1^u \subset TO_1$ is parameterized by $\hat{b}$, we can describe TO_1^u as a graph α in $(\hat{b}, \hat{a})$ space over the $\hat{b}$ variables. See Figure 11 for an example with $\ell = 1, k = 2$. Furthermore, for s a point in $\alpha = \{\hat{b}, \hat{a}(\hat{b})\}$, we can describe the tangent vectors that span $T_s\alpha$ as follows. Let $e_{\hat{b}_1}, \ldots, e_{\hat{b}_\ell}$ denote the unit vectors in the $\hat{b}$ directions in $(\hat{b}, \hat{a})$ space. Then $T_s\alpha$ is spanned by the ℓ tangent vectors given by $e_{\hat{b}_j} + \left(0_\ell, \frac{\partial \hat{a}_1}{\partial \hat{b}_j}(s), \ldots, \frac{\partial \hat{a}_k}{\partial \hat{b}_j}(s)\right)$, for $j = 1, \ldots, \ell$, where 0_ℓ denotes an ℓ-tuple of zeroes.

We now turn our attention to the k-dimensional manifold TD_1^s. By the same arguments as above, but in backward time, $TD_1^s \subset TD_1$ is parameterized by $\hat{a}$. Therefore, TD_1^s can be described as a graph β in $(\hat{b}, \hat{a})$ space, but this time over the $\hat{a}$ variables. Furthermore, for s a

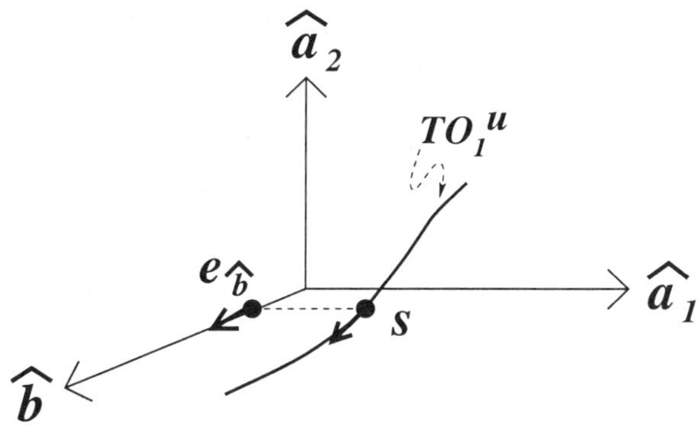

FIG. 11. *Representation of TO_1^u in the take off set $TO_1 = \{(\hat{a},\hat{b},Y_1)|Y_1 = Y_1(TO)\}$, for some $Y_1(TO) > 0$. Here, $k = 2, \ell = 1, p = 1$. TO_1^u is a graph over the $\hat{b}$ variables.*

point in $\beta = \{\hat{b}(\hat{a}), \hat{a}\}$, $T_s\beta$ is spanned by the following k tangent vectors: $\left(\frac{\partial \hat{b}_1}{\partial \hat{a}_1}(s), ..., \frac{\partial \hat{b}_l}{\partial \hat{a}_1}(s), 0_k\right) + e_{\hat{a}_i}$ for $i = 1, ...k$, where 0_k denotes a k-tuple of zeroes and $e_{\hat{a}_1}, ..., e_{\hat{a}_k}$ denote the unit vectors in the $\hat{a}$ directions. See Figure 12 for an illustration with $\ell = 1, k = 2$.

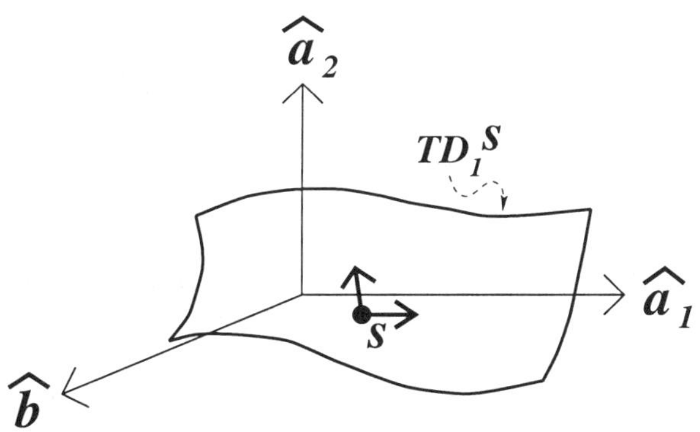

FIG. 12. *Representation of TD_1^s in the touch down set $TD_1 = \{(\hat{a},\hat{b},Y_1)|Y_1 = Y_1(TD)\}$, for some $Y_1(TD) < 0$. Here, $k = 2, \ell = 1, p = 1$. TD_1^s is a graph over the $\hat{a}$ variables.*

Next, we show that if a point p_1^u belongs to TO_1^u, its image in $f_f^L(TO_L^u)$ after the sequence of fast and slow maps has the same $\hat{a}$ and $\hat{b}$ coordinates as p_1^u does. Without loss of generality, we show this for $L = 2$. Consider $f_f^1(p_1^u) \in TD_2^s$. By definition of the fast map f_f^1, p_1^u and $f_f^1(p_1^u)$ have the same original global slow variables. Also, by Remark 6.3, we know that $f_f^1(p_1^u)$ and $f_s^2(f_f^1(p_1^u)) \in TO_2^u$ have the same $\hat{a}$ and $\hat{b}$ coordinates. In fact, only their Y_1 coordinates differ. Since $f_s^2(f_f^1(p_1^u))$ and

$f_f^2(f_s^2(f_f^1(p_1^u))) \in TD_1$ have the same global slow coordinates, we have that in local coordinates around the first slow manifold, p_1^u and $f_f^2(f_s^2(f_f^1(p_1^u)))$ have the same $\hat{a}$ and $\hat{b}$ coordinates, and only their Y_1 coordinates are different. So, one can think of $f_f^2(TO_2^u)$ as merely a translate of TO_1^u from TO_1 to TD_1, -*i.e*, only the Y_1 coordinate of a point in TO_1^u is different from its image in $f_f^2(TO_2^u)$. Completing the argument, the result holds for any $L \geq 2$, since the fast maps f_f^i leave the global slow variables unchanged, and the slow maps f_s^i only change the local Y_1 coordinates.

Finally, we will show that $f_f^L(TO_L^u)$ and TD_1^s intersect transversely. We can think of $f_f^L(TO_L^u) \subset TD_1$ as the graph α in $(\hat{a}, \hat{b})$ space, since every $p_1^u \in TO_1^u$ and its image in $f_f^L(TO_L^u)$ differ only in their Y_1-coordinates. Therefore, it suffices to show that the graphs α and β intersect transversely in $(\hat{a}, \hat{b})$ space. We first show that they intersect. This is a consequence from the following three facts. First, the domain of $\alpha = \{\hat{a}(\hat{b}), \hat{b}\}$ is I_b and its range is I_a, while the domain of $\beta = \{\hat{a}, \hat{b}(\hat{a})\}$ is I_a and its range is I_b. Second, the graph α is bounded away from the boundaries $\{a_i = a_{min}\}, \{a_i = a_{max}\}, i = 1, ..., k$ of $I_a \times I_b$. This is because since trajectories in $\pi_b I_\varepsilon$ obey $\dot{\hat{a}} = 0$, $\dot{\hat{b}} = 0$, points in α satisfy $a_{min} < |\hat{a}(\hat{b})| < a_{max}$. The inequalities hold because any trajectory starting on $\pi_b I_\varepsilon$ with initial a, $\hat{a}$ coordinates given by either a_{min} or a_{max} leaves the box $\mathcal{B}_{\mathcal{S}_1}$ with Y_1 coordinate different from the one corresponding to TO_1^u. Third, β is bounded away from the boundaries $\{b_j = b_{max}\}, \{b_j = b_{min}\}, j = 1, ..., \ell$, since for the same reasons as above points in β satisfy $b_{min} < |\hat{b}(\hat{a})| < b_{\max}$. Therefore, α and β must intersect inside the box $I_a \times I_b$ in $(\hat{a}, \hat{b})$ space.

We now show that the intersection of the graphs α and β is indeed transverse. Let s be a point in their intersection. By direct inspection of the ℓ tangent vectors that span $T_s\alpha$ and those k tangent vectors that span $T_s\beta$, we see that their union spans the whole $(\hat{a}, \hat{b})$ space. Therefore, the intersection of TO_1^u and TD_1^s in TD_1 is transverse, and it is a zero-dimensional manifold, since the dimension of TD_1 is $k + \ell$. See Figure 13 for an illustration with $k = 2, \ell = 1$. This concludes the proof of part i).

ii) Assume now that $p > 1$. The proof here consists of three main steps: showing that the intersection of $f_f^L(TO_L^u)$ and TD_1^s is nonempty, then showing that these manifolds are not coincident, and finally that their intersection is transverse. The extra complicating feature arising from the fact that $p > 1$ is that one needs to keep track of the $Y_r, \hat{\mathcal{Y}}_r$ coordinates of trajectories, for $r = 2, ..., p$. This complicating feature requires us to study the foliations of TO_1^u and TD_1^s over the variables $\hat{\mathcal{Y}}$, as a preliminary step.

We first study the foliation of TO_1^u. For any segment of a singular orbit in the neighborhood of the singular periodic orbit, let Y_{r,TO_j} and Y_{r,TD_j} denote, respectively, the (local) Y_r-values of this singular trajectory in the take off TO_j and touch down TD_j sets. Let $I_Y \equiv I_{Y_2} \times ... \times I_{Y_p}$. We will show that $TO_1^u = \cup_{\hat{\mathcal{Y}} \in I_Y} \left(TO_1^u|_{\hat{\mathcal{Y}}}\right)$, where the ℓ-dimensional manifolds

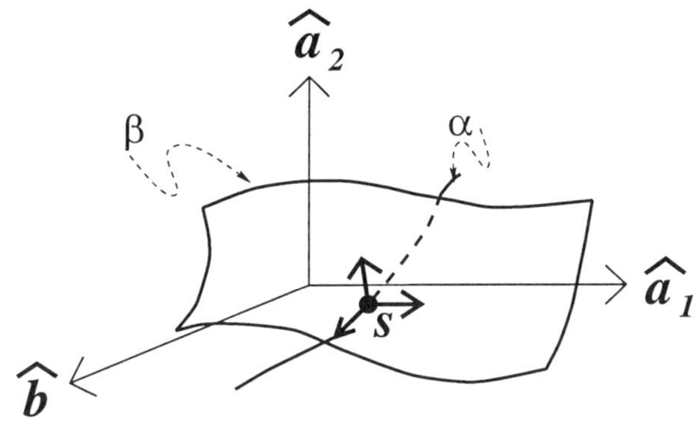

FIG. 13. *Illustration of the transverse intersection of* TD_1^s *and* $f_f^L(TO_L^u)$ *in the touch down set* TD_1*, for* $k = 2, \ell = 1, p = 1$. $f_f^L(TO_L^u)$ *and* TD_1^s *correspond, respectively, to the graphs* α *and* β.

$TO_1^u|_{\hat{\mathcal{Y}}}$ are constructed as follows. Let $\pi_b I_\varepsilon|_{\hat{\mathcal{Y}}} \cdot [0, \infty)$ denote the union of the forward-time trajectories with initial conditions in $\pi_b I_\varepsilon$ with fixed $Y_r, \hat{\mathcal{Y}}_r = \overline{Y}_r$, for $r = 2, ..., p$. By the same arguments as in part i) above, the $\varepsilon \to 0$ limit of the base points $\pi_a((\pi_b I_\varepsilon|_{\overline{\mathcal{Y}}} \cdot [0, \infty))|_{\{|a|=\Delta\}_L})$ transversely intersects the manifold $TO_1 \cap \{\hat{\mathcal{Y}}_r = \overline{Y}_r, Y = Y_{r,TO_1}, r = 2, ..., p\}$ in an ℓ-dimensional manifold. We will denote this intersection manifold by $TO_1^u|_{\overline{\mathcal{Y}}}$. Furthermore, also by the same arguments as in part i), $TO_1^u|_{\overline{\mathcal{Y}}}$ can be thought of as a graph $\alpha_{\hat{\mathcal{Y}}}$ in $(\hat{a}, \hat{b})$ space over the $\hat{b}$ variables. As we let $Y_r, \hat{\mathcal{Y}}_r$ vary in I_{Y_r}, with$Y_r = \hat{\mathcal{Y}}_r$ we obtain $\mathcal{V}_{\varepsilon,1}^u = \cup_{\overline{\mathcal{Y}} \in I_Y} \left(\pi_b I_\varepsilon|_{\overline{\mathcal{Y}}} \cdot [0, \infty)\right)$. Moreover, we know, by Proposition 6.2, that the $\varepsilon \to 0$ limit of the base points $\pi_a(\mathcal{V}_{\varepsilon,1}^u|_{\{|a|=\Delta\}_1})$, which is precisely BV_1^u, transversely intersects TO_1. This intersection has dimension $\ell + p - 1$, and has been denoted by TO_1^u. Therefore, TO_1^u is foliated by the $TO_1^u|_{\hat{\mathcal{Y}}}$.

The foliation of TD_1^s is obtained similarly, by considering the trajectories in $\pi_a I_\varepsilon|_{\overline{\mathcal{Y}}} \cdot [0, -\infty)$, the union of all backward-time trajectories that start in $\pi_a I_\varepsilon$ with fixed $Y_r, \hat{\mathcal{Y}}_r = \overline{Y}_r$, for $r = 2, ..., p$. That is, $TD_1^s = \cup_{\hat{\mathcal{Y}} \in I_Y} \left(TD_1^s|_{\hat{\mathcal{Y}}}\right)$, where the k-dimensional manifolds $TD_1^s|_{\hat{\mathcal{Y}}}$ are defined in a fashion analogous to the $TO_1^u|_{\overline{\mathcal{Y}}}$, but in backward time. Moreover, each $TD_1^s|_{\hat{\mathcal{Y}}}$ lies in the set $TO_1 \cap \{\hat{\mathcal{Y}}_r = \overline{Y}_r, Y_r = Y_{r,TD_1}, r = 2, ..., p\}$ and can be thought of as a graph $\beta_{\hat{\mathcal{Y}}}$ in $(\hat{a}, \hat{b})$ space over the $\hat{a}$ variables.

We now show that that $f_f^L(TO_L^u) \cap TD_1^s$ is nonempty. This will be a consequence of the existence of the periodic orbit. We shall need some extra notation. Consider first the slow flow on $\mathcal{S}_1^\varepsilon$ and the locally unique singular periodic orbit. For $r = 2, ..., p$, let $Y_r^* \in I_{Y_r}$ be the value of the coordinate Y_r such that the singular solution starting on the set $Y_1 = 0$, with $Y_r = Y_r^*$ corresponds to the first segment of the singular periodic orbit.

For $j = 1, ..., L$, let Y^*_{r,TO_j} be value of Y_r that the singular periodic orbit has at the take off set TO_j, and Y^*_{r,TD_1} its Y_r-coordinate in the touch-down set TD_1. Therefore, in global coordinates, $Y^*_{TD_1} = Y^*_{TO_L}$. Recall that for any other segment of a singular orbit in the neighborhood of the singular periodic orbit, Y_{r,TO_j} and Y_{r,TD_j} denote, respectively, the (local) Y_r-values of this singular trajectory in the take off TO_j and touch down TD_j sets.

Consider now $\pi_b I_\varepsilon|_{Y^*} \cdot [0, \infty)$ and $\pi_a I_\varepsilon|_{Y^*} \cdot [0, -\infty)$. Recall that the evolution of the slow coordinates of all these trajectories is the same as that of the coordinates of the singular periodic orbit. So, this is as though we had $p = 1$, since the concatenation of fast f^i_f and slow f^i_s maps leaves the $Y_r, \hat{\mathcal{Y}}_r$ variables fixed. Therefore, part i) applies and it follows that the manifolds $f^L_f(TO^u_L|_{Y^*})$ and $TD^s_1|_{Y^*}$ intersect transversely in $TD_1 \cap \{Y_r, \hat{\mathcal{Y}}_r = Y^*_r, r = 2, ..., p\}$ and also that $f^L_f(TO^u_L)$ and TD^s_1 have nonempty intersection, since $f^L_f(TO^u_L|_{Y^*}) \subset f^L_f(TO^u_L)$ and $TD^s_1|_{Y^*} \subset TS^s_1$.

Next, the fact that the singular periodic orbit is locally unique will give us that $f^L_f(TO^u_L)$ and TD^s_1 intersect, but are not coincident. Consider any slow trajectory starting on $\{Y_1 = 0\}$ in the neighborhood of the singular periodic orbit with $Y_2, ..., Y_p$ different from $Y^*_2, ..., Y^*_p$. Its slow global coordinates in TD_1 and TO_L cannot coincide, since by assumption this is not the singular periodic orbit. Thus, in global coordinates, $Y_{TD_1} \neq Y_{TO_L}$. This implies that the sets $TD^s_1|_{\hat{\mathcal{Y}}}$ and $f^L_f(TO^u_L|_{\hat{\mathcal{Y}}})$ cannot intersect, since points in $TD^s_1|_{\hat{\mathcal{Y}}}$ have Y_r coordinates given by Y_{r,TD_1} and points in $f^L_f(TO^u_L|_{\hat{\mathcal{Y}}})$ have Y_r coordinates given by Y_{r,TO_L}. Furthermore, for any two distinct $\overline{Y}'$ and $\overline{Y}''$ in I_Y, the sets of base points $TD^s_1|_{\overline{\mathcal{Y}}'}$ and $f^L_f(TO^u_L|_{\overline{\mathcal{Y}}''})$ cannot intersect, since points in $TD^s_1|_{\overline{\mathcal{Y}}'}$ have $\hat{\mathcal{Y}}_r$ coordinates given by $\hat{\mathcal{Y}}_r = \overline{Y}'_r$ while points in $f^L_f(TO^u_L|_{\overline{\mathcal{Y}}''})$ have $\hat{\mathcal{Y}}_r$ coordinates given by $\hat{\mathcal{Y}}_r = \overline{Y}''_r$. Therefore, the intersection $f^L_f(TO^u_L|_{\hat{\mathcal{Y}}'}) \cap (TO^s_1|_{\hat{\mathcal{Y}}''})$ is not empty if and only if $\hat{\mathcal{Y}}' = \hat{\mathcal{Y}}'' = \hat{\mathcal{Y}}^*$.

We finally show that the intersection of $f^L_f(TO^u_L)$ and TD^s_1 is transversal, using the fact that all eigenvalues of the derivative of the singular map F, as defined in Section 2 for the original system, have norm different from 1. We prove it for the case in which F is a contraction (*i.e.*, all eigenvalues lie inside the unit circle). The other case is proven similarly. Let q be a point in the intersection of $f^L_f(TO^u_L)$ and TD^s_1. We will show that the union of the two sets of tangent vectors that span, respectively, $T_q(f^L_f(TO^u_L))$ and $T_q(TD^s_1)$ spans the tangent space $T_q(TD_1)$, thus proving the proposition.

Now, since $TD^s_1 = \cup_{\hat{\mathcal{Y}} \in I_Y} (TD^s_1|_{\hat{\mathcal{Y}}})$, the tangent space $T_q(TD^s_1)$ is spanned by

(6.5) the k tangent vectors that span $T_q(TD^s_1|_{\hat{\mathcal{Y}}^*})$, and

(6.6) $p - 1$ tangent vectors in the directions normal to $T_q(TD^s_1|_{\hat{\mathcal{Y}}^*})$

Similarly, $T_q(f_f^L(TO_L^u))$ is spanned by

(6.7) the k tangent vectors that span $T_q(f_f^L(TO_L^u|_{\hat{\mathcal{Y}}^*}))$, and

(6.8) $p-1$ tangent vectors in the directions normal to $T_q(f_f^L(TO_L^u|_{\hat{\mathcal{Y}}^*}))$,

since $F_f^L(TO_L^u) = \cup_{\hat{\mathcal{Y}}\in I_Y} f_f^L\left(TO_L^u|_{\hat{\mathcal{Y}}}\right)$. Furthermore, we know that the vectors in (6.5) are given by the k tangent vectors to the graph $\beta|_{\hat{\mathcal{Y}}^*}$ at q, and that (6.7) are the ℓ tangent vectors to the graph $\alpha|_{\hat{\mathcal{Y}}^*}$ at q. Recall e_r denotes the unit vector, in Fenichel coordinates $(a, b, \hat{a}, \hat{b}, Y, \hat{\mathcal{Y}})$ along the r-coordinate, and e_{rs} denotes the unit vector along the diagonal of the plane spanned by the unit vectors e_r and e_s. For $r = 2, ..., p$, $e_{Y_r,\hat{\mathcal{Y}}_r}$ are $p-1$ vectors tangent to the sets $\pi_b I_\varepsilon$ and $\pi_a I_\varepsilon$ in the $Y_r, \hat{\mathcal{Y}}_r$ directions, see Figure 10.

The tangent vectors (6.6) are obtained by following the backward-time evolution of the $p-1$ tangent vectors $e_{Y_r,\hat{\mathcal{Y}}_r}$ under the linearization of (1.3) (*i.e.*, the variational equations) around the singular periodic orbit in $\mathcal{S}_1^\varepsilon$, in backward time, until the slow segment of the singular periodic orbit reaches the touch-down set TD_1^s. Since all the eigenvalues of dF have norm less than 1, the Y_r-coordinates of any of the singular trajectories in the neighborhood of the singular periodic orbit satisfy $|Y_{r,TO_j} - Y^*_{r,TO_j}| < |Y_{r,TD_j} - Y^*_{r,TD_j}|$. That is, in backward time, the Y_r coordinates of any if these singular trajectories get further away from those of the singular periodic orbit. Therefore, for $r = 2, ..., p$, the image of $e_{Y_r,\hat{\mathcal{Y}}_r}$ on $T_q(TD_1^s)$, which we shall denote by $n^-_{Y_r,\hat{\mathcal{Y}}_r}$, is a vector whose projection onto $(\hat{\mathcal{Y}}_r, Y_r)$ space is a line with slope greater than 1. Figure 14 illustrates the $p = 2$ case.

The $p-1$ tangent vectors (6.8) are obtained by the evolution of the same $e_{Y_r,\hat{\mathcal{Y}}_r}$ tangent vectors, but this time under the variational equations for the slow flow in forward time. The tangent vectors are followed until the singular periodic solution reaches TO_L^u. In this case, the image of $e_{Y_r,\hat{\mathcal{Y}}_r}$ on $T_q(f_f^L(TO_L^u))$, which we denote by $n^+_{Y_r,\hat{\mathcal{Y}}_r}$, is a vector whose projection onto $(\hat{\mathcal{Y}}_r, Y_r)$ space is a line with slope smaller than 1. Figure 15 illustrates $p = 2$.

We now complete the proof. $T_q(TD_1^s|_{\hat{\mathcal{Y}}^*})$ is spanned by the k vectors tangent to the graph $\beta|_{\hat{\mathcal{Y}}^*}$, and the $p-1$ tangent vectors $\{n^-_{Y_r,\hat{\mathcal{Y}}_r}, r = 2, ..., p\}$. The tangent space $T_q(f_f^L(TO_L^u|_{\hat{\mathcal{Y}}^*}))$ is spanned by the ℓ vectors tangent to $\alpha|_{\hat{\mathcal{Y}}^*}$, and the $p-1$ tangent vectors $\{n^+_{Y_r,\hat{\mathcal{Y}}_r}, r = 2, ..., p\}$. By part i), we know that the union of the vectors tangent to $\beta|_{\hat{\mathcal{Y}}^*}$ and $\alpha|_{\hat{\mathcal{Y}}^*}$ is a set of linearly independent tangent vectors that spans $T_q(TD_1|_{\{Y_r,\hat{\mathcal{Y}}_r=Y_r^*,r=2,\dots,p\}})$ (*i.e.*, the $\hat{a}$ and $\hat{b}$ directions in $T_q(TD_1)$). The complement of $T_q(TD_1)$ in the directions normal to $T_q(TD_1|_{\{Y_r,\hat{\mathcal{Y}}_r=Y_r^*,r=2,\dots,p\}})$ is spanned by the $2p-2$ linearly independent

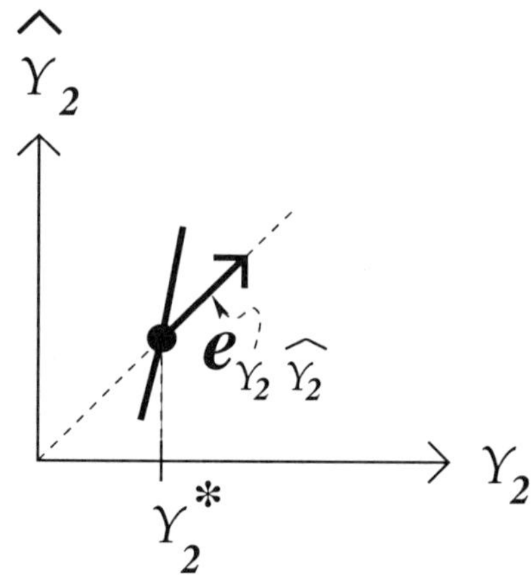

FIG. 14. *The line with positive slope corresponds to the projection of* $n^-_{y_2,\hat{\mathcal{Y}}_2}$ *onto* $(Y_2, \hat{\mathcal{Y}}_2)$ *space, where* $n^-_{y_2,\hat{\mathcal{Y}}_2}$ *is the backward-time image of* $e_{y_2,\hat{\mathcal{Y}}_2}$ *in* TD_1 *under the variational equations for the slow flow.*

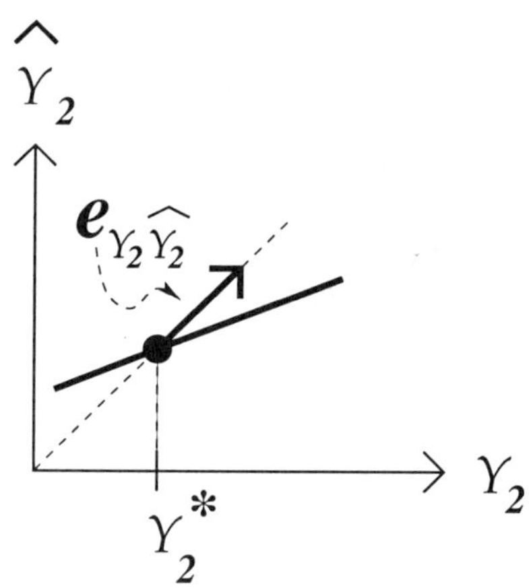

FIG. 15. *The line with negative slope corresponds to the projection of* $n^+_{y_2,\hat{\mathcal{Y}}_2}$ *onto* $(Y_2, \hat{\mathcal{Y}}_2)$ *space, where* $n^+_{y_2,\hat{\mathcal{Y}}_2}$ *is the forward-time image of* $e_{(y_2,\hat{\mathcal{Y}}_2)}$ *in* TD_1 *under the variational equations for the slow flow.*

vectors $\{n^-_{Y_r,\hat{\mathcal{Y}}_r}, n^+_{Y_r,\hat{\mathcal{Y}}_r}, r = 2, ..., p\}$, since for each $r = 2, ..., p$, the $(Y_r, \hat{\mathcal{Y}}_r)$ space is spanned by the projection of $n^-_{Y_r,\hat{\mathcal{Y}}_r}$ and $n^+_{Y_r,\hat{\mathcal{Y}}_r}$ onto that space. This proves that the intersection of TD^s_1 and $f^L_f(TO^u_L)$ is indeed transverse. It follows then that this intersection is zero-dimensional. □

REMARK 6.6. The existence of a locally unique singular periodic orbit in the original system (1.1) implies that the augmented system (3.1) possesses a $(k+\ell+p-1)$-dimensional manifold of singular periodic orbits $\mathcal{P}$.

The transverse intersection of TD_1^s and $f_f^L(TO_L^u)$ guaranteed by the above proposition picks out a point in TD_1 that belongs to a unique singular periodic orbit of $\mathcal{P}$. It is this singular periodic orbit that the actual one will be close to.

Proof of Theorems 1 and 2. By the discussion at the beginning of this section we know that if we show that $\mathcal{M}_\varepsilon^+$ transversely intersects $\mathcal{M}_\varepsilon^-$ in a one dimensional manifold, then this intersection manifold corresponds to the periodic orbit we seek. That this periodic orbit arises as a transverse intersection gives automatically its local uniqueness. Now, by Proposition 6.6 we have that $f_f^L(TO_L^u)$ transversely intersects TD_1^s in the touch-down set TD_1. Therefore, the hypothesis of Proposition 6.5 is satisfied and we conclude that $\mathcal{V}_{\varepsilon,1}^s$, a submanifold of $W^s(\mathcal{S}_1^\varepsilon)$, and $\mathcal{V}_L^u$, a submanifold of $W^u(\mathcal{S}_L^\varepsilon)$, intersect transversely. Now, recall that when trajectories in $\mathcal{M}_\varepsilon^-$ leave the neighborhood of $\mathcal{S}_1^\varepsilon$, $\mathcal{M}_\varepsilon^-$ is $\mathcal{C}^1$ exponentially close to $\mathcal{V}_{\varepsilon,1}^s$. Similarly, $\mathcal{M}_\varepsilon^+$ is $\mathcal{C}^1$ exponentially close to $\mathcal{V}_{\varepsilon,L}^u$, after trajectories in it have spent $\mathcal{O}(\frac{1}{\varepsilon})$ amounts of time in the neighborhood of each of the slow manifolds in the sequence $\{\mathcal{S}_j^\varepsilon\}$. Since $\mathcal{V}_{\varepsilon,1}^s$ and $\mathcal{V}_{\varepsilon,L}^u$ transversely intersect, so do $\mathcal{M}_\varepsilon^-$ and $\mathcal{M}_\varepsilon^+$. Finally, both $\mathcal{M}_\varepsilon^-$ and $\mathcal{M}_\varepsilon^+$ have dimension $k+\ell+p$, and their intersection occurs in $2k+2\ell+2p-1$ space, so their intersection is a one-dimensional manifold. □

7. Application to periodic orbits in the FitzHugh-Nagumo equations. The FitzHugh-Nagumo equations are a system of reaction diffusion equations:

$$\begin{aligned} \frac{\partial u}{\partial t} &= \frac{\partial^2 u}{\partial x^2} + f(u) - w \\ \frac{\partial w}{\partial t} &= \varepsilon(u - \gamma w), \end{aligned} \tag{7.1}$$

where $0 < \varepsilon \ll 1$, $f(u)$ is a cubic-shaped function and $\gamma \geq 0$ is such that the curves $w = f(u)$ and $w = \dfrac{u}{\gamma}$ intersect only at $(u, w) = (0, 0)$. These equations were first derived as a simplified system of the Hodgkin-Huxley equations, see Section 8.

When traveling-wave solutions of (7.1) are sought, that is solutions of the form $u = u(x + \theta t), w = w(x + \theta t)$, where θ corresponds to the wave speed, the following system of ordinary differential equations is obtained:

$$\begin{aligned} \dot{u} &= v \\ \dot{v} &= \theta v - f(u) + w \\ \dot{w} &= \frac{\varepsilon}{\theta}(u - \gamma w) \end{aligned} \tag{7.2}$$

where $\cdot \equiv \dfrac{d}{d\xi}$ and $\xi \equiv x + \theta t$. Since $0 < \varepsilon \ll 1$, (7.2) is of the form (1.1), with $k + \ell = 2$ and $p = 1$. For γ sufficiently small, the origin is the only fixed point of (7.2). It is well known [6, 33, 38] that in this case, there exists a value $\theta^*(\varepsilon)$ of the wave speed such that (7.2) has an orbit homoclinic to the fixed point. Using Theorem 1, in Section 7.1 we show that there is a range of values of θ for which there exist periodic orbits of 7.2, for ε sufficiently small. In Section 7.2 we study the behavior of these periodic orbits as $\theta \to \theta^*(\varepsilon)$. This makes possible the spectral analysis of the periodic orbits as solutions to 7.1. See [18] for details.

7.1. Existence of periodic orbits. The main result in this subsection is

THEOREM 3. *Let $\gamma \geq 0$ be such that $(u = 0, v = 0, w = 0)$ is the only fixed point of* (7.2). *Let θ^* denote the value of θ for which* (7.2) *has a singular homoclinic orbit to the point $(u = 0, v = 0, w = 0)$. Then, for each $\theta \in (0, \theta^*)$, there exists an ε_0 such that if $0 < \varepsilon < \varepsilon_0$, then the system* (7.2) *possesses a locally unique periodic orbit.*

REMARK 7.1. The existence of the homoclinic orbit to $(u = 0, v = 0, w = 0)$ was first proven in [6] using isolating blocks (topological methods). In [33] the existence and local uniqueness of this orbit was proven using geometric methods. The existence of periodic orbits for (7.2) was also proven in [6], and for the special case $(\gamma = 0)$ in [23] by topological shooting methods. We remark that $\varepsilon_0 \to 0$ as $\theta \to \theta^*$.

Proof. Theorem 3 will be proven by showing that (7.2) satisfies Assumptions 1 and 2 and by constructing a singular periodic orbit such that Assumption 3 holds. Then we will apply Theorem 1 to (7.2) to complete the proof. The reduced fast system

$$\begin{aligned} \dot{u} &= v \\ \dot{v} &= \theta v - f(u) + w \\ \dot{w} &= 0 \end{aligned} \tag{7.3}$$

possesses a pair of one-dimensional slow manifolds $\mathcal{S}_1^0 \equiv \{(u, v, w) : u = f_1^{-1}(w), v = 0, w \in [w_{min}, \infty]\}$ and $\mathcal{S}_2^0 \equiv \{(u, v, w) : u = f_2^{-1}(w), v = 0, w \in [-\infty, w_{max}]\}$, which correspond to the points on the cubic $w = f(u)$ in the $v = 0$ plane for which $f'(u) < 0$. (See Figure 16). The manifolds $\mathcal{S}_1^0$ and $\mathcal{S}_2^0$ have two-dimensional center unstable manifolds $W^u(\mathcal{S}_1^0)$ and $W^u(\mathcal{S}_2^0)$ and two-dimensional center stable manifolds $W^s(\mathcal{S}_1^0)$ and $W^s(\mathcal{S}_2^0)$. Therefore, (7.3) satisfies Assumption 1.

We now verify that Assumption 2 holds. It has been shown [6] that for every value of the wave speed θ in an interval $[0, \theta^*]$, there exist two heteroclinic connections, one from $\mathcal{S}_1^0$ to $\mathcal{S}_2^0$ and the other from $\mathcal{S}_2^0$ to $\mathcal{S}_1^0$. More precisely, for all $\theta \in [0, \theta^*]$, there exist unique values of the variable w, $w^f(\theta)$ and $w^b(\theta)$ (both away from the values of w at which $f'(u) = 0$) such that there is a heteroclinic connection $\Gamma_f(\theta)$ of (7.3) on the plane $w = w^f(\theta)$ from $\mathcal{S}_1^0$ to $\mathcal{S}_2^0$ and a heteroclinic connection $\Gamma_b(\theta)$ from $\mathcal{S}_2^0$ to $\mathcal{S}_1^0$ on the $w = w^b(\theta)$ plane. (See Figure 17). Furthermore, in [30, 33] it has been shown that Γ_f (resp. Γ_b) lies in the transverse intersection of $W^u(\mathcal{S}_1^0)$ and

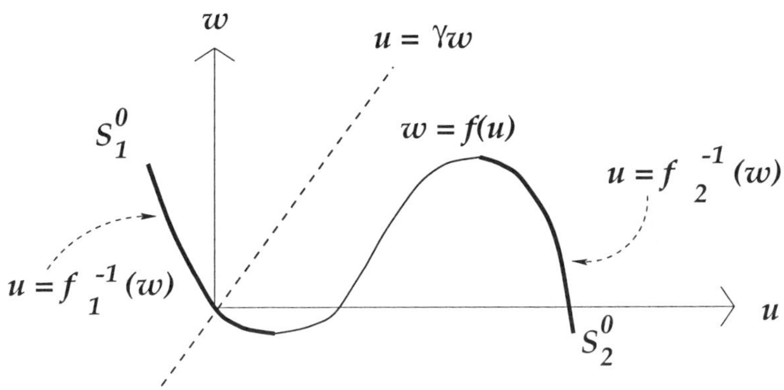

FIG. 16. *The one-dimensional slow manifolds $\mathcal{S}_1^0$ and $\mathcal{S}_2^0$ lie in the $\{v = 0\}$ plane.*

$W^s(S_2^0)$ (resp. of $W^u(\mathcal{S}_2^0)$ and $W^s(S_1^0)$). Therefore, Assumption 2 holds. Also, in this case the sets TO_j and TD_j, for $j = 1, 2$, are zero-dimensional, and the flow on $\mathcal{S}_2^0$ is monotone. The flow on $\mathcal{S}_1^0$, away from the fixed point is also monotone, so Assumption 3 holds.

We now construct the singular periodic orbit. Let $\theta \in (0, \theta^*)$ be such that the distance $|w^b(\theta) - w^f(\theta)|$ is $\mathcal{O}(1)$. This means that the points TD_i and TO_i are $\mathcal{O}(1)$ apart in $\mathcal{S}_i^0$. The first slow segment of the singular periodic orbit consists of the piece of S_1^0 from TD_1 to TO_1. The first fast segment corresponds to the heteroclinic connection $\Gamma_f(\theta)$. The second slow segment is the piece of $\mathcal{S}_2^0$ from TD_2 to TO_2. Finally, the last segment is the fast heteroclinic connection $\Gamma_b(\theta)$. (See Figure 17).

Now that we have the appropriate singular periodic orbit, Theorem 1 implies that for $0 < \varepsilon \ll 1$ there exists a locally unique periodic orbit of (7.2) that is $\mathcal{O}(\varepsilon)$ close to the singular one. □

7.2. Periodic orbits for $\varepsilon > 0$ fixed. Having shown the existence (and local uniqueness) of periodic orbits in (7.2) for $0 < \varepsilon < \varepsilon_0$, we wish to understand the relationship between these periodic orbits and the homoclinic solution. Since $\varepsilon_0 \to 0$ as $\theta \to \theta^*$, more work must be done. We will show in this subsection that for a small but fixed value of $\varepsilon > 0$, there exists an interval $(0, \theta^*(\varepsilon))$ of wave speeds, and for each value of θ in this interval, there is a periodic orbit of (7.2) with that wave speed. This family of periodic orbits limits (uniformly on compact intervals) on the homoclinic solution with $\theta = \theta^*(\varepsilon)$. Here, we also note that $\theta^*(\varepsilon)$ is a decreasing function of ε, see Figure 18a, and a comprehensive analysis on the behavior of the homoclinic solutions of (7.2) is given in [38].

Since we are interested in the periodic solutions as θ varies, we append $\dot{\theta} = 0$ to (7.2):

$$\dot{u} = v$$
$$\dot{v} = \theta v - f(u) + w$$

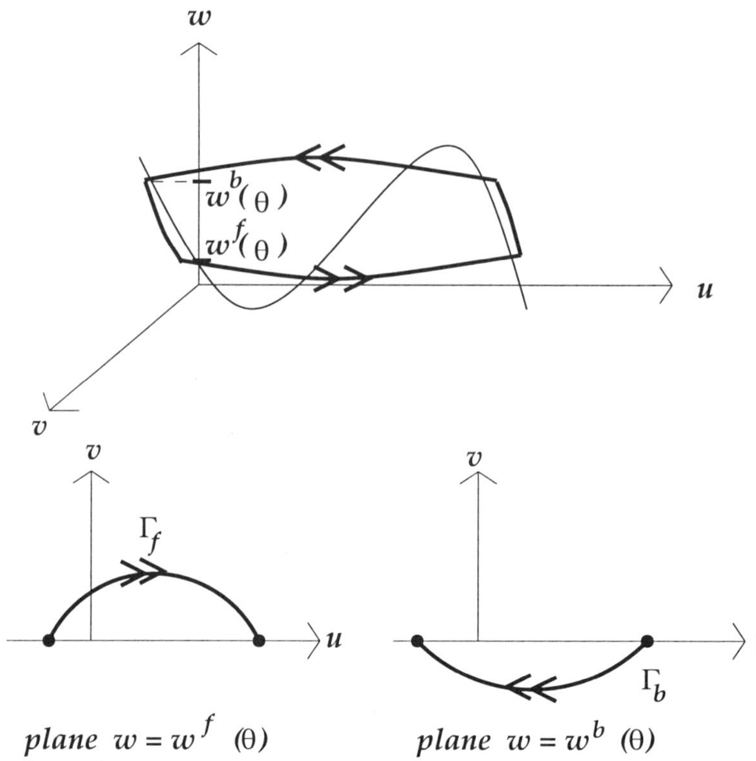

FIG. 17. *Above: the singular periodic orbit. Below: the fast heteroclinic connections for θ fixed.*

(7.4)
$$\begin{aligned} \dot{w} &= \frac{\varepsilon}{\theta}(u - \gamma w) \\ \dot{\theta} &= 0. \end{aligned}$$

The fast system associated to (7.4)

$$\begin{aligned} \dot{u} &= v \\ \dot{v} &= \theta v - f(u) + w \\ \dot{w} &= 0 \\ \dot{\theta} &= 0 \end{aligned}$$

possesses a pair of 2-dimensional slow manifolds, given by

(7.5)
$$\begin{aligned} S_1^0 &= \{(u, v, w, \theta) : u = f_1^{-1}(w), v = 0, w \in [w_{min}, w_{max}], \theta \in [0, \theta^* + \delta]\} \\ S_2^0 &= \{(u, v, w, \theta) : u = f_2^{-1}(w), v = 0, w \in [w_{min}, w_{max}], \theta \in [0, \theta^* + \delta]\}, \end{aligned}$$

where $\delta > 0$, $w_{min} < 0, w_{max} > 0$. Furthermore, $W^u(S_1^0)$ and $W^s(S_2^0)$ (resp. $W^u(S_2^0)$ and $W^s(S_1^0)$) intersect transversely in a two-dimensional

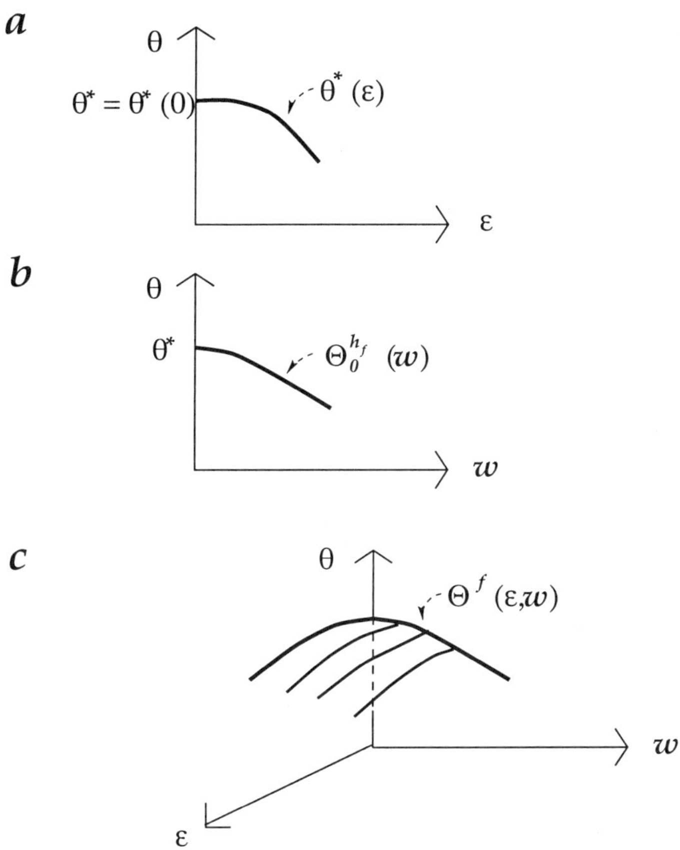

FIG. 18. **(a)** *Behavior of the wave speed of the homoclinic pulse with respect to* ε. **(b)** *Representation of* $\Theta_0^{h_f}(w)$. **(c)** *Illustration of the surface* $\theta = \Theta^f(\varepsilon, w)$ *in* (ε, w, θ) *space. It corresponds to the zero-set of the function* $d(\varepsilon, w, \theta)$, *which measures the distance between the unstable and the stable manifolds.*

heteroclinic manifold H_f (resp. H_b) formed by the union of all the fast heteroclinic connections $\Gamma_f(\theta)$ (resp. $\Gamma_b(\theta)$) over θ. The union of the fixed points in S_1^0 and S_2^0 that are connected via these heteroclinic orbits form four one-dimensional manifolds parameterized by w (or θ):

$$
\begin{aligned}
TO_1 &= \{u = f_1^{-1}(w), v = 0, w \in [w_{min}, w_{max}], \theta = \Theta_0^{h_f}(w)\} \subset S_0^1 \\
TD_2 &= \{u = f_2^{-1}(w), v = 0, w \in [w_{min}, w_{max}], \theta = \Theta_0^{h_f}(w)\} \subset S_0^2 \\
TO_2 &= \{u = f_2^{-1}(w), v = 0, w \in [w_{min}, w_{max}], \theta = \Theta_0^{h_b}(w)\} \subset S_0^2 \\
TD_1 &= \{u = f_1^{-1}(w), v = 0, w \in [w_{min}, w_{max}], \theta = \Theta_0^{h_b}(w)\} \subset S_0^1
\end{aligned}
\tag{7.6}
$$

where $\Theta_0^{h_{b,f}}(w)$ is smooth and $\dfrac{d\Theta_0^{h_f}}{dw}, \dfrac{d\Theta_0^{h_b}}{dw} < 0$. See Figure 18b. In the following proposition, we determine the effect of ε on the $\Theta_0^{h_f}(w)$ curve. A similar result holds for $\Theta_0^{h_b}(w)$.

PROPOSITION 7.1. *For $\varepsilon > 0$ fixed but small, there exists a curve $\theta = \Theta_\varepsilon^h(w)$ in (ε, w, θ) space such that $\frac{\partial \Theta_\varepsilon^h}{\partial w}(w) < 0$, and for every value of θ on this curve, $W^u(S_1^\varepsilon)$ and $W^s(S_2^\varepsilon)$ intersect transversely in a heteroclinic connection whose fast excursion lies close to $\Gamma_f(\theta)$.*

Proof. Let $q_0 = q_0(\theta)$ be a point in the heteroclinic connection $\Gamma_f(\theta)$. Let $d = d(\varepsilon, w, \theta)$ be the distance between $W^u(S_1^\varepsilon)$ and $W^s(S_2^\varepsilon)$ at this reference point q_0. We point out that d is smooth. We will find the zero-set of d in (ε, w, θ) space.

We know that $d(0, 0, \theta^*0) = 0$, and this corresponds to $\Gamma_f(\theta^*)$. From the implicit function theorem we have that, since $\frac{\partial d}{\partial \theta} \neq 0$, there exists a smooth function $\Theta^f(\varepsilon, w)$ such that $d(\varepsilon, w, \Theta^f(\varepsilon, w)) = 0$, for (ε, w) close to $(0, 0)$, see Figure 18c. Now, the geometric proof in [33] of the existence of the heteroclinic orbit of (7.2) implies that $\Theta^f(\varepsilon, 0) = \theta^*(\varepsilon)$. This gives $\frac{\partial \Theta^f}{\partial \varepsilon}(\varepsilon, 0) < 0$. By continuity, for w close to 0, we have $\frac{\partial \Theta^f}{\partial \varepsilon}(\varepsilon, w) < 0$. Also, we know that $\Theta^f(0, w) = \Theta_0^{h_f}(w)$, and thus $\frac{\partial \Theta^f}{\partial w}(0, w) < 0$. Therefore, by continuity, $\frac{\partial \Theta^f}{\partial w}(\varepsilon, w) < 0$ for $\varepsilon > 0$ close to 0.

Let $\varepsilon > 0$ be small and fixed. We define $\Theta_\varepsilon^h(w) \equiv \Theta^f(\varepsilon, w)$. Now, since every point on the curve $\theta = \Theta_\varepsilon^h(w) \equiv \Theta^f(\varepsilon, w)$ is a zero of the distance function d, we have that the claimed result holds. □

We now state and prove the main result in this subsection:

THEOREM 4. *For each small but fixed value of $\varepsilon > 0$, there exists an interval $(0, \theta^*(\varepsilon))$ of wave speeds such that the system (7.2) possesses a one-parameter family of periodic orbits, one for each value of θ in this interval.*

Proof. The proof follows from the previous proposition and the construction of the periodic orbits carried out on the proof of Theorem 1.

Fix $\varepsilon > 0$. Consider the curve $\theta = \Theta_\varepsilon^h(w)$. We know that $\Theta_\varepsilon^h(0) = \theta^*(\varepsilon)$, since the fixed point of (7.2) is in the $w = 0$ plane, and so $\Theta_\varepsilon^h(0)$ corresponds to the value of θ for which the the homoclinic orbit $\Gamma_f(\theta^*)$ persists for this value of ε.

Now, from Figure 18c and Proposition 7.1 we have that for $\theta < \Theta_\varepsilon^h(0)$, $W^u(S_1^\varepsilon)$ and $W^s(S_2^\varepsilon)$ intersect transversely, and similarly for $W^u(S_2^\varepsilon)$ and $W^s(S_1^\varepsilon)$. We can now construct a periodic orbit for each $\theta < \Theta_\varepsilon^h(0) = \theta^*(\varepsilon)$ in the same fashion as we did in the proof of Theorem 1. This concludes the proof of Theorem 4. □

For the spectral analysis of periodic traveling waves in a class of reaction-diffusion equations, we refer the reader to [12, 17, 18, 42, 43, 47].

8. Application to periodic orbits in the Hodgkin-Huxley equations. The Hodgkin-Huxley equations are a reaction-diffusion equation coupled to one fast and two slow ordinary differential equations:

$$\begin{aligned} v_t &= v_{xx} - \hat{G}(v,m,n,h) \\ m_t &= \frac{1}{\delta}\gamma_m(v)(m_\infty(v) - m) \\ n_t &= \varepsilon\gamma_n(v)(n_\infty(v) - n) \\ h_t &= \varepsilon\gamma_h(v)(h_\infty(v) - h), \end{aligned}$$

where the subscripts t and x denote partial derivatives with respect to t and x, respectively. These equations were first derived to study the propagation of a voltage impulse along the giant squid axon, where $v = v(x,t)$ represents the transmembrane potential, m and h correspond to the activation and inactivation of the sodium channels, respectively, and n is the activation of the potassium channels. The function $\hat{G}(v,m,n,h)$ is the sum of the ionic currents, and typically is of the form $\hat{G}(v,m,n,h) = g_{Na}m^3h(v - V_{Na}) + g_K n^4(v - V_K) + g_L(v - V_L)$. Here, $g_{Na}, g_K, g_L > 0$ are parameters that correspond to the maximal conductances of the various channels. The parameters V_{Na}, V_K, V_L are also constants, corresponding to the sodium, potassium and leak reversal potentials; in general $V_K < V_L < V_{Na}$ (see [29] for examples of these values). The functions $m_\infty(v), n_\infty(v)$ and $h_\infty(v)$ are usually determined experimentally; and, in general, they are sigmoidal functions of the voltage, bounded between 0 and 1, see Figure 19. In most cases, δ is small, since the m process is very fast compared to the other ones. Thus, for a leading order approximation, one substitutes $m = m_\infty(v)$ in the equations above, effectively reducing the dimension of the system:

$$\begin{aligned} v_t &= v_{xx} - G(v,n,h) \\ n_t &= \varepsilon\gamma_n(v)(n_\infty(v) - n) \\ h_t &= \varepsilon\gamma_h(v)(h_\infty(v) - h), \end{aligned} \tag{8.1}$$

where

$$G(v,n,h) \equiv \hat{G}(v, m_\infty(v), n, h). \tag{8.2}$$

Traveling-wave solutions of (8.1) are solutions of the following system of ordinary differential equations:

$$\begin{aligned} \dot{v} &= w \\ \dot{w} &= \theta w + G(v,n,h) \\ \dot{n} &= \frac{\varepsilon}{\theta}\gamma_n(v)(n_\infty(v) - n) \\ \dot{h} &= \frac{\varepsilon}{\theta}\gamma_h(v)(h_\infty(v) - h), \end{aligned} \tag{8.3}$$

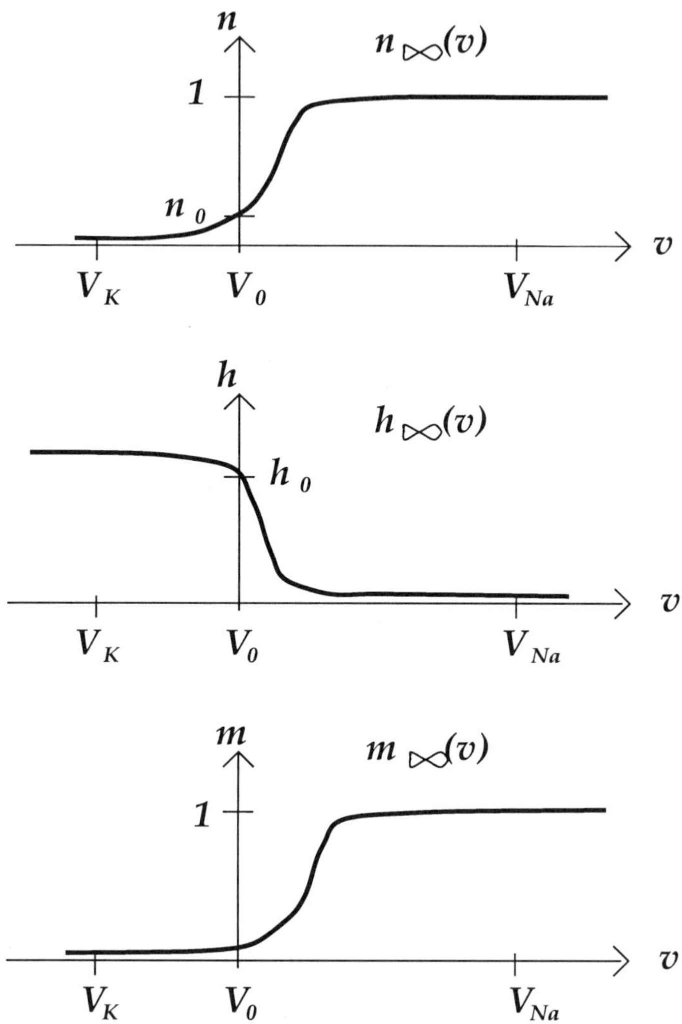

FIG. 19. *The functions* $n_{\infty(v)}, h_\infty(v), m_\infty(v)$.

where $\dot{r} \equiv \dfrac{dr}{d\xi}$ and $\xi = x + \theta t$, and θ is the constant $\mathcal{O}(1)$ wave speed. It is easy to see that (8.3) is of the form (1.1) with $p = 2$, since n and h are the slow variables. We will require (8.3) to have the following properties:

(G1) *For all* $y \equiv (n, h)$ *in a set of the form* $I_n \times I_h$*, where* I_n *and* I_h *are intervals of* n *and* h *values, respectively,* $G(v, n, h) = G(v, y)$ *is a cubic-like function of* v*, with three zeroes.* See Figure 20.

(G2) $\dfrac{\partial G}{\partial h} < 0$ *for* $v < V_{Na}$ *and* $\dfrac{\partial G}{\partial n} > 0$ *for* $v > V_K$.

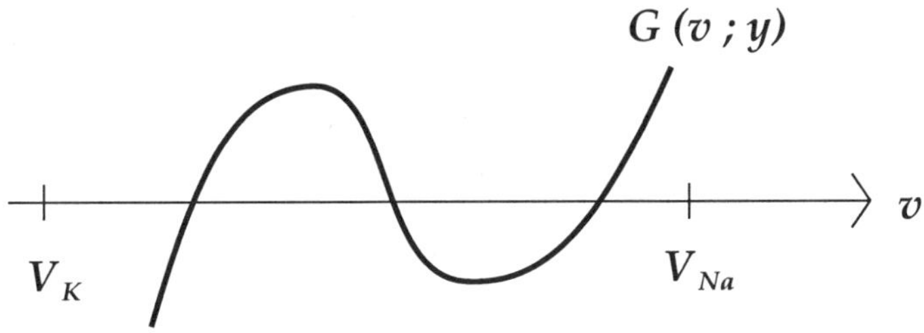

FIG. 20. *Behavior of* $G(v;y)$ *for fixed* $y = (n, h)$.

(G3) *For each* h *fixed,* $G(v, n, h) = 0$ *is a cubic-like one-dimensional curve in* (v, n) *space, that can be written as* $n = N(v; h)$. *As* h *increases,* N *rises.* See Figure 21.

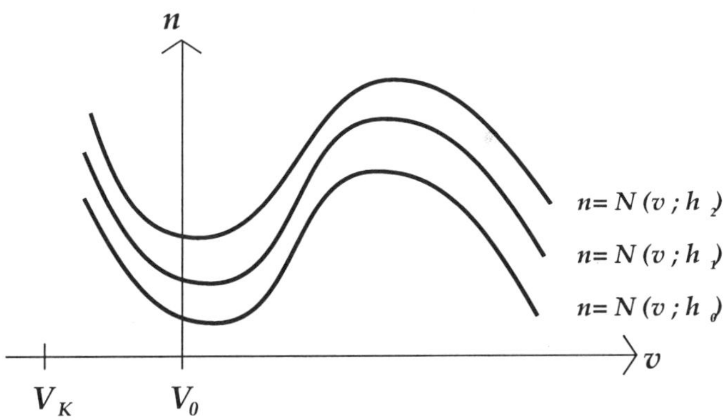

FIG. 21. *The zero set of* $G(v, n, h)$ *for fixed* h *is a cubic-like graph* $n = N(v; h)$ *in* (v, n) *space.*

For G of the form (8.2), $N(v, h) = \left(-\frac{g_{Na} m_\infty^3(v) h(v - V_{Na}) + g_L(v - V_L)}{g_K(v - V_K)}\right)^{\frac{1}{4}}$. Conditions **(G1)–(G3)** are generally satisfied for most physiologically relevant parameter regimes. The next two conditions we require refer to the detailed behavior of the functions $n_\infty, h_\infty, \gamma_n$ and γ_h. These conditions are also satisfied for many physiologically relevant parameter ranges.

(NH1) $\gamma_n(v) > 0, \gamma_h(v) > 0$ *and* (8.3) *has a unique fixed point* $(v = V_0, w = 0, n = n_0, h = h_0)$, *where* $n_0 = n_\infty(V_0), h_0 = h_\infty(V_0)$ *and* $V_0 > V_K$. *Furthermore,* V_0 *is such that* $n_\infty(V_0), h_\infty(V_0)$ *are already at the saturating (i.e. almost constant) parts of the sigmoidal curves, with* n_0 *close to* 0 *and* h_0 *close to* 1. V_0 *is to the left of the smallest value of* v *for which* $N_v(v, h) = 0$. See Figures 19 and 21.

(NH2) *For values of* v *satisfying* $N_v < 0$, $\frac{\partial n_\infty}{\partial v}, \frac{\partial h_\infty}{\partial v} \ll 1$ *and* $|\frac{\partial n}{\partial v}| < |\gamma_n \frac{\partial n_\infty}{\partial v}|, |\frac{\partial h}{\partial v}| < |\gamma_h \frac{\partial h_\infty}{\partial v}|$.

Our main result for (8.3) concerns the existence and local uniqueness of periodic orbits of this system. In [6] it was shown that a homoclinic solution to the fixed point $(V_0, 0, n_0, h_0)$ of (8.3) exists for a value $\theta^*(\varepsilon) > 0$ of the wave speed. In [7], the existence of periodic orbits for wave speed values in the interval $(0, \theta^*)$ was shown using topological isolating blocks.

THEOREM 5. *Let* (8.3) *satisfy conditions* **G1–G3, NH1** *and* **NH2.** *For every $\theta \in (0, \theta^*)$, where θ^* corresponds to the wave-speed value for which* (8.3) *has a singular homoclinic solution to $(V_0, 0, n_0, h_0)$, system* (8.3) *has a locally unique periodic orbit, provided* $0 < \varepsilon \ll 1$.

The proof of Theorem 5 follows from Theorem 2. In order to show that we may apply Theorem 2 to (8.3), we first verify that Assumptions 1–4 are satisfied by (8.3) under the above conditions. Then, for every $\theta \in (0, \theta^*)$, we construct a two-dimensional singular periodic manifold of (8.3). The rest of this section is devoted to carrying out this strategy.

Let $\theta > 0$ be fixed. The fast system of (8.3) is

$$\begin{aligned} \dot{v} &= w \\ \dot{w} &= \theta w + G(v; n, h) \\ \dot{n} &= 0 \\ \dot{h} &= 0. \end{aligned} \tag{8.4}$$

Since n and h are now parameters, the fixed points of (8.4) are given by $w = 0, G(v; n, h) = 0$. The stability of each fixed point is determined by the sign of the determinant of the Jacobian J, where

$$J = \begin{pmatrix} 0 & 1 \\ G_v & \theta \end{pmatrix}.$$

For fixed points at which $G_v > 0$, J has one positive and one negative eigenvalue and hence the phase portrait has a saddle. Whereas at saddle-node fixed points, $G_v = 0$. In (v, w, n, h) space, the fixed points form a two-dimensional manifold, parameterized by n and h. By **(G3)**, this manifold is given by $w = 0, n = N(v; h)$. Thus, in the subspace $w = 0$, this manifold is described by the graph of $N(v; h)$ in (v, n, h) space, and for each constant h slice, it has a cubic-like shape (see Figures 21 and 22). On the manifold given by the graph $n = N(v; h)$, fixed points with $G_v = 0$ are those at which $N_v < 0$. The saddle-nodes correspond to the folds, or "knees" of $n = N(v; h)$; that is the points that satisfy $N_v = 0$. Therefore, the two-dimensional manifold $n = N(v; h)$ has two normally hyperbolic components (*i.e.* where the fixed points have eigenvalues with real part bounded away from zero), $\mathcal{S}_1$ and $\mathcal{S}_2$. $\mathcal{S}_1$ (resp. $\mathcal{S}_2$) corresponds to the points to the left (resp. right) of the left fold (resp. right fold) of $n = N(v, h)$. We can parameterize $\mathcal{S}_1$ and $\mathcal{S}_2$ by (n, h) as $v = V_1(n, h)$ and $v = V_2(n, h)$, respectively. Although the $\mathcal{S}_i$ are not overflowing invariant as defined, they can be made so by using a bump-function to modify of the flow outside their boundaries. See [32] for details. Therefore, (8.3) satisfies Assumption 1.

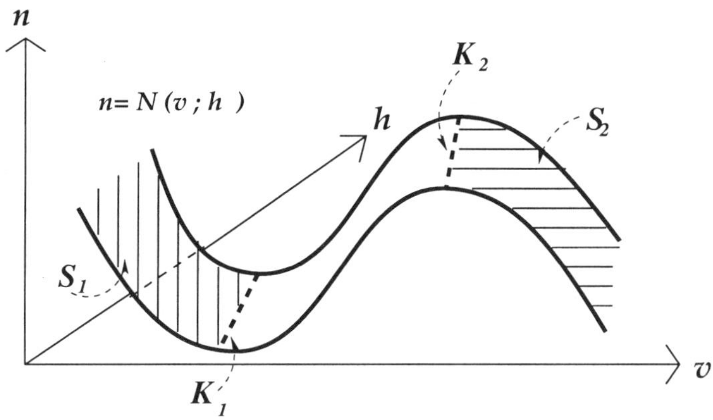

FIG. 22. *The graph of* $n = N(v;h)$ *in* (v,h,n) *space. The knees (or folds)* $\mathcal{K}_1$ *and* $\mathcal{K}_2$ *are curves of saddle-node fixed points of the fast system. The shaded surfaces, labeled as* $\mathcal{S}_1$ *and* $\mathcal{S}_2$ *correspond to the normally hyperbolic invariant manifolds.*

We now verify that Assumption 2 holds, even when $\varepsilon = 0$. In particular, we show that for all $\theta \in [0, \theta^*]$, there are one-dimensional submanifolds $TO_1(\theta)$ of $\mathcal{S}_1$ and $TD_2(\theta)$ of $\mathcal{S}_2$, such that every point in the submanifold $TO_1(\theta)$ is connected to a point in $TD_2(\theta)$, via a heteroclinic orbit. There are also a one-dimensional submanifolds $TO_2(\theta)$ of S_2 and $TD_1(\theta)$ of S_1, such that each point in $TO_2(\theta)$ is connected to $TD_1(\theta)$. We first work with the $\theta = 0$ case, then we carry out a more involved analysis to obtain the result for $\theta > 0$.

Consider (8.4) with $\theta = 0$:

$$\dot{v} = w$$
$$\dot{w} = G(v; n, h).$$

For each vector of slow variables $y = (n, h)$, this is a Hamiltonian system, with Hamiltonian

$$H(v, w; y) = \frac{1}{2}w^2 - \int_0^v G(r; y)dr.$$

Let $(v = v_{fp}(y), w = 0)$ be a fixed point of (8.4); then the value of the Hamiltonian there is $H(v_{fp}(y), 0; y) = -\int_0^{v_{fp}(y)} G(r; y)dr$. Thus, for $\theta = 0$, the two fixed points $(v^1_{fp}(y), 0; y) \in \mathcal{S}_1$ and $(v^2_{fp}(y), 0; y) \in \mathcal{S}_2$ are (doubly) connected if and only if they have the same energy, that is, if $\int_0^{v^1_{fp}(y)} G(r; y)dr = \int_0^{v^2_{fp}(y)} G(r; y)dr$ or $\int_{v^1_{fp}(y)}^{v^2_{fp}(y)} G(r; y)dr = 0$, for some y. (See Figure 23a). By continuous dependence on parameters, and by (**G2**), it is easy to see that $TO_1(0)$ is a one-dimensional curve that can be parameterized by h as $\{v = V_1(h, T(h)\}$, with $n = T(h)$ increasing with h. This automatically gives us $TD_2(0) = \{v = V_2(h, T(h))\}$. Note that for the $\theta = 0$ case, $TO_i(0) = TD_i(0)$, $i = 1, 2$. See Figure 24.

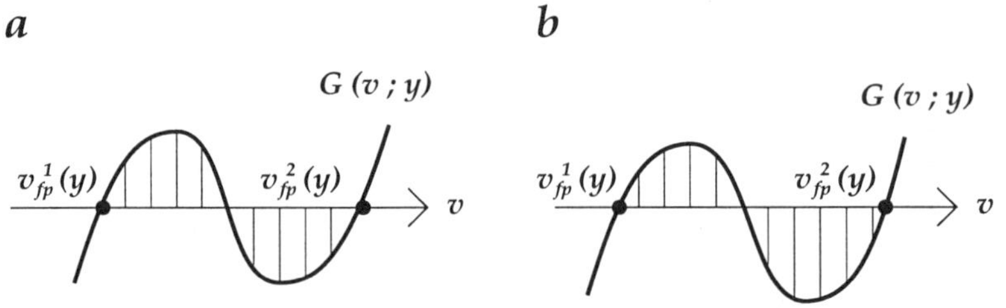

FIG. 23. **(a)** *Equal area rule. For $\theta = 0$, the fixed points $v^1_{fp}(y)$ and $v^2_{fp}(y)$ are (doubly) connected.* **(b)** *For $\theta > 0$, the area shaded must be < 0 for there to be a heteroclinic connection between the fixed points $v^1_{fp}(y)$ and $v^2_{fp}(y)$.*

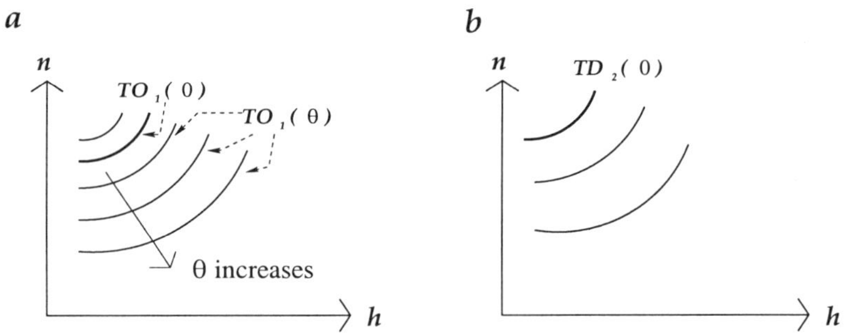

FIG. 24. **(a)** *Projection of the take off sets $TO_1(\theta) \subset S^1$ onto parameter space (n, y).* **(b)** *Projections of the touch down sets $TD_2(\theta) \subset S^2$ onto parameter space (n, y).*

Now, when $\theta \neq 0$, the system (8.4) is no longer Hamiltonian. Along trajectories, the rate of change of H is given by $\dot{H} = \theta w^2$. For $\theta > 0$, a heteroclinic connection going from a fixed point $(v^1_{fp}(y), 0; y) \in \mathcal{S}_1$ to a fixed point $(v^2_{fp}(y), 0; y) \in \mathcal{S}_2$ must increase its energy, so that $\int_{v^1_{fp}(y)}^{v^2_{fp}(y)} G(r; y)dr < 0$. It has been shown [2, 15] that if this condition is satisfied, there is a $\theta(y) > 0$ such that there is a heteroclinic connection between these two fixed points. Similarly, if $\int_{v^1_{fp}(y)}^{v^2_{fp}(y)} G(r; y)dr > 0$ there exists $\theta(y) > 0$ such that a heteroclinic connection from $(v^2_{fp}(y), 0; y) \in \mathcal{S}_2$ to $(v^1_{fp}(y), 0; y) \in \mathcal{S}_1$ exists. As in the $\theta = 0$ case above, using continuous dependence on parameters and **(G2)**, it can be shown that for fixed $\theta > 0$, there are one-dimensional manifolds $TO_1(\theta) \subset \mathcal{S}_1, TD_2(\theta) \subset \mathcal{S}_2$ whose points are connected via heteroclinic orbits of (8.4) going from $\mathcal{S}_1$ to $\mathcal{S}_2$. Similarly, there are one-dimensional manifolds $TO_2(\theta) \subset \mathcal{S}_2$ and $TD_1(\theta) \subset \mathcal{S}_1$, with the heteroclinic orbits going from $\mathcal{S}_2$ to $\mathcal{S}_1$ (See Figure 24 and the remark following Proposition 8.1).

An argument identical to that of Lemma 2 in [33] shows that the heteroclinic manifold formed by the union of these heteroclinic connections, arises from the transverse intersection of $W^u(S_i)$ and $W^s(\mathcal{S}_j)$. Thus, we have verified that Assumption 2 is satisfied by (8.3).

In order to verify Assumption 3, we will describe the slow flow on $\mathcal{S}_1$ and $\mathcal{S}_2$. As mentioned before, $\mathcal{S}_1$ can be parameterized as $v = V_1(n,h)$ in (v,n,h) space. Thus, the reduced slow flow is given by

$$(8.5)\qquad \begin{aligned} n' &= \frac{\gamma_n(V_1(n,h))}{\theta}\Big[n_\infty(V_1(n,h)) - n\Big] \\ h' &= \frac{\gamma_h(V_1(n,h))}{\theta}\Big[h_\infty(V_1(n,h)) - h\Big]. \end{aligned}$$

By condition (**NH1**), we know that the global fixed point $(V_0, 0, n_0, h_0)$ lies in $\mathcal{S}_1$. Thus, for $(n = n_0, h = h_0)$, $n' = 0, h' = 0$ in (8.5). The stability of this fixed point with respect to the slow flow is given by the sign of the determinant of the linearization of (8.5) at (n_0, h_0):

$$\begin{aligned} det J(n_0,h_0) = \ & -\frac{\gamma_n(V_0)\gamma_h(V_0)}{\theta^2}\left(\frac{\partial n_\infty}{\partial v}\bigg|_{V_0}\frac{\partial V_1}{\partial n}\bigg|_{(n_0,h_0)}\right. \\ & \left. + \frac{\partial h_\infty}{\partial v}\bigg|_{V_0}\frac{\partial V_1}{\partial h}\bigg|_{(n_0,h_0)} - 1\right). \end{aligned}$$

By (**NH1**), $\dfrac{\partial n_\infty}{\partial v}, \dfrac{\partial h_\infty}{\partial v} \ll 1$; so the determinant is positive, and since the trace is close to -2, the fixed point is stable.

There are two nullclines in $\mathcal{S}_1$, given by $n = n_\infty(V_1(n,h))$ and $h = h_\infty(V_1(n,h))$. They correspond to the intersection of the surfaces $n = n_\infty(v)$ and $h = h_\infty(v)$ with $\mathcal{S}_1$. Figure 25 shows the projection of $\mathcal{S}_1$ onto (n,h) space, together with the nullclines, the fixed point, and $\mathcal{K}_1$, the manifold of "knees" or saddle-nodes (for which $G_v(V_1(n,h),n,h) = 0$). A solution on $\mathcal{S}_1$ starting with $n > n_0, h < h_0$ evolves towards the fixed point; the magnitude of the vector field in the n direction (resp. the h direction) depends on the distance between the n-coordinate (resp. h-coordinate) and the n-nullcline (resp. h-nullcline).

One can perform a similar analysis for the flow on $\mathcal{S}_2$. The main difference is that there is no fixed point on $\mathcal{S}_2$. The n-nullcline is now located at values of n close to 1, while the h-nullcline is located at values of h close to 0. A solution on $\mathcal{S}_2$ evolves with its n-coordinate getting closer to the n-nullcline, and its h-coordinate getting closer to the h-nullcline. Solutions leave $\mathcal{S}_2$ through the curve of knees $\mathcal{K}_2$. See Figure 25.

The following proposition completes the picture for the slow flow on $\mathcal{S}_1$ and $\mathcal{S}_2$. It shows that Assumption 3 holds.

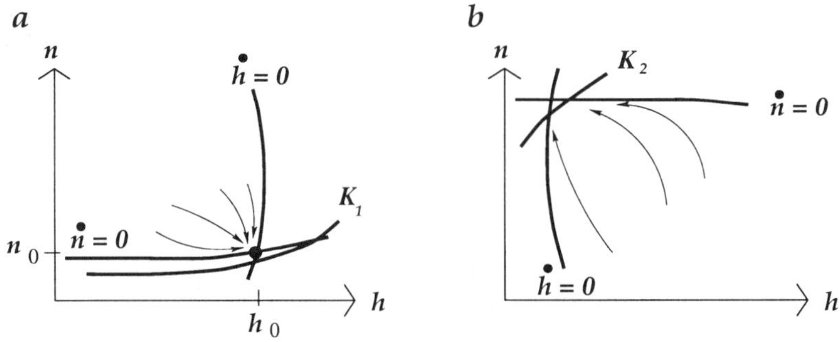

FIG. 25. **(a)** *Projection of the slow flow on S^1 onto (n,h) space. the curve labeled $\mathcal{K}_1$ corresponds to the projection of the curve of saddle-nodes that is the boundary of $\mathcal{S}^1$. The other two curves are the projections of the nullclines. Their intersection corresponds to the projection of the fixed point on $\mathcal{S}^1$.* **(b)** *Projection of the slow flow on $\mathcal{S}^2$ onto (n,h) space.*

PROPOSITION 8.1. *For $\theta \geq 0$, the one-dimensional manifolds $TO_i(\theta)$ and $TD_i(\theta)$, for $i = 1,2$ are transverse to the slow flow on the manifold $\mathcal{S}_i$.*

Proof. We will show this for $TO_1(\theta)$. The results for the others are similarly shown. Fix θ and let $y = (n,h)$, with $n > n_0, h < h_0$, be such that there exists a heteroclinic connection between the fixed points $(v^1_{fp}(y), 0; y) \in \mathcal{S}_1$ and $(v^2_{fp}(y), 0; y) \in \mathcal{S}_2$. See Figure 26a. Take $y_1 = y + \Delta y = (n + \Delta n, h + \Delta h)$ infinitesimally close to y, along the solution of (8.5) through y. Thus, $\Delta h > 0, \Delta n < 0$. We will show that for y_1 and this fixed value of θ, there cannot be a heteroclinic connection from $(v^1_{fp}(y_1), 0; y_1) \in \mathcal{S}_1$ to $(v^2_{fp}(y_1), 0; y_1) \in \mathcal{S}_2$.

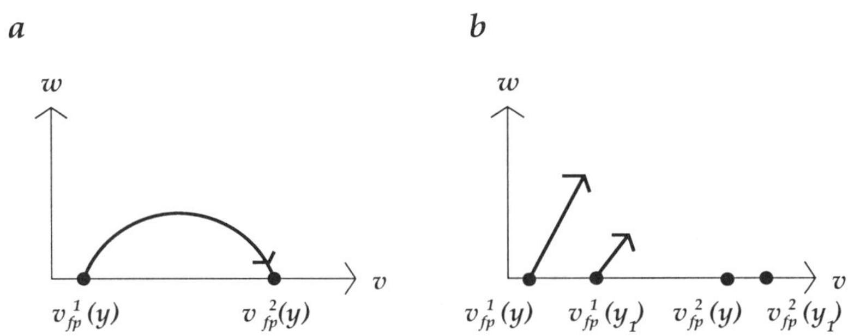

FIG. 26. **(a)** *A heteroclinic connection between the fixed points $v^1_{fp}(y) \in \mathcal{S}^1$ and $v^2_{fp}(y) \in \mathcal{S}^2$.* **(b)** *For $i = 1,2$ the fixed points $v^i_{fp}(y_1)$ must lie to the right of the fixed points $v^i_{fp}(y)$, respectively. The unstable eigenvector of $v^1_{fp}(y)$ has a larger slope than that of $v^1_{fp}(y_1)$.*

We first identify the relative location of the fixed points for y and y_1. We can write $v^1_{fp}(y_1) = V_1(y_1) \approx V_1(y) + \dfrac{\partial V_1}{\partial n}(y)\Delta n + \dfrac{\partial V_1}{\partial h}(y)\Delta h$. Since on $\mathcal{S}_1$, $\dfrac{\partial V_1}{\partial n} < 0$, $\dfrac{\partial V_1}{\partial h} > 0$, we have $v^1_{fp}(y_1) = v^1_{fp}(y) + \Delta_1 v$, with $\Delta_1 v > 0$. Similarly, $v^2_{fp}(y_1) = v^2_{fp}(y) + \Delta_2 v$ with $\Delta_2 v > 0$. See Figure 26b.

We next compute the unstable eigenvectors of $(v^1_{fp}(y), 0; y)$ and $(v^1_{fp}(y_1), 0; y_1)$. They are, respectively, $(1, \lambda_u(y))$ and $(1, \lambda_u(y_1))$, where $\lambda_u(y) = \dfrac{\theta + \sqrt{\theta^2 + 4G_v(v;y)}}{2}$. Since $\dfrac{\partial G_v}{\partial h} < 0$, $\dfrac{\partial G_v}{\partial n} > 0$ and $\dfrac{\partial G_v}{\partial v} < 0$ on $\mathcal{S}_1$, it follows that $\lambda_u(y_1) < \lambda_u(y)$. Therefore, the unstable eigenvector of $(v^1_{fp}(y), 0; y)$ has a larger slope than that of $(v^1_{fp}(y_1), 0; y_1)$. See Figure 26b.

Now, let $w^u(y;t)$ denote the w-coordinate of a point on $W^u_{loc}\Big(v^1_{fp}(y), 0; y\Big)$ and $w^u(y_1;t)$ denote the w-coordinate of a point on $W^u_{loc}\Big(v^1_{fp}(y_1), 0; y_1\Big)$. From the evolution equation for w, $\dot{w} = \theta w + G(v;y)$, we see that $|\dot{w}|$ depends on the distance to the cubic $w = -\dfrac{G(v;y)}{\theta}$ in the (v,w) plane. Since the cubic $w = -\dfrac{G(v;y_1)}{\theta}$ is above the cubic $w = -\dfrac{G(v;y)}{\theta}$ (see Figure 27a), for fixed v, $w^u(y_1;t)$ has smaller $\dot{w}$ than $w^u(y;t)$. This in turn implies that the graph of $W^u(v^1_{fp}(y_1), 0; y_1)$ is always below the graph of $W^u(v^1_{fp}(y), 0; y)$, at least while points on them have w-coordinate greater than or equal to zero (see Figure 27b). Therefore $W^u(v^1_{fp}(y_1), 0; y_1)$ cannot be a heteroclinic connection for this value of θ. This completes the proof of the proposition. □

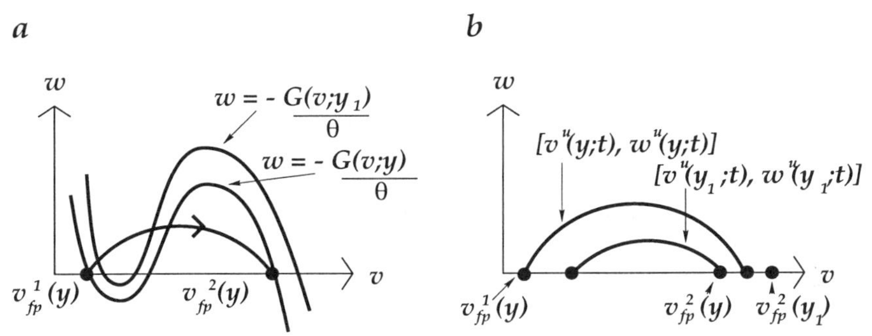

FIG. 27. **(a)** *The cubic* $w = -\frac{G(v;y_1)}{\theta}$ *is above the cubic* $= -\frac{G(v;y)}{\theta}$. **(b)** *The unstable manifold of* $v^1_{fp}(y_1)$,$W^u(v^1_{fp}(y_1), 0; y_1) = (v^u(y_1;t), w^u(y_1;t))$ *is always below to that of* $v^1_{fp}(y)$.

REMARK 8.1. An argument similar to the proof of this proposition shows that the manifolds $TO_i(\theta), TD_i(\theta)$ can be parameterized by h, with their n coordinate increasing as h increases, as is illustrated in 24.

So far, we have shown that Assumptions 1–3 hold for (8.3). The last hypothesis necessary for the use of Theorem 1 is the construction of the two-dimensional singular manifold. Recall θ^* denotes the value of θ for which there exists a singular heteroclinic connection of (8.4) between the global fixed point $(V_0, 0; n_0, h_0) \in \mathcal{S}_1$ and a point in $\mathcal{S}_2$. Pick $0 < \theta < \theta^*$ such that the distance between the submanifolds $TD_1(\theta)$ and $TO_1(\theta)$ is of $\mathcal{O}(1)$. The following proposition shows that every solution of the slow flow (8.5) that starts in a sufficiently small segment of $TD_1(\theta)$ reaches $TO_1(\theta)$. This follows because the slow flow, in addition of being transverse to the manifolds $TD_1(\theta), TO_1(\theta)$, is contracting, in the sense that the distance between trajectories starting at two nearby points in $TD_1(\theta)$ decreases with time.

PROPOSITION 8.2. *Let $p \in TD_i(\theta)$ and q be a tangent vector of $TD_i(\theta)$ at p. Then $|q(t)| < |q(t')|$ for $t > t'$.*

Proof. We prove this for $i = 1$. The proof for $i = 2$ is similar. The result follows from the variational equations of (8.5):

$$\begin{aligned}
\delta n' &= \frac{\gamma_n}{\theta}\left(\frac{\partial n_\infty}{\partial v}\frac{\partial V_1}{\partial n}\delta n + \frac{\partial n_\infty}{\partial v}\frac{\partial V_1}{\partial h}\delta h - \delta n\right) \\
&\quad + [n_\infty - n]\frac{1}{\theta}\left(\frac{\partial \gamma_n}{\partial v}\frac{\partial v_1}{\partial n}\delta n + \frac{\partial \gamma_n}{\partial v}\frac{\partial v_1}{\partial h}\delta h\right) \\
\delta h' &= \frac{\gamma_h}{\theta}\left(\frac{\partial h_\infty}{\partial v}\frac{\partial V_1}{\partial n}\delta n + \frac{\partial h_\infty}{\partial v}\frac{\partial V_1}{\partial h}\delta h - \delta h\right) \\
&\quad + [h_\infty - h]\frac{1}{\theta}\left(\frac{\partial \gamma_h}{\partial v}\frac{\partial v_1}{\partial n}\delta n + \frac{\partial \gamma_h}{\partial v}\frac{\partial v_1}{\partial h}\delta h\right)
\end{aligned}$$

Using (**NH2**) to compare the relative contribution of each of the terms, we can see that the dominant ones are $-\frac{\gamma_n}{\theta}\delta n$ in the first equation and $-\frac{\gamma_h}{\theta}\delta h$ in the second one. The variational system can therefore be written as

$$\begin{aligned}
\delta n' &= -a(h,n)\delta n + \eta_1(n,h)\delta h \\
\delta h' &= -b(h,n)\delta h + \eta_2(n,h)\delta n
\end{aligned}$$

where $a \approx \frac{\gamma_n}{\theta} > 0, b \approx \frac{\gamma_h}{\theta} > 0$ and $|\eta_i| \ll |a|, |b|$ for $i = 1, 2$. By the theory of linear systems [21] we have that δn and δh decay exponentially. If we let $q(t) = (\delta n(t), \delta h(t))$, with $\delta n(0), \delta h(0) > 0$, we obtain the claimed result.

The construction of the two-dimensional singular periodic manifold now follows. Let $\hat{T}D_1(\theta)$ be the largest one-dimensional subset of $TD_1(\theta)$ such that its image under the slow map f_s^1 satisfies $f_s^1(\hat{T}D_1(\theta)) \subset TO_1$. By Proposition 8.2, the set $\hat{T}D_1(\theta)$ exists. By the properties of the slow flow on $\mathcal{S}_1$ and $\mathcal{S}_2$, we have that the image of $\hat{T}D_1(\theta)$ under the map F, where $F(\hat{T}D_1(\theta)) = f_f^2 f_s^2 f_f^1 f_s^1(\hat{T}D_1(\theta))$, is a one-dimensional subset of $\hat{T}D_1(\theta)$. Furthermore, by Proposition 8.2, F is contracting. Therefore, for each value of $\theta \in (0, \theta^*)$, we have a locally unique periodic orbit of (8.3) by Theorem 2. Thus, Theorem 5 is proven. □

9. Proof of MELESE. In this section, we prove the modified version of the Exchange Lemma with Exponentially Small Error (MELESE), stated as Lemma 3 in Section 5. We concentrate on showing the $\mathcal{C}^1$ closeness of the tracked manifold and the appropriate submanifold of the slow manifold's unstable manifold, since the proof of the $\mathcal{C}^0$ closeness is essentially identical to that of Proposition 3.1 in [31] or Proposition 3.2 in [56].

We are dealing with the tracking a $(k+\sigma)$-dimensional invariant manifold $\mathcal{L}_\varepsilon$, where $1 \le \sigma \le n$, of a singularly perturbed system of the form (5.1) in the neighborhood of a slow manifold $\mathcal{S}^\varepsilon$. We use the Fenichel normal form in this neighborhood:

$$\begin{aligned} \dot{a} &= \Lambda(a, b, y; \varepsilon)a \\ \dot{b} &= \Gamma(a, b, y; \varepsilon)b \\ \dot{y} &= \varepsilon\left(m(y, \varepsilon) + h(a, b, y; \varepsilon)ab\right), \end{aligned} \tag{9.1}$$

where $a \in \mathbb{R}^{\mathbf{k}}, \mathbf{b} \in \mathbb{R}^{\ell}, \mathbf{y} \in \mathbb{R}^{\mathbf{n}}$ on the compact set $\mathcal{B} = \{(a, b, y) : |a| \le \Delta, |b| \le \Delta, y \in U\}$. Here, Λ and Γ are square matrices of dimension k and ℓ, respectively, m is a vector in $\mathbb{R}^{\mathbf{n}}$ and h is a tensor of rank three. The i-th component of $h(a, b, y; \varepsilon)ab$ is $\sum_{m=1}^{\ell} \sum_{j=1}^{k} h_{ijm} a_j b_m$.

To obtain the $\mathcal{C}^1$ closeness results, we need to follow the $(k+\sigma)$-dimensional tangent space of $\mathcal{L}_\varepsilon$. Tangent vectors of $\mathcal{L}_\varepsilon$ evolve according to the variational equations of (9.1):

$$\begin{aligned} (\delta a_i)' &= \Lambda_i \cdot \delta a + a \cdot L_i \\ (\delta b_i)' &= \Gamma_i \cdot \delta b + b \cdot G_i \\ (\delta y_i)' &= \varepsilon\Big[\nabla_y m_i \cdot \delta y + \sum_{j=1}^{k} \sum_{m=1}^{\ell} \Big\{h_{ijm} \delta a_j b_m + h_{ijm} a_j \delta b_m \\ &\qquad + (\nabla_z h_{ijm} \cdot \delta z) a_j b_m\Big\}\Big], \end{aligned} \tag{9.2}$$

where $z \equiv (a, b, y)$, Λ_i, Γ_i are the i-th rows of Λ and Γ, respectively, $L_{ij} \equiv \nabla \Lambda_{ij} \cdot \delta z$ and $G_{ij} \equiv \nabla \Gamma_{ij} \cdot \delta z$. Then, the evolution of the tangent spaces can be studied using Plücker coordinates (see Section 2 of [56]). In general, if $\mathcal{W}$ is a $(k+\sigma)$ dimensional subspace of $\mathbb{R}^{\mathbf{k}+\ell+\mathbf{n}}$, and if $v_1, ..., v_{k+\sigma}$ a set of spanning vectors of $\mathcal{W}$, the Plücker coordinates of $\mathcal{W}$ are the maximal minors of the matrix whose columns are the vectors v_i. Geometrically, these minors correspond to the projected "volumes" of the parallelepiped formed by the $v_1, ..., v_{k+\sigma}$ onto the different $(k+\sigma)$-dimensional subspaces of $\mathbb{R}^{\mathbf{k}+\ell+\mathbf{n}}$.

We obtain Plücker coordinates for the tangent spaces of $\mathcal{L}_\varepsilon$ in system (9.1) as follows. Let $\{e_1^a, ..., e_k^a, e_1^b, ..., e_\ell^b, e_1^y, ..., e_n^y\}$ be an ordered basis for $\mathbb{R}^{k+\ell+n}$ in Fenichel coordinates. Let $\{r_1, ..., r_\sigma\}$ be a subset of $\{1, ..., n\}$, containing σ elements of this set in an ordered fashion. Let $\hat{\mathcal{Z}}$ contain all the $(k+\sigma)$ forms corresponding to the k expanding (a) directions and σ of the slow (y) directions: $\delta a_1 \wedge ... \wedge \delta a_k \wedge \delta y_{r_1} \wedge ... \wedge \delta y_{r_\sigma}$. $\hat{\mathcal{X}}$ will contain all the other $(k+\sigma)$ forms. Let $q \in \mathcal{L}_\varepsilon$ and $V(q) = \{v_1, ..., v_{k+\sigma}\}$ be a set of vectors that span $T_q\mathcal{L}_\varepsilon$. The $\hat{\mathcal{Z}}$ part of the Plücker coordinates of the $(k+\sigma)$-dimensional linear subspace $T_q\mathcal{L}_\varepsilon$, $\hat{\mathcal{Z}}(T_q\mathcal{L}_\varepsilon)$, is obtained by evaluating all the $(k+\sigma)$ forms in $\hat{\mathcal{Z}}$ on the set $V(q)$. Each $\delta a_1 \wedge ... \wedge \delta a_k \wedge \delta y_{r_1} \wedge ... \wedge \delta y_{r_\sigma}(V(q))$ is obtained from the matrix whose column vectors are the elements of $V(q)$ by computing the $(k+\sigma)$ minor resulting from retaining only the $k+\sigma$ rows corresponding to the $a_1, ..., a_k, y_{r_1}, ..., y_{r_\sigma}$ rows. Similarly for $\hat{\mathcal{X}}(T_q\mathcal{L}_\varepsilon)$.

For a trajectory $q(t)$ of (9.1) inside $\mathcal{B}$, the evolution of the tangent space $T_{q(t)}\mathcal{L}_\varepsilon$ is given by the evolution of the $(k+\sigma)$ forms. For brevity, we will only state the evolution equations for the $\hat{\mathcal{Z}}$ and $\hat{\mathcal{X}}$ components of the Plücker coordinates. Their derivation can be checked by hand, or obtained in a manner similar to [31, 32, 56]:

$$\begin{aligned} \hat{\mathcal{Z}}' &= (Tr\Lambda + \Phi + D^{(\sigma)}m + \Theta_1)\hat{\mathcal{Z}} + \Theta_2\hat{\mathcal{X}} \\ \hat{\mathcal{X}}' &= (B + \Psi_2)\hat{\mathcal{X}} + \Psi_1\hat{\mathcal{Z}} \end{aligned} \tag{9.3}$$

where $\Phi \equiv \sum_{i,s=1}^k a_s \dfrac{\partial \Lambda_{is}}{\partial a_i}$ and each element of $D^{(\sigma)}m(y,\varepsilon)$ is of the form $\varepsilon \sum_{t=1}^\sigma \dfrac{\partial m_{rt}}{\partial y_{r_t}}$. Each of the terms Θ_i, Ψ_i are matrices of appropriate order; in particular, every term of Θ_1 is multiplied by a term of the form $\varepsilon a_i b_j$ with $i \in \{1, ..., k\}, j \in \{1, ..., \ell\}$, every term of Θ_2 is multiplied by a term a_i with $i \in \{1, ..., k\}$. Every term of Ψ_1 is multiplied by a b_j for $j \in \{1, ..., \ell\}$, and every term of Ψ_2 is multiplied by an a_i, or a b_j, or a sum of them, for $i \in \{1, ..., k\}, j \in \{1, ..., \ell\}$. The term $B = B(a, b, y; \varepsilon)$ is a matrix that satisfies $|| \exp\{\int_s^t (B - Tr\Lambda) d\xi\} || \le Me^{-\mu(t-s)}$, for some $M \ge 1$ and $\mu > 0$. Furthermore, for ε and Δ sufficiently small, the following bounds hold:

$$\begin{aligned} |\Theta_1| &\le C_{\theta_1}\varepsilon|a||b| \\ |\Theta_2| &\le C_{\theta_2}|a| \\ |\Phi| &\le C_\phi|a| \\ |\Psi_1| &\le C_{\psi_1}|b| \\ |\Psi_2| &\le C_{\psi_2}(|a| + |b|). \end{aligned} \tag{9.4}$$

We now perform a time-dependent rescaling in equations (9.3), see [56]. Let

$$\mathcal{Z} = \exp\left(-\int_0^t (Tr\Lambda + \Phi)(z(\xi, \varepsilon))d\xi\right)\hat{\mathcal{Z}}$$

$$\mathcal{X} = \exp\left(- \int_0^t (Tr\Lambda + \Phi)(z(\xi, \varepsilon))d.\xi \right) \hat{\mathcal{X}},$$

Then (9.3) becomes:

$$\begin{aligned} \mathcal{Z}' &= (D^{(\sigma)}m + \Theta_1)\mathcal{Z} + \Theta_2 \mathcal{X} \\ \mathcal{X}' &= (B - Tr\Lambda - \Phi + \Psi_2)\mathcal{X} + \Psi_1 \mathcal{Z}. \end{aligned} \tag{9.5}$$

The solution of (9.5) will provide us with the information on the evolution of the tangent spaces of $\mathcal{L}_\varepsilon$ we require. Our goal is to show that for a trajectory $q(t)$ in $\mathcal{L}_\varepsilon$ that exits the box $\mathcal{B}$ after an $\mathcal{O}(\frac{1}{\varepsilon})$ time T, $\mathcal{X}(T) \equiv \mathcal{X}(T_{q(T)}\mathcal{L}_\varepsilon)$ is exponentially small. This implies that $T_{q(T)}\mathcal{L}_\varepsilon$ has exponentially small projections onto any of the $(k+\sigma)$-dimensional subspaces of the tangent space of $W^s_{loc}(\mathcal{S}^\varepsilon)$. Furthermore, we will also show that the $\mathcal{Z}$ component of the Plücker coordinates of $T_{q(T)}\mathcal{L}_\varepsilon$ is exponentially close to the $\mathcal{Z}$ component of the Plücker coordinates of a tangent space of a submanifold $\tilde{\mathcal{L}}_\varepsilon$ of $W^u_{loc}(\mathcal{S}^\varepsilon)$. $\tilde{\mathcal{L}}_\varepsilon$ corresponds to the forward trajectories that start from the projection of the set of initial conditions K_ε onto $W^u_{loc}(\mathcal{S}^\varepsilon)$. To achieve this goal, in Section 9.1 we obtain several estimates concerning this system and a related, uncoupled system. In Section 9.2 we obtain a system similar to (9.5), but designed to track the tangent spaces of $\tilde{\mathcal{L}}_\varepsilon$. We also compare the Plücker coordinates of $\tilde{\mathcal{L}}_\varepsilon$ and those of $\mathcal{L}_\varepsilon$. Finally, in Section 9.3 we interpret these results in terms of the tracked manifold $\mathcal{L}_\varepsilon$ and its projection onto $W^u_{loc}(\mathcal{S}^\varepsilon)$, thus proving MELESE.

9.1. Basic estimates. The first estimates we need concern trajectories $z(t) = (a(t), b(t), y(t))$ that lie in $\mathcal{B}$ during the interval $[t_0, t_1]$ of system (9.1). In particular, Proposition 9.1 below gives a lower bound on the exponential growth of the a-component and an upper bound on the exponential decay of the b-component. We define:

$$\begin{aligned} \lambda_0 &= \inf_{y \in U} \mathrm{Re}(\mathrm{spec}\Lambda(0,0,y;0)) \\ \gamma_0 &= \sup_{y \in U} \mathrm{Re}(\mathrm{spec}\Gamma(0,0,y;0)), \end{aligned} \tag{9.6}$$

and note that both λ_0, γ_0 are bounded away from zero since $\mathcal{S}^\varepsilon$ is normally hyperbolic. We have the following

PROPOSITION 9.1. *For $\lambda < \lambda_0, \gamma > \gamma_0$, there exist $\varepsilon_0, \Delta_0 > 0, K_a$, $K_b > 0$ such that if $\varepsilon < \varepsilon_0, \delta < \Delta_0$, then for $t \in [t_0, t_1]$*
(i) $|a(t)| \geq K_a |a(t_0)| e^{\lambda(t-t_0)}$, *and* (ii) $|b(t)| \leq K_b |b(t_0)| e^{\gamma(t-t_0)}$.

This is Lemma 3.1 in [31] or Corollary 2.4 in [56], so we omit its proof.

Next, we turn our attention to system (9.5). Since we would like to know the evolution of the $\mathcal{Z}$ component when the trajectory $z(t)$ evolves entirely on $W^u_{loc}(\mathcal{S}^\varepsilon)$, we know $b(t) = 0$. This forces $\Theta_1 = 0$ in the first equation above. So let $\mathcal{P}$ denote the $\mathcal{Z}$ component of the Plücker coordinates for a trajectory on an $(k+\sigma)$ dimensional manifold that evolves on $W^u_{loc}(\mathcal{S}^\varepsilon)$. We append the equation of $\mathcal{P}$ to (9.5) to obtain:

$$\begin{aligned}\mathcal{Z}' &= (D^{(\sigma)}m + \Theta_1)\mathcal{Z} + \Theta_2\mathcal{X}\\ \mathcal{X}' &= (B - Tr\Lambda - \Phi + \Psi_2)\mathcal{X} + \Psi_1\mathcal{Z}\\ \mathcal{P}' &= (D^{(\sigma)}m)\mathcal{P} + \Theta_2\mathcal{X}.\end{aligned}$$

We first study an uncoupled, linear system that is simpler than the one above:

$$\begin{aligned}\mathcal{Z}' &= (D^{(\sigma)}m + \Theta_1)\mathcal{Z}\\ \mathcal{X}' &= (B - Tr\Lambda - \Phi + \Psi_2)\mathcal{X}\\ \mathcal{P}' &= (D^{(\sigma)}m)\mathcal{P}\end{aligned}\tag{9.7}$$

The estimates obtained from (9.7) will be useful later on when we study the full system (9.5). Now let

$$\mu_0 = \inf_{y\in U}\left(\operatorname{spec}\left(Tr\Lambda(0,0,y;0) - B(0,0,y;0)\right)\right)$$

and note that $\operatorname{spec}(D^{(\sigma)}m(y;0)) = \{0\}$. The following proposition gives estimates on solutions of (9.7):

PROPOSITION 9.2. *For $\alpha > 0, \mu < \mu_0$ there exist $\Delta_0, \varepsilon_0, K_z, K_p, K_x > 0$ such that if $\Delta < \Delta_0, \varepsilon < \varepsilon_0$ and $z(t), t \in [0,T]$ is a trajectory of (9.1) inside $\mathcal{B}$, then solutions of (9.7) satisfy:*
(i) $|\mathcal{Z}(t)| \le K_z|\mathcal{Z}(0)|e^{\alpha t}$, *and* (ii) $|\mathcal{X}(t)| \le K_x|\mathcal{X}(0)|e^{-\mu t}$.
Furthermore, if $z(t)$ evolves on $W^u_{loc}(\mathcal{S}^\varepsilon)$, then (iii) $|\mathcal{P}(t)| \le K_p|\mathcal{P}(0)|e^{\alpha t}$.

This is essentially Proposition 2.7 in [56], so we omit its proof. We will now show that the behavior of the solutions of (9.5) is similar to that of the solutions of (9.7):

PROPOSITION 9.3. *For any $\alpha > 0, \mu < \mu_0$, $|\gamma| > \beta \ge \alpha$ and $-\kappa < \min\{\beta - |\gamma|, -\mu\}$, there exist $\Delta_0 > 0, \varepsilon_0 > 0$ such that for fixed $\Delta < \Delta_0$, there exist constants $\tilde{B}_1, \tilde{B}_2 > 0$ such that if $t \in [0,T]$, where $T = \mathcal{O}(\frac{1}{\varepsilon})$ and $\mathcal{Z}, \mathcal{X}$ evolve according to (9.5) then, as long as $\mathcal{Z}(0)$ and $\mathcal{X}(0)$ are $\mathcal{O}(1)$ and the trajectory $z(t)$ stays in $\mathcal{B}$*
(i) $|\mathcal{Z}(t)| \le \tilde{B}_1 e^{\alpha t}$, *and* (ii) $|\mathcal{X}(t)| \le \tilde{B}_2 e^{-\kappa t}$.

Proof. Consider first the solution for the $\mathcal{X}$ equation:

$$\begin{aligned}\mathcal{X}(t)\exp&\left\{-\int_0^t (B - Tr\Lambda - \Phi + \Psi_2)ds\right\}\\ &= \mathcal{X}(0) + \int_0^t \Psi_1\mathcal{Z}\exp\left\{-\int_0^r (B - Tr\Lambda - \Phi + \Psi_2)ds\right\}dr.\end{aligned}$$

From estimate (ii) in Proposition 9.2, we get a bound on the integrating factor; from (9.4) we have $|\Psi_1| \le C_{\psi_1}|b|$, and we can bound b by (ii) of

Proposition 9.1. Thus, we have

$$|\mathcal{X}(t)| \leq K_x \left\{ |\mathcal{X}(0)|e^{-\mu t} + C_{\psi_1} K_b |b(0)| \int_0^t |\mathcal{Z}| e^{-\gamma r} e^{-\mu(t-r)} dr \right\}.$$

Now, assume that $\mathcal{Z}$ satisfies the following a-priori bound:

$$|\mathcal{Z}(t)| \leq B_1 e^{\beta t} \tag{9.8}$$

for some $B_1 > 0$ and β such that $|\gamma| > \beta \geq \alpha > 0$, where α is as in Proposition 9.2. Substituting (9.8) in the equation above we have

$$\begin{aligned} |\mathcal{X}(t)| &\leq K_x \left\{ |\mathcal{X}(0)|e^{-\mu t} + B_1 C_{\psi_1} K_b |b(0)| e^{-\mu t} \int_0^t e^{(\mu+\beta-|\gamma|)r} dr \right\} \\ &= K_x \left\{ |\mathcal{X}(0)|e^{-\mu t} + B_1 C_{\psi_1} K_b |b(0)| e^{-\mu t} \frac{e^{(\mu+\beta-|\gamma|)t} - 1}{\mu + \beta - |\gamma|} \right\} \\ &= K_x \left\{ \left(|\mathcal{X}(0)| - \frac{B_1 C_{\psi_1} K_b |b(0)|}{\mu + \beta - |\gamma|} \right) e^{-\mu t} + \frac{B_1 C_{\psi_1} K_b |b(0)|}{\mu + \beta - |\gamma|} e^{(\beta - |\gamma|)t} \right\}. \end{aligned}$$

Now, recalling the definition of κ, we get

$$|\mathcal{X}(t)| \leq C e^{-\kappa t} \tag{9.9}$$

for some $C > 0$, provided $|\mathcal{X}(0)|$ is $\mathcal{O}(1)$.

Next, we will show that the a-priori bound (9.8) holds. We solve the equation for $\mathcal{Z}$ in (9.5) and use the bounds (i) of Proposition 9.2 for the integrating factor $e^{\int_0^t (D^{(\sigma)} m + \Theta_1) ds}$, as well as the estimate $|\Theta_2| \leq C_{\theta_2} |a|$ to get

$$|\mathcal{Z}(t)| \leq K_z \left\{ |\mathcal{Z}(0)| e^{\alpha t} + e^{\alpha t} C_{\theta_2} \int_0^t |a| |\mathcal{X}(s)| e^{-\alpha s} ds \right\}.$$

Substituting (9.9) and $|a| \leq \Delta$ into the above equation gives:

$$|\mathcal{Z}(t)| \leq K_z \left\{ |\mathcal{Z}(0)| e^{\alpha t} + \underline{C} \Delta e^{\alpha t} \int_0^t e^{-(\kappa+\alpha)s} ds \right\},$$

which upon integration yields

$$\begin{aligned} |\mathcal{Z}(t)| &\leq K_z \left\{ |\mathcal{Z}(0)| e^{\alpha t} + \frac{\underline{C} \Delta}{-(\kappa + \alpha)} e^{\alpha t} (e^{-(\kappa+\alpha)t} - 1) \right\} \\ &\leq B_1 e^{\beta t} \end{aligned}$$

for some $B_1 > 0$, since $\beta > \alpha$ and $|\mathcal{Z}(0)|$ is $\mathcal{O}(1)$. Hence, the a-priori bound (9.8) holds, and the Proposition is proven. □

9.2. Tracking of tangent planes. Consider two trajectories inside the box $\mathcal{B}$: $q(t)$ and $\tilde{q}(t)$. Furthermore, assume that $\tilde{q}(t)$ evolves entirely on $W^u_{loc}(\mathcal{S}^\varepsilon)$. If $\tilde{q}(t)$ is on a $(k+\sigma)$-dimensional invariant manifold, we can obtain the evolution equations for the tangent planes of this manifold along $\tilde{q}(t)$ by making $b=0$ in Θ_1 and Ψ_1 in equations (9.5). Recall $\mathcal{P}$ denotes the $\mathcal{Z}$ part of the Plücker coordinates of this invariant manifold. Let $\mathcal{Q}$ denote the $\mathcal{X}$ part. By (9.5), the evolution of $\mathcal{P}$ and $\mathcal{Q}$ is given by:

$$\begin{aligned} \mathcal{P}' &= (D^{(\sigma)}m)\mathcal{P} + \Theta_2\mathcal{Q} \\ \mathcal{Q}' &= (B - Tr\Lambda - \Phi + \Psi_2)\mathcal{Q}. \end{aligned} \tag{9.10}$$

The uncoupled, linear part

$$\begin{aligned} \mathcal{P}' &= (D^{(\sigma)}m)\mathcal{P} \\ \mathcal{Q}' &= (B - Tr\Lambda - \Phi + \Psi_2)\mathcal{Q}. \end{aligned} \tag{9.11}$$

satisfies the estimates (ii) and (iii) of Proposition 9.2, since the equation for Q in (9.10) is identical to that for $\mathcal{X}$ in (9.7). Thus, we have that $|\mathcal{Q}(t)| \le K_q|\mathcal{Q}(0)|e^{-\mu t}$ for $K_q > 0$ and $\mathcal{Q}$ satisfying the above equation. Combining this estimate with estimate (ii) of Proposition 9.3, we have that as long as $|\mathcal{Q}(0)|$ and $|\mathcal{X}(0)|$ are $\mathcal{O}(1)$ and $\mathcal{Q}$, $\mathcal{X}$ obey (9.10) and (9.5) respectively, then

$$|\mathcal{Q}(t) - \mathcal{X}(t)| \le B_3 e^{-\kappa t} \tag{9.12}$$

where $B_3 > 0$ and κ, t are as in Proposition 9.3.

We will now estimate $|\mathcal{Z}(t) - \mathcal{P}(t)|$. In order to do so, assume that the initial y-coordinates of trajectories $q(t)$ and $\tilde{q}(t)$ are exponentially close: $|y(q(0)) - y(\tilde{q}(0))| = \mathcal{O}(e^{-\hat{\lambda}T})$, where $\hat{\lambda} > \hat{\kappa} + \hat{\alpha}, \hat{\alpha} > 0$, $\hat{\kappa} < \min\{|\gamma_0|, \lambda_0\}$, and $T = \mathcal{O}(\frac{1}{\varepsilon})$. Recall that $\tilde{q}(t)$ evolves entirely on $W^u_{loc}(S^\varepsilon)$. We will now compare the evolution of $\mathcal{Z}$ along the trajectory $q(t)$ with that of $\mathcal{P}$ along the trajectory $\tilde{q}(t)$. From (9.5) and (9.10) we have:

$$\begin{aligned} \mathcal{Z}' - \mathcal{P}' = &\left(D^{(\sigma)}m(y;\varepsilon)\right)(\mathcal{Z} - \mathcal{P}) + (D^{(\sigma)}m(y;\varepsilon) - D^{(\sigma)}m(\tilde{y};\varepsilon))\mathcal{P} \\ &+ \Theta_1(q,\varepsilon)\mathcal{Z} + \Theta_2(q,\varepsilon)(\mathcal{X} - \mathcal{Q}) + (\Theta_2(q,\varepsilon) - \Theta_2(\tilde{q},\varepsilon))\mathcal{Q} \end{aligned} \tag{9.13}$$

Using this equation, we prove in the following lemma that

$$|\mathcal{Z}(t) - \mathcal{P}(t)| = |\mathcal{Z}(0) - \mathcal{P}(0)|K_p e^{\alpha t} + \tilde{B}_4 e^{-\tilde{\kappa} t} \tag{9.14}$$

where $-\tilde{\kappa} > \max\{-|\gamma| + \alpha, \kappa, \lambda + \alpha, -\mu + \alpha, -\hat{\lambda} + \alpha\}$ and $\alpha > 0$ can be made as small as we want.

LEMMA 4. *Let $|\mathcal{X}(0)|, |\mathcal{Z}(0)|, |\mathcal{P}(0)|$ and $|\mathcal{Q}(0)|$ be the $\mathcal{O}(1)$ initial conditions of the Plücker coordinates $\mathcal{X}, \mathcal{Z}$ that evolve according to (9.5) for a trajectory $q(t)$ inside the box $\mathcal{B}$, and $\mathcal{P}, \mathcal{Q}$ that evolve according to (9.10)* *for a trajectory* $\tilde{q}(t)$ *evolving on* $W^u_{loc}(\mathcal{S}^\varepsilon)$ *inside* $\mathcal{B}$ *. Let* $|y(0) - \tilde{y}(0)| \leq \hat{K}e^{-\hat{\lambda}T}$ *where* $\hat{\lambda} > \hat{\kappa} + \alpha$*,* $\alpha > 0$ *and* $\hat{\kappa} < \min\{\lambda_0, |\gamma_0|\}$*. If* $|\mathcal{Z}(0) - \mathcal{P}(0)|$ *is exponentially small, then for* $T = \mathcal{O}(\frac{1}{\varepsilon})$ *we have:* $|\mathcal{X}(T)|, |\mathcal{Q}(T)|$ *and* $|\mathcal{Z}(T) - \mathcal{P}(T)|$ *are exponentially small.*

Proof. We have already shown that $|\mathcal{X}(T)|$ and $\mathcal{Q}(T)|$ are exponentially small. To show $|\mathcal{Z}(T) - \mathcal{P}(T)|$ is exponentially small, we use equation (9.14). We obtain bounds for the solution of this equation from the bound on the integrating factor ($|e^{-\int_0^t D^{(\sigma)}m ds}| \leq K_p e^{\alpha t}$, from (iii) of Proposition 9.2, and the bounds given in (9.4):

$$|\mathcal{Z}(t) - \mathcal{P}(t)| \leq |\mathcal{Z}(0) - \mathcal{P}(0)| K_p e^{\alpha t} + e^{\alpha t} \int_0^t \Big\{ R|\mathcal{P}|y - \tilde{y}| + C_{\theta_1}\varepsilon|a||b||\mathcal{Z}| + C_{\theta_2}|a||\mathcal{X} - \mathcal{Q}| + C_{\theta_2}|a||\mathcal{Q}| + C_{\theta_2}|\tilde{a}||\mathcal{Q}| \Big\} e^{-\alpha r} dr$$

where we have used $|D^{(\sigma)}m(y;\varepsilon) - D^{(\sigma)}m(\tilde{y};\varepsilon)| \leq R|y - \tilde{y}|$ for $R > 0$. Now, we are assuming that the initial y-coordinates of the trajectories are exponentially close $|y(0) - \tilde{y}(0)| = \mathcal{O}(e^{-\hat{\lambda}T})$, and one can show that as long as they stay in the box $\mathcal{B}$ they remain exponentially close: $|y(t) - \tilde{y}(t)| \leq \tilde{R}e^{-\hat{\lambda}T}e^{(\hat{\lambda}-\hat{\kappa})t}$, for $\hat{\kappa} < \min\{\lambda_0, |\gamma_0|\}$ and $\hat{\lambda} > \hat{\kappa} + \hat{\alpha}$, for any $\hat{\alpha} > 0$. See Proposition 3.2 in [56]. Using this in the first term of the integrand above, the bounds from Propositions 9.1 and 9.3, and the fact that $|\mathcal{P}| \leq \tilde{B}_3 e^{\alpha t}$, we get

$$\begin{aligned} |\mathcal{Z}(t) - \mathcal{P}(t)| \leq\ & |\mathcal{Z}(0) - \mathcal{P}(0)| K_p e^{\alpha t} \\ & + e^{\alpha t} \hat{R}\tilde{B}_3 K_p \int_0^t e^{-\hat{\lambda}T} e^{-(\hat{\lambda}-\hat{\kappa})r} dr \\ & + e^{\alpha t} C_{\theta_1} \varepsilon \tilde{B}_1 K_p \int_0^t \frac{\Delta}{K_a} e^{-\lambda(t-r)} K_b e^{-|\gamma| r} dr \\ & + e^{\alpha t} C_{\theta_2} B_3 K_p \int_0^t \frac{\Delta}{K_a} e^{-\lambda(t-r)} e^{-\kappa r} e^{\alpha r} dr \\ & + 2e^{\alpha t} C_{\theta_2} K_q |\mathcal{Q}(0)| \int_0^t \frac{\Delta}{K_a} e^{-\lambda(t-r)} e^{-\mu r} e^{\alpha r} dr, \end{aligned}$$

which upon integration yields

$$\begin{aligned} |\mathcal{Z}(t) - \mathcal{P}(t)| \leq\ & |\mathcal{Z}(0) - \mathcal{P}(0)| K_p e^{\alpha t} + e^{\alpha t} + \overline{\overline{R}} e^{-(\hat{\lambda}+\alpha)t} \\ & + \varepsilon \overline{\overline{C}}_{\theta_2} \left(\frac{e^{(-|\gamma|+\alpha)t} - e^{(\alpha-\lambda)t}}{-|\gamma| + \lambda} \right) \\ & + \overline{\overline{\overline{C}}}_{\theta_2} \left(\frac{e^{-\kappa t} - e^{-(\alpha-\lambda)t}}{-(\kappa + \alpha - \lambda)} \right) \\ & + 2\overline{\overline{\overline{\overline{C}}}}_{\theta_2} |\mathcal{Q}(0)| \left(\frac{e^{-\mu t} - e^{(-\lambda+\alpha)t}}{-(\mu + \alpha - \lambda)} \right). \end{aligned}$$

Thus, provided $|\mathcal{Q}(0)|$ is $\mathcal{O}(1)$, we have

$$|\mathcal{Z}(t) - \mathcal{P}(t)| = |\mathcal{Z}(0) - \mathcal{P}(0)|K_p e^{\alpha t} + \tilde{B}_4 e^{-\tilde{\kappa} t}$$

where $-\tilde{\kappa} > \max\{-|\gamma| + \alpha, \kappa, \lambda + \alpha, -\mu + \alpha, -\hat{\lambda} + \alpha\}$ and $\alpha > 0$. □

9.3. Final step in the proof of MELESE. In this final section, we show how Lemma 4 enables us to complete the proof of the Modified ELESE (Lemma 3). We first give an interpretation of Lemma 4 in terms of the tracked manifold $\mathcal{L}_\varepsilon$. Recall that $\mathcal{L}_\varepsilon = \cup_{t \in [0,\infty)} K_\varepsilon \cdot t$, where K_ε has dimension $k + \sigma - 1$. Let $q(t)$ be a trajectory in $\mathcal{L}_\varepsilon$ with $q(0) \in K_\varepsilon$. The evolution of $T_{q(t)}\mathcal{L}_\varepsilon$ is given by the Plücker coordinates $\mathcal{Z}(q(t)) \equiv \mathcal{Z}(t), \mathcal{X}(q(t)) \equiv \mathcal{X}(t)$, where the $\mathcal{Z}$ component denotes the projection of $T_{q(t)}\mathcal{L}_\varepsilon$ onto the $(k+\sigma)$-dimensional subspaces of $TW^u(\mathcal{S}^\varepsilon)$. The $\mathcal{X}$ component denotes the projection onto all the other $(k+\sigma)$-dimensional subspaces of $\mathbb{R}^{\mathbf{k}+\ell+\mathbf{n}}$.

The fact that $\mathcal{X}(T)$ is exponentially small after an $\mathcal{O}(\frac{1}{\varepsilon})$ time T means that, upon exiting the box $\mathcal{B}$, the components of $T_{q(t)}\mathcal{L}_\varepsilon$ in the stable directions are exponentially small. Let $\tilde{q}(0)$ denote the projection of $q(0)$ onto $W^u(\mathcal{S}^\varepsilon)$ along the stable directions obtained using the Fenichel stable fibration, and let πK_ε denote the projection of K_ε onto $W^u(\mathcal{S}^\varepsilon)$. Let $\tilde{q}(t)$ denote the trajectory through $\tilde{q}(0)$, and $\tilde{\mathcal{L}}_\varepsilon \equiv \cup_{t \in [0,\infty)} (\pi K_\varepsilon) \cdot t$. IF $\tilde{\mathcal{L}}_\varepsilon$ is also of dimension $k + \sigma$, then we can also determine the evolution of its tangent spaces along the trajectory $\tilde{q}(t)$: $T_{\tilde{q}(t)}\tilde{\mathcal{L}}_\varepsilon$ by looking at its Plücker coordinates. Thus, $\mathcal{P}(t)$ and $\mathcal{Q}(t)$ correspond, respectively, to the $\mathcal{Z}$ and $\mathcal{Q}$ components of the Plücker coordinates of $T_{\tilde{q}(t)}\tilde{\mathcal{L}}_\varepsilon$.

If the hypotheses of Lemma 4 are satisfied, then one can conclude that for times T of $\mathcal{O}(\frac{1}{\varepsilon})$, $\mathcal{Z}(T)$ and $\mathcal{P}(T)$ are exponentially close, which in turn means that the tangent spaces $T_{q(t)}\mathcal{L}_\varepsilon$ and $T_{\tilde{q}(t)}\tilde{\mathcal{L}}_\varepsilon$ are exponentially close.

Let $\pi\mathcal{L}_\varepsilon$ denote the projection of $\mathcal{L}_\varepsilon$ onto $W^u(\mathcal{S}^\varepsilon)$ (assuming $\pi\mathcal{L}_\varepsilon$ is of dimension $k + \sigma$). The hypotheses of Lemma 4, regarding the initial values of the $\mathcal{P}$ and $\mathcal{Q}$ components of the Plücker coordinates of $T_{\tilde{q}(t)}\tilde{\mathcal{L}}_\varepsilon$, and the exponentially small difference $|\mathcal{Z}(0) - \mathcal{P}(0)|$ can be verified from $\pi\mathcal{L}_\varepsilon$, as is shown in Lemma 5. Note that Lemma 5 is equivalent to Lemma 3 (MELESE) in Section 5.

LEMMA 5. *Let $\mathcal{L}_\varepsilon$ be a $(k+\sigma)$ dimensional invariant manifold, obtained by flowing a $(k+\sigma-1)$-dimensional manifold K_ε. Let $q \in K_\varepsilon$ be such that its trajectory leaves $\mathcal{B}$ after an $\mathcal{O}(\frac{1}{\varepsilon})$ amount of time T. Let $\overline{q}$ be the exit point from $\mathcal{B}$ of this trajectory. Let π denote the Fenichel projection along the stable directions. If $\pi\mathcal{L}_\varepsilon$ is of dimension $(k+\sigma)$ and πK_ε of dimension $(k+\sigma-1)$, and if the following conditions hold:*
(i) $|\mathcal{X}(T_q\mathcal{L}_\varepsilon)|, |\mathcal{Q}(T_{\pi q}(\pi\mathcal{L}_\varepsilon))|, |\mathcal{Z}(T_q\mathcal{L}_\varepsilon)|$, and $|\mathcal{P}(T_{\pi q}(\pi\mathcal{L}_\varepsilon))|$ are $\mathcal{O}(1)$
(ii) $|\mathcal{Z}(T_q\mathcal{L}_\varepsilon) - \mathcal{P}(T_{\pi q}(\pi\mathcal{L}_\varepsilon))|$ are exponentially close,
then $\mathcal{L}_\varepsilon$ at $\overline{q}$ is C^1 exponentially close to $W^u(\mathcal{S}^\varepsilon|_{\pi K_\varepsilon \cdot [T-\gamma, T+\gamma]})$ at $\pi q \cdot T$, for some $\gamma > 0$ independent of ε.

Proof. First, note that q and πq have the same y-coordinates, so the requirement in Lemma 4 that $|y(0) - \tilde{y}(0)|$ is exponentially small is automatically satisfied.

Next, we verify that $|\mathcal{Q}(T_{\pi q}(\pi\mathcal{L}_\varepsilon))|$ and $|\mathcal{Q}(T_{\pi q}(\tilde{\mathcal{L}}_\varepsilon))|$ are exponentially close, and similarly for $|\mathcal{P}(T_{\pi q}(\pi\mathcal{L}_\varepsilon))|$ and $|\mathcal{P}(T_{\pi q}(\tilde{\mathcal{L}}_\varepsilon))|$. Let $V \equiv \{v_1, ..., v_{k+\sigma-1}\}$ be a set of tangent vectors that span $T_q K_\varepsilon$. Then $T_q\mathcal{L}_\varepsilon$ is spanned by $V \cup \{\mathbf{v}(q)\}$, where $\mathbf{v}(q) = \big(\Lambda a, \Gamma b, \varepsilon m(y;\varepsilon) + \varepsilon h(a,b,y;\varepsilon)\, ab\big)$ is the vector field at q. The space $T_{\pi q}(\pi\mathcal{L}_\varepsilon)$ is spanned by $\pi V \cup \{\pi(\mathbf{v}(q))\}$, where $\pi V \equiv \{\pi v_1, ..., \pi v_{k+\sigma-1}\}$ and $\pi(\mathbf{v}(q)) = (\Lambda a, 0, \varepsilon m(y;\varepsilon) + \varepsilon h(a,b,y;\varepsilon)ab)$. The space $T_{\pi q}(\tilde{\mathcal{L}}_\varepsilon)$ is spanned by $\pi V \cup \{\mathbf{v}(\pi q)\}$, where $\mathbf{v}(\pi q) = (\Lambda a, 0, \varepsilon m(y;\varepsilon))$. Since the a-coordinate of q must be exponentially small, then $\mathbf{v}(\pi \mathbf{q})$ and $\pi\mathbf{v}(q)$ are exponentially close. This implies that the respective $\mathcal{Q}$ and $\mathcal{P}$ coordinates are also exponentially close.

Now, assume $|\mathcal{Q}(T_{\pi q}(\pi\mathcal{L}_\varepsilon))|$ and $|\mathcal{X}(T_q\mathcal{L}_\varepsilon)|$ are $\mathcal{O}(1)$. This implies $|\mathcal{Q}(T_{\pi q}(\tilde{\mathcal{L}}_\varepsilon))|$ is also $\mathcal{O}(1)$. In terms of the notation used in Lemma 4, $|\mathcal{Q}(T_{\pi q}(\tilde{\mathcal{L}}_\varepsilon))| \equiv |\mathcal{Q}(0)|$ and $|\mathcal{X}(T_q\mathcal{L}_\varepsilon)| \equiv |\mathcal{X}(0)|$. Similarly, assume $|\mathcal{P}(T_{\pi q}(\pi\mathcal{L}_\varepsilon))|$ and $|\mathcal{Z}(T_q\mathcal{L}_\varepsilon)| \equiv |\mathcal{Z}(0)|$ are each $\mathcal{O}(1)$. Then $|\mathcal{P}(T_{\pi q}(\tilde{\mathcal{L}}_\varepsilon))| \equiv |\mathcal{P}(0)|$ is also $\mathcal{O}(1)$. Finally if $|\mathcal{P}(T_{\pi q}(\pi\mathcal{L}_\varepsilon)) - \mathcal{Z}(T_q\mathcal{L}_\varepsilon)|$ is exponentially close, then so is $|\mathcal{P}(0) - \mathcal{Z}(0)|$. Therefore, the hypotheses of Lemma 4 are satisfied, and we can conclude that at the exit time T, $\mathcal{Z}(T_{\bar{q}}\mathcal{L}_\varepsilon)$ is exponentially close to $\mathcal{P}(T_{\pi q \cdot T}\tilde{\mathcal{L}}_\varepsilon)$, and the conclusion of Lemma 5 then follows.

Finally, it is straightforward to verify that Lemma 5 is equivalent to MELESE in Section 5.2, since the condition $(M2)$ of MELESE is equivalent to conditions (i) and (ii) of Lemma 5. This concludes the proof of MELESE. □

Acknowledgement. This work constitutes Chapter 4 of the author's Ph.D. thesis [53]. The author thanks T. Kaper for his guidance throughout this work. The author also thanks C. Jones for having suggested the use of an augmented system and for bringing [17] to our attention.

REFERENCES

[1] S.B. Angenent, J. Mallet-Paret, and L.A. Peletier. Stable transition layers in semilinear boundary value problems. *J. Diff. Eq.*, **67**:212–242, 1987.

[2] D. Aronson and H. Weinberger. Nonlinear diffusions in population genetics, combustion and nerve propagation. In J. Goldstein, editor, *Partial Differential Equations and related topics*, Volume **446** of *Lecture Notes in Math.* Springer Verlag, 1975.

[3] T.B. Benjamin. Instability of periodic wavetrains in dispersive systems. *Proc. Roy. Soc. London, Series A*, **299**:59–75, 1967.

[4] C. Bonet. Singular perturbation of relaxed periodic orbits. *J. Diff. Eq.*, **66**:301–339, 1987.

[5] R. Camassa, G. Kovačič, and S.K. Tin. A Melnikov method for homoclinic orbits with many pulses. *Arch. Rat. Mech. Analysis*, **143**:105–193, 1998.

[6] G. Carpenter. A geometric approach to singular perturbation problems with applications to nerve impulse equations. *J. Diff. Eq.*, **23**:335–367, 1977.

[7] G. Carpenter. Periodic solutions of nerve impulse equations. *J. Math. Anal. Appl.*, **58**:152–173, 1977.

[8] S.N. Chow and X.-B. Lin. Bifurcation of a homoclinic orbit with a saddle-node equilibrium. *Diff. Int. Eq.*, **3**:435–466, 1990.

[9] A. Doelman, T.J. Kaper, and P. Zegeling. Pattern formation in the 1-d Gray-Scott model. *Nonlinearity*, **10**:523–563, 1997.

[10] S.R. Dunbar. Traveling waves in diffusive predator-prey equations: periodic orbits and point-to-periodic heteroclinic orbits. *SIAM J. Appl. Math.*, **46**:1057–1078, 1986.

[11] W. Eckhaus. *Studies in nonlinear stability theory.* Springer-Verlag, New York, 1965.

[12] E. Eszter. *On the stability of periodic traveling waves solutions associated with the FitzHugh-Nagumo system.* PhD thesis, Department of Mathematics, University of Massachusetts, 1999.

[13] N. Fenichel. Geometrical singular perturbation theory for ordinary differential equations. *J. Diff. Eq.*, **31**:53–98, 1979.

[14] P.C. Fife. Boundary and interior transition layer phenomena for pairs of second order differential equations. *J. Math. Anal. Appl..*, **54**:497–521, 1976.

[15] P.C. Fife. *Mathematical aspects of reacting and diffusing systems.* Springer Verlag, New York, 1979.

[16] R. FitzHugh. Impulses and physiological states in models of nerve membrane. *Biophys. J.*, **1**:445–466, 1961.

[17] R.A. Gardner. On the structure of the spectra of periodic traveling waves. *J. Math. Pures Appl.*, **72**:415–439, 1993.

[18] R.A. Gardner. Spectral analysis of long wavelength periodic waves and applications. *J. reine u angew. Math.*, **491**:149–181, 1997.

[19] R.A. Gardner and J. Smoller. The existence of periodic travelling waves for singularly perturbed predatory-prey equations via the conley index. *J. Diff. Eq.*, **47**:133–161, 1983.

[20] J. Guckenheimer and P. Holmes. *Nonlinear oscillations, dynamical systems and bifurcations of vector fields.* Springer Verlag, New York, 1983.

[21] J. Hale. *Ordinary differential equations.* Robert E. Krieger Publishing Company, Inc., Malabar,FLA, 1980. Ch. 3.

[22] G. Haller and S. Wiggins. N-Pulse homoclinic orbits in perturbations of resonant Hamiltonian systems. *Arch. Rational Mech. Anal.*, **130**:25–101, 1995.

[23] S. Hastings. The existence of periodic solutions to Nagumo's equation. *Quart. J. Math. Oxford*, **3**(25):369–378, 1974.

[24] S. Hastings. On traveling wave solutions of the Hodgkin-Huxley equations. *Arch. Rat. Mech. Anal.*, **60**:229–257, 1976.

[25] S.P. Hastings and J.B. McLeod. Periodic solutions of a forced second order differential equation. *J. Nonlinear Science*, **1**:225–245, 1991.

[26] S.P. Hastings and J.B. McLeod. Chaotic motion of a pendulum with oscillatory forcing. *Am. Math. Monthly*, **100**:563–572, 1993.

[27] M. Hayes, K. Ono, T.J. Kaper, and N. Kopell. On the application of geometric singular perturbation theory to some classical two-point boundary value problems. *Intl. J. Bif. Chaos*, **8**:189–209, 1998.

[28] A.L. Hodgkin and A.F. Huxley. A quantitative description of membrane current and its application to conduction and excitation in nerve. *J. Physio. (London)*, **117**:500–542, 1952.

[29] D. Johnston and S.M. Wu. *Foundations of cellular neurophysiology.* The MIT Press, Cambridge,MA; London,England, 1995.

[30] C.K.R.T. Jones. Geometric singular perturbation theory. In R. Johnson, editor, *Dynamical Systems, Montecatibi Terme*, Volume **1609** of *Lecture Notes in Mathematics*. Springer Verlag, 1994.

[31] C.K.R.T. Jones, T.J. Kaper, and N. Kopell. Tracking invariant manifolds up to exponentially small errors. *SIAM J. Math. Anal.*, **27**(2):558–577, 1996.

[32] C.K.R.T. Jones and N. Kopell. Tracking invariant manifolds with differential forms in singularly perturbed systems. *J. Diff. Eq.*, **108**:64–88, 1994.

[33] C.K.R.T. Jones, N. Kopell, and R. Langer. Construction of the FitzHugh-Nagumo pulse using differential forms. In G. Aris H. Swinney and D. Aronson, editors, *Patterns and dynamics in reactive media*, Volume **37** of *IMA Volumes in Mathematics and its Applications*, New York, 1991. Springer Verlag.

[34] T.J. Kaper and G. Kovačič. Multi-bump orbits homoclinic to resonance bands. *Trans. A.M.S.*, **348**(10):3835–3887, 1996.

[35] T.J. Kaper and S. Wiggins. A commentary on "Periodic solutions of a forced second-order equation" by S.P. Hastings and J.B. McLeod. *J. Nonlinear Sci.*, **1**:247–253, 1991.

[36] J. Kevorkian and J.D. Cole. *Perturbation methods in applied mathematics.* Springer-Verlag, New York, 1981.

[37] N. Kopell and L. Howard. Target patterns and horseshoes from a perturbed central-force problem: some temporally-periodic solutions to reaction-diffusion equations. *Stud. in Appl. Math.*, **64**:1–56, 1981.

[38] M. Krupa, B. Sandstede, and P. Szmolyan. Fast and slow waves in the fitzhugh-nagumo equation. *J. Diff. Eq.*, **133**:49–97, 1997.

[39] J.J. Levin and N. Levinson. Singular perturbations of nonlinear systems of differential equations and an associated boundary layer equation. *J. Rat. Mech. Analysis*, **3**:247–270, 1954.

[40] N. Levinson. A second order equation with singular solutions. *Ann. Math.*, **50**:127–153, 1949.

[41] N. Levinson. Perturbations of discontinuous solutions of nonlinear systems of differential equations. *Acta Math.*, **82**:71–106, 1950.

[42] K. Maginu. Existence and stability of periodic traveling wave solutions to Nagumo's equation. *J. Math. Bio.*, **10**:133–153, 1980.

[43] K. Maginu. Stability of periodic traveling wave solutions with large spatial periods in reaction diffusion systems. *J. Diff. Eq.*, **39**:73–99, 1981.

[44] E.F. Mischenko and N. Rozov. *Differential equations with small parameters and relaxation oscillations.* Plenum Press, New York, 1980.

[45] D. Morgan, A. Doelman, and T.J. Kaper. Stationary periodic patterns in the 1-d Gray-Scott model. *Technical Report* **13***, Center for BioDynamics, Boston University*, November 1998.

[46] H. Ockendon, J.R. Ockendon, and A.D. Johnson. Resonant sloshing in shallow water. *J. Fluid Mech.*, **167**:465–479, 1986.

[47] J. Rinzel and J.B. Keller. Traveling wave solutions of a nerve conduction equation. *Biophys. J.*, **13**:1313–1337, 1973.

[48] C. Robinson. Sustained resonance for a nonlinear system with slowly varying coefficients. *SIAM J. Math. Anal.*, **14**(5):847–860, 1983.

[49] K. Sakamoto. Invariant manifolds in singular perturbation problems for ordinary differential equations. *Proc. Roy. Soc. Edin*, **116A**:45–78, 1990.

[50] J. Smoller. *Shock waves and reaction-diffusion equations.* Springer Verlag, New York, 1983.

[51] C. Soto-Treviño and T.J. Kaper. Periodic orbits in singularly perturbed systems. In A. Doelman and A.V. Harten, editors, *Pattern formation in the natural environment*, Pitman Research Notes in Mathematics, 295–314, 1995.

[52] C. Soto-Treviño and T.J. Kaper. Higher-order Melnikov theory for adiabatic systems. *J. Math. Phys.*, **37**(12):6220–6249, 1996.

[53] C. Soto-Treviño. *Geometric methods for periodic orbits in singularly perturbed systems.* PhD thesis, Department of Mathematics, Boston University, 1998.

[54] D. Terman. Chaotic spikes arising from a model for bursting in excitable membranes. *SIAM J. Appl. Math*, **51**:1418–1450, 1991.

[55] A.N. TIKHONOV. On the dependence of the solutions of differential equations on a small parameter. *Mat. Sb.*, **31**:575–586, 1948.

[56] S.K. TIN. *On the dynamics of tangent spaces near normally hyperbolic invariant manifolds.* PhD thesis, Division of Applied Mathematics, Brown University, 1994.

[57] S.K. TIN, N. KOPELL, AND C.K.R.T. JONES. Invariant manifolds and singularly perturbed boundary value problems. *SIAM J. Numer. Anal*, **31**(6):1558–1576, 1994.

[58] A.M. TURING. The chemical basis of morphogenesis. *Phil. Trans. Roy. Soc. Lond., Series B*, **237**:37–72, 1952.

THE PHENOMENON OF DELAYED BIFURCATION AND ITS ANALYSES

JIANZHONG SU*

Abstract. In this note, a general survey of the interesting phenomenon of delayed bifurcation is given. These delayed bifurcations arise in particular fast-slow systems under various situations, and their mathematical justification and application are discussed.

Key words. Fast-slow systems, Delay of bifurcation, Periodic forcing, Hopf bifurcation, Simple eigenvalue bifurcation.

AMS(MOS) subject classifications. 34C35, 34D20, 58F14, 92C30.

1. Introduction. Delayed Bifurcations (or delays of bifurcation) are dynamical phenomena typically arising in some dynamical systems containing a parameter that is a slowly varying function of time t or, more generally in fast-slow systems where the parameter is determined by a slow equation such as

$$\frac{\partial u}{\partial t} = f(u, I) , \qquad u \in \mathbb{X} , \tag{1a}$$

$$\frac{\partial I}{\partial t} = \varepsilon g(u, I) , \qquad I \in \mathbb{R}^1 . \tag{1b}$$

For each fixed value of the "parameter" I, there is a "static" solution (for example, an equilibrium, or a periodic solution) which is in fact a singular solution of Eq. (1) by setting $\varepsilon = 0$. Further, we assume that there is a critical value (which is usually also a bifurcation point) in the parameter I so that the linearized orbital stability (in the classical sense) of these singular solutions in Eq. (1a) changes as the parameter moves across the critical value.

The interesting phenomenon of delayed bifurcations occurs for the solutions of initial value problems of Eq. (1) (where the parameter I is slowly varying rather than fixed) in a scenario as follows. The solutions remain close to the unstable singular solutions mentioned earlier, and only move away from them at some point of the parameter which is above the bifurcation point at a distance depending only on the initial parameter value i.e., independent of ε. The limits of these solutions as ε goes to zero fit into the pattern that the solutions of the initial value problems are attracted immediately to the "static" solutions (which generally are not solutions of Eq. (1)) and then stay close to them up until certain points above the bifurcation point, then move away.

---

*Department of Mathematics, University of Texas at Arlington, Box 19408, Arlington, TX 76019. Email: su@uta.edu.

This phenomenon is different from what is understood as a bifurcation, and is named as a Delayed Bifurcation. Further, for a certain range of the initial parameter, the separation has a memory effect in such a way that if the initial parameter is further away from the critical value, then the separation would be further above the critical value.

These delayed bifurcations are counter-intuitive, and their mathematical structure is rather complicated. The theory, however, offers a good explanation to a number of natural phenomena.

2. Background. Delayed bifurcations are of interest for their importance in applications to various scientific fields. We start by exposing some previous work in the literature to show the connection with physiology, physics and others.

The theory of the reverse accommodation phenomenon in the experiments of membranes of the giant axons of squids was studied by Jakobsson and Guttman [15]. If a membrane is connected with a constant electric current, then there exists a threshold with the property that that the membrane potential can accommodate any current below the threshold, and when the current is above the threshold, the potential starts to oscillate (burst) periodically. The discovery of Jakobsson and Guttman in [15] concerns the response of the membrane potential with respect to a continuously increasing current, and it was found by them that if the rate of increasing is slow, the potential would accommodate until the current reaches a point which is substantially higher than the threshold. More significantly, the magnitude of delay is independent of the slowness of the increase.

Baer, Erneux and Rinzel [3, 30] considered the corresponding mathematical problem by studying the FitzHugh Nagumo equation and made extensive asymptotic analyses and numerical computations for the slow passage problems in the FitzHugh Nagumo equation. In particular, by using exponential asymptotic expansions, Baer, Erneux and Rinzel [3, 30] showed that the initial current $I_i < I_-$ and the current I_q, where a large amplitude bursting starts, must satisfy the relationship

$$\int_{I_i}^{I_q} Re\lambda_1(\tau)d\tau = 0 \tag{2}$$

where λ_1 is the first eigenvalue of FHN system. The relationship in Eq. (2) was later proven rigorously in [33, 34].

Similar phenomena are observed in experiments of lasers with saturable absorbers (LSA) [10–12]. The LSA have two interesting modes of operation: optical bistability and time-periodic intensity modulation. With the LSA equations, these two modes are stable stationary solutions and stable periodic solutions respectively. The bifurcation diagram indicates that the emergence of a bifurcating branch of the time-periodic solutions occurs at the bifurcation point through an exchange of instability of stationary solutions [20]. Experimental results in [10, 12, 20] were obtained under the physical settings where the control parameter was slowly swept across the domain to be studied. Delayed bifurcations were among the resulting pictures that cannot be directly explained by the bifurcation diagram.

We also note that applications of delayed bifurcations in other fields are available in the literature [1, 2, 5–7, 16, 17, 21, 25, 26, 30, 40, 41, and their references]. The theory of delayed bifurcations would also help in the formulation of mathematical equations to various biological models, particularly in the area of modeling impulses and nerve electrical waves that have a relaxation oscillation character [6, 7, 40].

3. Delayed Hopf-bifurcations in the FitzHugh Nagumo model. Let us start with a mathematical interpretation of the experimental results of Jakobsson and Guttman [15] which is a typical example of a delayed bifurcation. Using the Hodgkin-Huxley theory of nerve electrical activities, we take a qualitative mathematical model FitzHugh Nagumo equation (FHN) to describe the reverse accommodation phenomenon. It has been shown (in [3, 33, 34] for the case of $\phi \equiv 1$ and in [28, 29, 35, 36] for the more general case of $\phi(x) > 0$) that the FHN system

$$
\begin{aligned}
&\frac{\partial v}{\partial t} = Dv_{xx} - f(v) - w + I\phi(x) \qquad \text{for} \quad 0 < x < 1\,, \quad t > 0\,,\\
&\frac{\partial w}{\partial t} = bv - b\gamma w \qquad \text{for} \quad 0 < x < 1\,, \quad t > 0\,,\\
&\frac{\partial v}{\partial x}(x,t) = 0 \qquad \text{for} \quad x = 0, 1\,, \quad t > 0\,,\\
&\frac{\partial w}{\partial x}(x,t) = 0 \qquad \text{for} \quad x = 0, 1\,, \quad t > 0
\end{aligned}
\tag{3}
$$

where I is a constant parameter, has the following properties: (1) there exist a family of equilibria $u_0(x, I) = (V_0(x, I), W_0(x, I))$ for Eq. (3) for $I \in (-\infty, \infty)$; (2) there exists a critical value $I = I_-$ and some $\sigma > 0$ such that when $I_- - \sigma < I < I_-$, the eigenvalues λ_j of

$$
A = \begin{pmatrix} D\partial_x^2 - f'(V_0(x, I)) & -1 \\ b & -b\gamma \end{pmatrix}
$$

are all in the negative half plane i.e., $\text{Re}\lambda_j(I) < 0$, and $\lambda_2(I) = \bar{\lambda}_1(I)$; and when $I_- < I < I_- + \sigma$, the eigenvalues $\lambda_1(I)$ and $\lambda_2(I) = \bar{\lambda}_1(I)$ are in the positive half plane i.e., $Re\lambda_1(I) > 0$, while for $j \geq 3$, $\text{Re}\lambda_j(I) < 0$.

Efforts have been made to understand the dynamical behavior of solutions of Eq.(3) when I is replaced by $I(t) \equiv I_i + \varepsilon t$ which is a slowly varying parameter passing through the critical value $I = I_-$ corresponding to Jakobsson-Guttman's experiments where the current is slowly changed to study the reverse accommodation phenomenon. Change coordinates so that the new independent variable is $I = I_i + \varepsilon t$, and denote $u(x, I) = (v(x, I), w(x, I))^T$, $y = u(x, I) - u_0(x, I)$. Eq. (3) then becomes

$$\text{(4)} \qquad \begin{aligned} \varepsilon \frac{\partial y}{\partial I} &= \begin{pmatrix} D\frac{\partial^2}{\partial x^2} - f'(V_0(x, I)) & -1 \\ b & -b\gamma \end{pmatrix} y + F_1(y, I, \varepsilon) + \varepsilon Q_1(I, \varepsilon) , \\ \frac{\partial y}{\partial x} &= 0 \qquad \text{for} \quad x = 0, 1, \end{aligned}$$

where $F_1(y, I, \varepsilon)$ involves small linear terms along with second and third order homogeneous polynomials of y with the property $|F_1(y, I, \varepsilon)| \leq K_1(\varepsilon|y| + |y|^2 + |y|^3)$, $Q_1 = -\frac{\partial u_0}{\partial I}$ is the nonhomogeneous term. For the moment, let us assume $\phi \equiv 1$ and consider only the spatially uniform solutions. Then Eq. (4) becomes a two dimensional ordinary differential equation.

Eq.(4) motivates a basic result of delayed Hopf bifurcations that holds under more general conditions. Let $u(t)$ be the solution of the initial value problem

$$\text{(5a)} \qquad \frac{\partial u}{\partial t} = f(u, I_i + \varepsilon t) ,$$

$$\text{(5b)} \qquad u(t)|_{t=0} = u_0(I_i) + O(\varepsilon)$$

satisfying the following assumptions: (A1) The function $f(u, I) : \mathbb{R}^2 \times \mathbb{R} \to \mathbb{R}^2$ has an analytic extension $f(U, z)$ for $|U| < \sigma$ in the complex U-plane and $|z| < r_a$ in the z-complex plane for some $r_a > 0$, (A2) For each fixed I, the system $V_t = f(V, I)$ has an equilibrium solution $u_0(I)$ which is also analytic in I for $|I| < r_a$, (A3) The variational system of Eq. (5a) about $u_0(I)$:

$$W_t = f_u(u_0(I), I)W$$

is a linear system with coefficients depending on the parameter I. Let $A(I) = f_u(u_0(I), I)$. Assume that two eigenvalues of $A(I)$, $\lambda_1(I)$ and $\lambda_2(I)$, are conjugate to each other i.e., $\lambda_2(I) = \bar{\lambda}_1(I)$ for I on the real axis and $|I| < r_a$. Further, we assume that there exists a real number I_- such that $Re\lambda_j(I) < 0$ when $I < I_-$; $Re\lambda_j(I) > 0$ when $I > I_-$; $Im\lambda_1(I_-) < 0$. Also the initial parameter $I_i < I_-$ starts below the critical value.

Let us change the coordinates so that $I = I_i + \varepsilon t$ is the independent variable, and $y = u(t) - u_0(I_i + \varepsilon t)$. Eq. (5) then becomes

$$\text{(6a)} \qquad \varepsilon \frac{\partial y}{\partial I} = f_u(u_0(I), I)y + f_2(u_0(I), I, y) + \varepsilon G_1(I) ,$$

$$\text{(6b)} \qquad y|_{I=I_i} = O(\varepsilon)$$

where f_2 involves small linear terms along with homogeneous polynomials of y in second order and higher with coefficients depending on I. The nonhomogeneous term $G_1(I) = -\frac{\partial}{\partial I} u_0(I)$ is a bounded vector function of I.

THEOREM 1. *Let $y(I, \varepsilon)$ be a family of solutions of Eq. (6) with initial conditions that satisfy $|y(I_i, \varepsilon)| \leq M_1 \varepsilon$. Then there exist $M = M(M_1)$, $I_q = I_q(M_1, M) > I_-$, $\varepsilon_0 = \varepsilon_0(M_1, M)$ such that*

$$|y(I, \varepsilon)| \leq M\varepsilon$$

whenever $I_i \leq I \leq I_q, \varepsilon \leq \varepsilon_0$. Further, if I_i is close enough to I_-, then I_q and I_i satisfy the relationship

$$\int_{I_i}^{I_q} Re\lambda_1(\tau) d\tau = 0.$$

Proofs of Theorem 1 can be found in [22–26, 33–35]. We demonstrate the behavior of $u(I)$ in Figure 1.

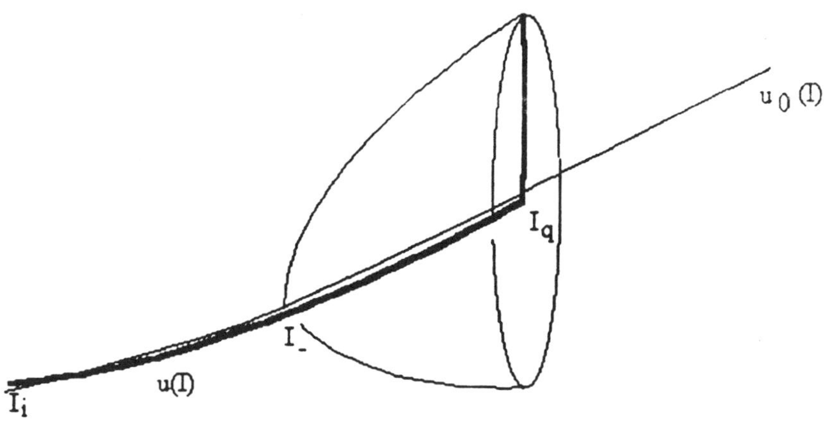

FIG. 1. *The behavior of the solutions $u(I)$ fits the pattern that $u(I)$ stay close to $u_0(I)$ for $I_i \leq I \leq I_q$ where $I_q > I_-$ is independent of ε.*

Rigorous studies of delayed Hopf bifurcation date at least from the 1970's. Shishkova [32] discovered delayed bifurcations in a particular ordinary differential equation system. Later, Neishtadt [22–26] considered general systems having delayed Hopf bifurcations. Su [33, 34] gave a rigorous proof of delayed Hopf bifurcation for the spatially uniform FitzHugh Nagumo equation. The case of the non-spatially uniform FitzHugh Nagumo system where $\phi \equiv \phi(x) > 0$ was also studied for delayed Hopf bifurcations [35]. A quite different approach was utilized in order to deal with the parabolic system. Candelpergher, F. Diener and M. Diener [4] also attempted to explain the delayed Hopf bifurcation phenomenon from non-standard analysis point of view. Most discussions have been concerned with the cases $\frac{\partial I}{\partial t} = \varepsilon$ in Eq. (1b), but the general cases involving nontrivial g are similar.

One particular feature of the FHN equation is that there are two critical values I_- and I_+ where linear stability changes, and $I_- < I_+$. The results of [23–26, 33–35] indicated that the delays depend heavily upon the structure of the level curves of $Re \int_{I_-}^{z} \lambda_1(\tau)\, d\tau$ in the complex z-plane. The different parameter settings in the FHN result in different Riemann surfaces of the analytic function $\int_{I_-}^{z} \lambda_1(\tau)\, d\tau$. We note that some of the parameter settings do allow level curves to connect the points below I_- to I_+. See Figure 2, details were provided in [33, 34]. Such a structure implies the existence of solutions of Eq. (4) which may never burst at all. These phenomena are considered interesting in mathematical biology [30].

4. Related problems in delayed bifurcation. These mathematical studies have provided a solid theoretical foundation for giving mathematical justifications of natural phenomena such as reverse accommodations. A question was raised by experimentalists on the reason why the reverse accommodation phenomenon was observed in some experiments but not in others, or how sensitive these persistent unstable solutions would be with respect to perturbations. Further, reverse accommodations were found on the tissues under laboratory circumstances, and it would be interesting to know whether this type of behavior happens to live tissues where other disturbances exist.

Baer, Erneux and Rinzel [3] studied this problem numerically and observed that if the right-hand side of Eq. (4) had added a small perturbation $g_1 = \delta \sin \omega t$, then the unstable solutions persisted when $\omega \neq \frac{1}{3}|\omega_0|, \frac{1}{2}|\omega_0|, |\omega_0|, 2|\omega_0|$ where $|\omega_0| = |Im \lambda_1(I_-)|$ is the frequency at Hopf bifurcation point. Near those resonance frequencies, the delay amounts $I_q - I_-$ were significantly reduced. It is conceivable, that due to the high sensitivity of the delayed bifurcation phenomenon with respect to the roundoff errors in computations, it might be extremely difficult to observe what really happened exactly at those resonant frequencies. We provide an example below by using a modified Shishkova-Wallet equation to show that, in fact, when the forcing frequency $\omega = \omega_0$, the delay amount can be reduced to $O(\sqrt{\varepsilon} |\ln(\varepsilon)|)$ rather than $O(1)$ as in Eq. (4).

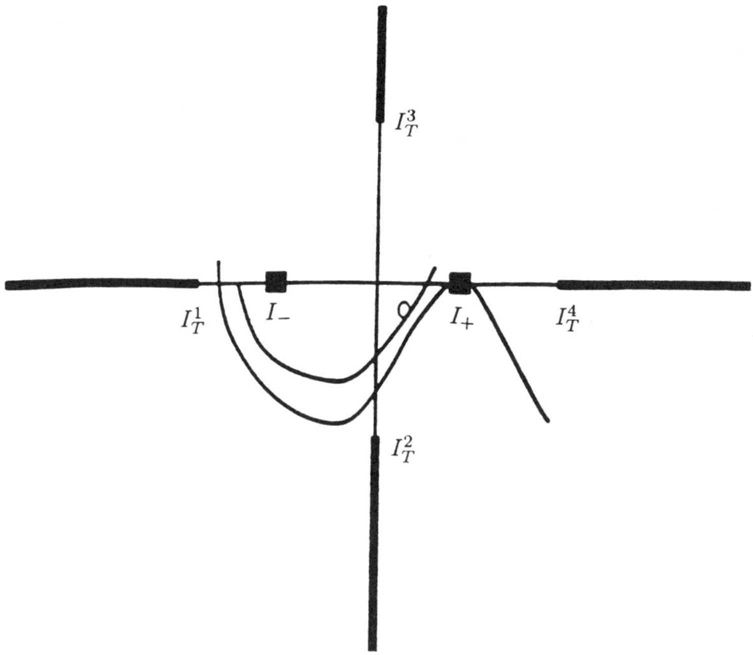

FIG. 2. *Some parameter settings of FHN equation allow the level curves of* $\int_{I_-}^{z} \lambda_1(\tau)\, d\tau$ *in the z-plane to connect points below the critical value* I_- *to another critical value* I_+ *in the real axis.* $I_T^1, I_T^2, I_T^3, I_T^4$ *are branching points in the Riemann surface of the analytic function* $\int_{I_-}^{z} \lambda_1(\tau)\, d\tau$.

Consider the following modified Shishkova-Wallet equation:

$$\varepsilon \frac{\partial u}{\partial I} = (I + i\omega_0)u + \varepsilon e^{\frac{i\omega I}{\varepsilon}}, \tag{7a}$$

$$u\big|_{I=I_i} = O(\varepsilon) \tag{7b}$$

for $I_i < I_- \equiv 0$. We denote u_+ the solution of Eq. (7a) with the initial condition $u_+(1) = 0$, and let u_- be the solution of Eq. (7a) with the initial condition $u_-(-1) = 0$. From [33, 34, 36–39], the distance $|u_+(0) - u_-(0)|$ determines the points of bifurcation. If $|u_+(0) - u_-(0)| \leq e^{-\frac{c}{\varepsilon}}$ for some $c > 0$, then the delays are O(1). If $|u_+(0) - u_-(0)| \geq \varepsilon^r$ for some $r > 0$, then the delays are $O(\sqrt{|\varepsilon \log \varepsilon|})$. It can be easily shown through a direct calculation that

$$|u_+(0) - u_-(0)| = \left| \int_{-1}^{1} e^{-\frac{1}{2\varepsilon}(s^2 + 2i(\omega_0 - \omega)s)} ds \right|. \tag{8}$$

When $\omega = \omega_0$, the distance $|u_+(0) - u_-(0)| = \int_{-1}^{1} e^{-\frac{1}{2\varepsilon}s^2} ds \geq K\sqrt{\varepsilon}$ for some $K > 0$, provided $\varepsilon \leq \varepsilon_0$. Thus, u_- jumps away from u_+ (in the sense of $|u_+(I) - u_-(I)| = O(1)$) at the point $I_q \equiv \inf\left\{I > 0 | \sqrt{\varepsilon} e^{\frac{1}{2\varepsilon}I^2} = O(1)\right\}$ $= O(\sqrt{|\varepsilon \log \varepsilon|})$. In other words, the bifurcation of $u_-(I)$ from $u_0(I) \equiv 0$ occurs near the critical value $I_- = 0$ at $I_q = O(\sqrt{|\varepsilon \log(\varepsilon)|})$. Hence the bifurcation has not been delayed substantially. The key element that prevents the delayed bifurcation is the interference between the frequency at the critical value ω_0 and the frequency of the periodic forcing ω. However if $\omega \neq \omega_0$, then we can derive the estimate $|u_+(0) - u_-(0)| \leq e^{-\frac{c}{\varepsilon}}$ for some $c > 0$ which implies delayed bifurcations. This example can be further modified to

$$\varepsilon \frac{\partial u}{\partial I} = (I + i\omega_0)u + \varepsilon \left(\sum_{n \in \mathbb{Z}} \frac{1}{1 + n^2} e^{\frac{in\omega I}{\varepsilon}} \right), \tag{9a}$$

$$u\big|_{I=I_i} = O(\varepsilon) \tag{9b}$$

to show $I_q = O(\sqrt{|\varepsilon \log(\varepsilon)|})$ if $\omega = \dfrac{1}{n}\omega_0$. Thus the resonance effect is a vital factor in problems.

The problems of delayed Hopf bifurcation under the influence of periodic perturbations were studied in [36, 39]. The existence of delayed bifurcations was shown under the nonresonance conditions $\omega \neq \dfrac{2|\omega_0|}{n}, n \in \mathbb{N}$. The situation near the resonances was also considered in [39]. The amounts of delays there are indeed proportional to the gap between the forcing frequency and the resonance frequencies of unperturbed systems. If the systems are added with a periodic perturbation εg_2, which has a slowly changing frequency $\omega = \omega(I_i + \varepsilon t)$, then the phenomenon of a shifted interference can be observed. Such processes are commonly seen in real physical experiments where various parameters usually drift slowly with time. In these slowly varying frequency cases, the resonance occurs when $\omega(I_-) + \omega'(I_-)(I_- - I_i) = \dfrac{2|\omega_0|}{n}$ [39]. Therefore if $\omega'(I) > 0$, then resonance frequencies $\omega(I_-)$ which people observe in the experiments should be lower than $\dfrac{2|\omega_0|}{n}$. For $\omega'(I) < 0$, the result goes in the other direction. Thus interferences are shifted because of the frequency change. Neishtadt gave out his insight on the cause of this shift by introducing a dynamic frequency [in a personal communication]. Let the fast scale time be $\theta = \int_{I_i}^{I} \gamma(\tau)\, d\tau = \omega(I)(I - I_i)$ with the dynamic frequency $\gamma(I)$. Then $\omega(I_-) + \omega'(I_-)(I_- - I_i) = \gamma(I_-) = \dfrac{2|\omega_0|}{n}$ is the resonance frequency.

In other directions, there are results [23, 27, 37, 38] in more general cases where delayed bifurcations occur to a family of periodic motions whose critical exponents moved across the imaginary axis as the parameter slowly moves past a critical value under some nonresonance conditions. Also delayed bifurcations on analytic mappings were studied in [13, 23]. The theory of maximal delays was developed in [9, 26].

5. Delayed simple eigenvalue bifurcations. Both the theory and some applications relating to delayed simple eigenvalue bifurcations are available in the literature [1, 5, 8, 14, 16–19, 31, 39]. Many studies are concerned with the dynamical behavior of solutions of dynamical systems that have a simple eigenvalue slowly moving across zero along the real axis. Such systems had been studied in various cases by F. Diener and M. Diener [8], Schecter [31], Lebovitz and Schaar [18, 19] and Haberman [14] with particular attention to whether there is such a substantial delay in the bifurcation at a magnitude of O(1).

Under a restrictive condition that $u \equiv 0$ is a constant solution of $u_t = f(u, I_i + \varepsilon t)$, Haberman [14], F. Diener and M. Diener [8] and Schecter [31] independently proved the existence of delayed simple eigenvalue bifurcations. Without the restrictive condition stated above, the problems are rather difficult and open. The reason is that, generally, a nonhomogeneous equation cannot be changed into a homogeneous equation by finitely many transformations, even in the linear system case. Although n-step iterations of some transformations can reduce the nonhomogeneous terms to $O(\varepsilon^n)$, the resulting sequence of the nonhomogeneous terms is divergent as $n \to \infty$ [23].

Let us consider a one dimensional equation:

$$\frac{\partial u}{\partial t} = f(u, I_i + \varepsilon t) \ , \tag{10a}$$

$$u(t)|_{t=0} = u_0(I_i) + O(\varepsilon) \tag{10b}$$

where $f : \mathbb{R} \times \mathbb{R} \to \mathbb{R}$ has analytic extensions for both variables. Assume that, for each fixed I, the equation $U_t = f(U, I) = 0$ has an equilibrium $u_0(I)$ which is also analytic on I. Further, assume that there exists $I = I_-$ such that when $I < I_-$, $f_u(u_0(I), I) < 0$ and when $I > I_-$, $f_u(u_0(I), I) > 0$. Also $I_i < I_-$. Under the assumption that $u_0(I) \equiv 0$, Haberman [14], Diener [8] and Schecter [31] showed that $|u(I) - u_0(I)| \leq M\varepsilon$ for $I_i \leq I \equiv I_i + \varepsilon t \leq I_q$ where $I_q > I_-$ satisfies $\int_{I_i}^{I_q} f_u(0, I)dI = 0$. There, the assumption $u_0 \equiv 0$ is very crucial in the sense that if the system is perturbed by a constant with its magnitude larger than $\varepsilon^{-\frac{c}{\varepsilon}}$, (e.g. ε^n for any $n > 0$) for ε sufficiently small, then the delay vanishes. Lebovitz and Schaar [18, 19] gave some conditions under which no delay of bifurcation would occur.

A different approach of delayed simple eigenvalue bifurcation problems is motivated by the following consideration. The system

$$\varepsilon \frac{\partial u}{\partial I} = Iu + \varepsilon \sum_{n \neq 0} a_n e^{\frac{in\omega I}{\varepsilon}} \ , \tag{11a}$$

$$u(I_i) = O(\varepsilon) \tag{11b}$$

with $\sum_{n\neq 0}|a_n| \leq M$ can be shown to have a delayed bifurcation pattern. Indeed, if we let u_+ be the solution of Eq. (11a) with the initial condition $u_+(1) = 0$, and u_- be the solution of Eq. (11a) with the initial condition $u_-(-1) = 0$, then it can be easily shown, through a direct calculation, that

$$(12) \qquad |u_+(0) - u_-(0)| = \left| \sum_{n\neq 0} a_n \int_{-1}^{1} e^{-\frac{1}{2\varepsilon}(s^2+2in\omega s)} ds \right| \leq M e^{-\frac{c(\omega)}{\varepsilon}} .$$

The latter inequality can be derived by using a simple contour integration. Thus we find that, by avoiding certain particular resonant frequencies, delayed bifurcations persist even with perturbations at a magnitude of $O(\varepsilon)$.

We now consider a generalization of Eq. (11). Consider the system:

$$(13a) \qquad \frac{\partial u}{\partial t} = f(u, I_i+\varepsilon t) + \varepsilon \left(A a_0(I_i+\varepsilon t, \varepsilon) + \sum_{n\neq 0} a_n(I_i+\varepsilon t, \varepsilon)\, e^{in\omega t} \right) ,$$

$$(13b) \qquad u(t)|_{t=0} = u_0(I_i) + O(\varepsilon)$$

where the function f is the same as in Eq. (10), the perturbations are real functions in the form of a series, $\omega > 0$ is a constant, A is a real parameter, and $a_n = \bar{a}_{-n}$. For $|z - I_-| \leq r_a$ in the complex z-plane, $\varepsilon \leq \varepsilon_0$, the coefficients have analytic extensions $a_n(z, \varepsilon)$ that satisfy $\sum_{n\in\mathbb{Z}} ||a_n(z,\varepsilon)|| \leq M$. Also assume that $|a_0(I_-, \varepsilon)| \geq \delta_{10} > 0$ for $\varepsilon \leq \varepsilon_0$. Let us also use $I = I_i + \varepsilon t$ as the new independent variable. The general theorem for delayed simple eigenvalue bifurcations in [39] is as follows:

THEOREM 2. *Let the above assumptions hold. Suppose that* $u(I, I_i, M_1, \varepsilon)$ *are a family of solutions of Eq. (13) with initial conditions satisfying:* $|u(I, I_i, M_1, \varepsilon) - u_0(I_i)||_{I=I_i} \leq M_1\varepsilon$ *for some* $I_i < I_-$. *Given a set of analytic coefficients* $\{a_n(I,\varepsilon), n \in \mathbb{Z}\}$, *there exists a parameter value* $A_0 \equiv A_0(\{a_n, n \in \mathbb{Z}\}, \varepsilon)$ *such that when* $A = A_0$, *the corresponding Eq. (13) presents a delayed bifurcation pattern, i.e., for a large enough* M, *there exist* $r_{10} = r_{10}(M) > 0$, $\varepsilon_0 = \varepsilon_0(M) > 0$, *such that for* $|I_i - I_-| \leq r_{10}$, $\varepsilon \leq \varepsilon_0$,

$$(14) \qquad |u(I, I_i, M_1, \varepsilon) - u_0(I)| \leq M\sqrt{\varepsilon}$$

when and only when $I_i \leq I \leq I_q$ *where* $I_q = I_i^* + O(\varepsilon) > I_-$ *is a point above the critical value, and* I_i^* *satisfies* $Re \int_{I_i}^{I_i^*} \lambda_1(\tau)d\tau = 0$.

Thus for each scalar ordinary differential equation which has a pitchfork (simple eigenvalue) bifurcation, there exists a co-dimension one family of periodic perturbations with moderately large norms such that delayed bifurcation occurs under these perturbations.

REFERENCES

[1] AHLERS, G.M., CROSS, C., HOHENBERG, P.C. AND SAFRAN, S., *The amplitude equation near the convective threshold: Application to time-dependent heating experiments*, J. Fluid Mech., **110** (1981), 297–334.

[2] ARNOLD, V.I., Bifurcations and singularities in mathematics and mechanics, XVIIth *International Conference in Theoretical and Applied Mechanics* (1988).

[3] BAER, S.M., ERNEUX, T. AND RINZEL, J., *The slow passage through a Hopf bifurcation: Delay, Memory effects and resonance*, SIAM Appl. Math., **49** (1989), 55–71.

[4] CANDELPERGHER, B., DIENER, F. AND DIENER, M., Retard a la bifurcation: Du local au global, *Bifurcations of planar vector fields (edited by J.P. Francoise and R. Roussarie)*, Springer-Verlag, 1990, 1–19.

[5] CROSS, M.C., HOHENBERG, P.C. AND LUCKE, M., *Forcing of convection due to time-dependent heating near threshold*, J. Fluid Mech., **136** (1983), 269–276.

[6] DESTEXHE, A., BABLOYANTZ, A. AND SELNOWSKI, T.J., *Ionic mechanisms for intrinsic slow oscillations in thalamic relay neurons*, Biophy. J., **65** (1993), 1538–1552.

[7] DESTEXHE, A., MCCORMICK, D.A. AND SELNOWSKI, T.J., *A model for 8-10 Hz spindling in interconnected thalamic relay and reticularis neorons*, Biophy. J., **65** (1993), 2473–2477.

[8] DIENER, F. AND DIENER, M., *Sept formules relatives aux canards*, C.R. Acad. Sci. Paris, **297** (1983), 577–580.

[9] DIENER, F. AND DIENER, M., *Maximal delay*, Dynamical bifurcation (E. Beniot, editor), Springer, Lecture notes in math., **1493** (1993), 71–86.

[10] ERNEUX, T. AND MANDEL, D., *Stationary, harmonic and pulsed operations of an optically bistable laser with saturable absorber, II*, Phy. Rev. A, **30** (No. 4) (1984), 1902–1909.

[11] ERNEUX, T. AND MANDEL, D., *Imperfect bifurcation with a slowly-varying control parameter*, SIAM J. Appl. Math., **46** (1986), 1–16.

[12] ERNEUX, T. AND MANDEL, D., *Slow passage through the laser first threshold: Influence of the initial condition*, Optics Comm., **85** (1991), 43–46.

[13] FRUCHARD, *Canards des systems dynamiques discrets*, These de doctorat, U.F.R. de mathematiques de l'Universite Paris 7, F75251 Paris Cedex 05 (1991).

[14] HABERMAN, R., *Slowly-varying jump and transition phenomena associated with algebraic bifurcation problems*, SIAM J. Appl. Math., **37** (1979), 69–105.

[15] JAKOBSSON, E. AND GUTTMAN, R., *Continuous stimulation and threshold of axons: The other legacy of Kenneth Cole*, "Biophysical approach to excitable system" (eds., W. Adelman and D. Goldman) (1981), Plenum, 197–211.

[16] KAPILA, A.K., *Arrhenius system, dynamics of jump due to slow passage through criticality*, SIAM J. Appl. Math., **41** (1981), 29–42.

[17] LEBOVITZ, N.R. AND PESCI, A.I., *Dynamic bifurcation in Hamiltonian systems with one degree of freedom*, SIAM J. Appl. Math., **55** (No. 4) (1995), 1117–1133.

[18] LEBOVITZ, N.R. AND SCHAAR, R.J., *Exchange of stabilities in autonomous systems*, Stud. in Applied Math., **54** (1975), 229–260.

[19] LEBOVITZ, N.R. AND SCHAAR, R.J., *Exchange of stabilities in autonomous systems, II, vertical bifurcation*, Stud. in Applied Math., **56** (1977), 1–50.

[20] MANDEL, D. AND ERNEUX, T., *Stationary, harmonic and pulsed operations of an optically bistable laser with saturable absorber, I*, Phy. Rev. A, **30** (No. 4) (1984), 1893–1901.

[21] MENON S.V.G., *Integral-equation approach to delayed bifurcation in noisy dynamical system*, Phy. Rev. A, **43** (No. 8) (1991), 4216–4223.

[22] NEISHTADT, A.I., *Asymptotical study of stability loss of equilibrium under slow transition of two eigenvalues through imaginary axes*, Uspehi Math Nayk, **40:5** (1985), 300–301.

[23] NEISHTADT, A.I., *On delayed stability loss under dynamical bifurcations I*, Differential Equations, **23** (1981), 1385–1390.

[24] NEISHTADT, A.I., *On delayed stability loss under dynamical bifurcations II*, Differential Equations, **24** (1988), 171–176.

[25] NEISHTADT, A.I., *Averaging and passage through resonances*, Proceedings of the International Congress of Mathematicians, Kyoto, Japan, 1990 (1991), The Mathematical Society of Japan, 1271–1283.

[26] NEISHTADT, A.I., *On calculation of stability loss delay time for dynamical bifurcations*, Proceedings of the International Congress of Mathematical Physics, 1994 (ed. by D. Iagoluitzer) (1995), International press, 280–287.

[27] NEISHTADT, A.I., SIMO, C. AND TRESCHEV, D.V., *On stability loss delay for a periodic trajectory*, Progress in nonlinear diff. equations and their appl., **19** (1996), 253–278.

[28] RINZEL, J., *Repetitive activity and Hopf bifurcation under point-stimulation for a simple FitzHugh-Nagumo nerve condition model*, J. Math. Biology, **5** (1978), 363–382.

[29] RINZEL, J. AND KEENER, J.P., *Hopf bifurcation to repetitive activity in nerve*, SIAM J. Appl. Math., **43** (1983), 907–922.

[30] RINZEL, J. AND BAER, S.M., *Firing threshold of the Hodgkin-Huxley model for a slow current ramp: A memory effect and its dependence on fluctuations*, Biophys. J., **54** (1988), 551–555.

[31] SCHECTER, S., *Persistent unstable equilibria and closed orbits of a singularly perturbed equation*, Journal of Differential Equations, **60** (No. 1) (1985), 131–141.

[32] SHISHKOVA, M.A., *Examination of a system of differential equations with a small parameter in the highest derivatives*, Soviet Math. Dokl., **14** (No. 2) (1973), 384–387.

[33] SU, J., Delayed oscillation phenomena in the FitzHugh Nagumo Equation, *Ph.D. Thesis*, University of Minnesota, 1990.

[34] SU, J., *Delayed oscillation phenomena in the FitzHugh Nagumo equation*, Journal of Differential Equations, **105 (1)** (1993), 180–215.

[35] SU, J., *On delayed oscillations in nonspatially uniform FitzHugh Nagumo equation*, Journal of Differential Equations, **110 (1)** (1994), 38–52.

[36] SU, J., *Delayed Bifurcation properties in the FitzHugh Nagumo equation with periodic forcing*, Differential and Integral Equations, **9** (No. 3) (1996), 527–539.

[37] SU, J., *Persistent unstable periodic motions, I*, Jour. of Math. Anal. Appl., **198** (1996), 796–825.

[38] SU, J., *Persistent unstable periodic motions, II*, Jour. of Math. Anal. Appl., **199** (1996), 88–119.

[39] SU, J., *Effects of periodic forcing on delayed bifurcations*, Jour. of Dynamics and Differential Equations, **9 (4)** (1997), 561–625.

[40] WANG, X.-J. AND RINZEL, J., *Osicllatory and bursting properties of neurons*, Handbook of Brain Theory and Neural networks (ed., M.A. Arbib) (1995), MIT press, Cambridge, Mass., 686–691.

[41] ZEGHLACHE, H., MANDEL, D. AND VAN DEN BROECK, C., *Influence of noise on delayed bifurcations*, Phy. Rev. A, **40** (1989), 286–294.

SYNCHRONY IN NETWORKS OF NEURONAL OSCILLATORS

D. TERMAN*

Abstract. We review numerous recent results in which geometric singular perturbation methods have been used to analyze the population rhythms of neuronal networks. The neurons are modeled as relaxation oscillators and the coupling between neurons is modeled in a way that is motivated by properties of chemical synapses. The results give conditions for when excitatory or inhibitory synaptic coupling leads to either synchronized or desynchronized rhythms. Applications to models for sleep rhythms, image segmentation and wave propagation in inhibitory networks are also discussed.

Key words. oscillations, inhibition, synchronization, synaptic coupling, singular perturbation.

AMS(MOS) subject classifications. 34C15, 34C35, 34D15, 92C20.

1. Introduction. Oscillations arise throughout the central nervous system [12, 14, 18, 30, 38]. The thalamus, for example, is centrally important in the generation of sleep rhythms [7, 28, 30, 39]. It has also been implicated in the generation of epilepsy, Parkinson tremors, and 40 hz oscillations [10, 15, 29, 30]. Oscillatory behavior also arises in such areas as respiration, movement and secretion [4, 22].

These neuronal networks often exhibit a rich structure of dynamic behavior. The behavior of even a single cell can be quite complicated [3, 11, 35, 36]; it may, for example, fire repetitive action potentials or bursts of action potentials that are followed by a silent phase of near quiescent behavior [22, 24, 41]. The bursting behavior may wax and wane on a slower time scale [1, 5]. Examples of population rhythms include synchronous behavior, in which every cell in the network fires at the same time, and clustering [9, 17], in which the entire population of cells breaks up into subpopulations or blocks; every cell within a single block fires synchronously and different blocks are desynchronized from each other. Of course, much more complicated population rhythms are also possible [33, 37, 38]. The activity can also propagate through the network in a wave-like manner [6, 8, 16, 23].

A network's population rhythm results from interactions between three separate components: the intrinsic properties of individual neurons, the synaptic properties of coupling between neurons, and the architecture of coupling (i.e., which neurons communicate with each other). These components typically involve numerous parameters and multiple time scales. The synaptic coupling, for example, can be excitatory or inhibitory, and its possible turn on and turn off rates can vary widely. Neuronal systems may include several different types of cells as well as different types of coupling.

*Department of Mathematics, Ohio State University, Columbus, Ohio 43210.

An important and typically very challenging problem is to determine the role each component plays in shaping the emergent network behavior.

In this paper, we review numerous recent results in which geometric singular perturbation methods have been used to analyze the population rhythms of neuronal networks. The neurons that we analyze are modeled as relaxation oscillators and the synaptic coupling between neurons is modeled as in [31]. We begin by considering two mutually coupled cells and find conditions for when excitatory or inhibitory synaptic coupling leads to either synchronized or desynchronized rhythms. All four cases are possible and we demonstrate how geometric methods help to understand how each case depends on interactions between the intrinsic and synaptic properties. We then briefly discuss extensions of the results for two mutually coupled cells to larger networks.

An outline of the paper is as follows. In the next section we present the explicit equations for the networks to be considered. In Section 3, we give conditions for when excitatory coupling leads to synchronized rhythms and inhibitory coupling leads to desynchronized rhythms. These will be the case if the neurons have a sufficiently long active phase and the synapses turn on and turn off sufficiently fast. In Section 4, we review results in [31] where explicit conditions are given for when mutually coupled inhibitory neurons can synchronize. As we shall see, this is possible if the synapses turn off at a sufficiently slow rate. In Section 5, we review results in [2] for when excitatory coupling leads to desynchronized rhythms. Here the synapses may be fast; however, the active phase of the neurons must be sufficiently short. This corresponds to a spiking, instead of a bursting, neuron. There is a discussion of these results, with consequences for larger networks, in Section 6.

We note that there have been a tremendous number of papers related to the activity of neuronal networks. No attempt is made here to give an exhaustive list of such papers. A more complete list of papers related to issues considered here is given in [25].

2. The model.

2.1. The single neuron.

We model each individual neuron, without any coupling, as the relaxation oscillator

$$\begin{aligned} v' &= f(v,w) \\ w' &= \epsilon g(v,w) \end{aligned} \tag{2.1.1}$$

Here ϵ is assumed to be small. We assume that the $v-$nullcline $f(v,w) = 0$ defines a cubic-shaped curve and the $w-$nullcline $g(v,w) = 0$ is an increasing graph in the v,w plane that intersects $f = 0$ at a unique point p_0. We also assume that $f > 0$ ($f < 0$) below (above) the $v-$nullcline, and $g > 0$ ($g < 0$) below (above) the $w-$nullcline.

If the intersection p_0 of the nullclines lies on the left or the right branch of $f = 0$, then p_0 corresponds to a stable fixed point of (2.1.1) providing that ϵ is small enough. If p_0 is on the left branch, we say that (2.1.1) is *excitable*. If p_0 lies on the middle branch of $f = 0$, then (2.1.1) is *oscillatory*; there is a stable limit cycle, again if ϵ is sufficiently small. In the limit $\epsilon \to 0$, the periodic solution lies close to a singular periodic orbit. This singular orbit consists of four pieces, two slow and two rapid transitions between the slow pieces. We refer to the parts of the trajectory on the left and right branches as the *silent* and *active* phases. These phases end when the trajectory reaches either the left knee or the right knee of $f = 0$, initiating a rapid *jump up* to the active phase or *jump down* to the silent phase.

2.2. Coupling. We model a pair of mutually inhibitory neurons by the following system of differential equations

$$\begin{aligned} v_1' &= f(v_1, w_1) - s_2 g_{syn}(v_1 - v_{syn}) \\ w_1' &= \epsilon g(v_1, w_1) \\ v_2' &= f(v_2, w_2) - s_1 g_{syn}(v_2 - v_{syn}) \\ w_2' &= \epsilon g(v_2, w_2) \end{aligned} \tag{2.2.1}$$

Here (v_1, w_1) and (v_2, w_2) correspond to the two cells and g_{syn} is positive. The reversal potential v_{syn} determines whether the synapse is excitatory or inhibitory. If $v > v_{syn}$ along each bounded singular solution, then the synapse is inhibitory; while, if $v < v_{syn}$ along each bounded singular solution, then the synapse is excitatory. The terms s_i, $i = 1, 2$, measure how the postsynaptic conductance depends on the presynaptic potentials v_i. We consider two choices for the s_i. For a "direct" synapse, we assume that each s_i satisfies a first order equation of the form

$$s_i' = \alpha(1 - s_i)\, H(v_i - \theta_{syn}) \; - \; \beta s_i \tag{2.2.2}$$

where α and β are positive constants, H is the Heaviside step function, and θ_{syn} is a threshold above which one cell can influence the other. Note that α and β are related to the rates at which the synapses turn on or turn off. For *fast synapses*, we assume that both of these constants are $O(1)$ with respect to ϵ. For a *slow synapse*, we assume that $\alpha = O(1)$ and $\beta = O(\epsilon)$; hence, a slow synapse activates on the fast time scale, but turns off on the slow time scale.

Synapses may be very complicated, involving, for example, secondary processes such as G-proteins. We shall refer to these as *indirect synapses*. We model these by introducing new dependent variables x_1 and x_2, which satisfy first order equations. Each x_i and s_i satisfies the equations

$$\begin{aligned} x_i' &= \epsilon\alpha_x(1 - x_i)H(v_i - \theta_v) - \epsilon\beta_x x_i \\ s_i' &= \alpha(1 - s_i)H(x_i - \theta_x) - \beta s_i \end{aligned} \tag{2.2.3}$$

The constants α_x and β_x are assumed to be independent of ϵ. The effect of the indirect synapses is to introduce a delay from the time one oscillator jumps up until the time the other oscillator feels the synaptic input. For example, if the first oscillator jumps up, the secondary process is turned on when v_1 crosses the threshold θ_v. The synapse s_1 does not turn on until x_1 crosses θ_x; this takes a finite amount of (slow) time since x_1 evolves on the slow time scale, like the w_i.

2.3. Geometric approach. All of the networks in this paper are analyzed by treating ϵ as a small, singular perturbation parameter. The first step in the analysis is to identify the fast and slow variables. We then dissect the full system of equations into fast and slow subsystems. The fast subsystem is obtained by simply setting $\epsilon = 0$ in the original equations. This leads to a reduced set of equations for the fast variables with each of the slow variables constant. The slow subsystems are obtained by first introducing the slow time scale $\tau = \epsilon t$ and then setting $\epsilon = 0$ in the resulting equations. This leads to a reduced system of equations for just the slow variables, after solving for each fast variable in terms of the slow ones. The slow subsystems determine the evolution of the slow variables while the cells are in either the active or silent phases. During this time, each cell lies on either the left or right branch of some "cubic" determined by the total synaptic input which the cell receives. This continues until one of the cells reaches the left or right "knee" of its corresponding cubic. The cell may either jump up from the silent to the active phase or jump down from the active to the silent phase. The jumping up or down process is determined by the fast equations.

We note that the geometric approach used here is somewhat different from that used in many dynamical systems studies. All of the networks considered here consist of many differential equations, especially for larger networks. Traditionally, one would interpret the solution of this system as a single trajectory evolving in a very large dimensional phase space. We consider several trajectories, one corresponding to each cell, moving around in a much lower dimensional phase space. After reducing the full system to a system for just the slow variables, the dimension of the lower dimensional phase space equals the number of slow intrinsic variables and slow synaptic variables corresponding to each cell. In the worst case considered here, there is only one slow intrinsic variable for each cell and one slow synaptic variable; hence, we never have to consider phase spaces with dimension more than two. Of course, the particular phase space we need to consider may change, depending on whether the cells are active or silent and also depending on the synaptic input that a cell receives.

In what follows, it will be convenient to introduce the following notation. Let $\Phi(v,w,s) \equiv f(v,w) - g_{syn}s(v - v_{syn})$. We assume that $\mathcal{C}_s \equiv \{(v,w) : \Phi(v,w,s) = 0\}$ is a cubic-shaped curve for each $s \in [0,1]$.

3. Fast threshold modulation and post inhibitory rebound.

3.1. Synchronization with excitatory synapses. We now consider two mutually coupled cells with excitatory synapses. The synapses are assumed to be fast and direct, and each cell is assumed to be oscillatory for each fixed level of the synaptic input. There is then no problem in proving the existence of the synchronous solution. This orbit lies on the left branch of $\mathcal{C}_0$ during the silent phase and along the right branch of $\mathcal{C}_1$ during the active phase.

We next show that this solution is stable. To do this, we begin with both cells close to each other on the left branch of $\mathcal{C}_0$. We further assume that cell 1 is at the left knee of $\mathcal{C}_0$ about ready to jump up. We then follow the cells around in phase space and demonstrate that after one complete cycle they are closer to each other than when they started. The singular solution consists of several pieces.

The first piece of the singular solution begins when cell 1 jumps up. When $v_1(t)$ crosses θ_{syn}, $s_1(t) \to 1$. This 'raises' the cubic corresponding to cell 2 from $\mathcal{C}_0$ to $\mathcal{C}_1$. If $|w_1(0) - w_2(0)|$ is sufficiently small, then cell 2 lies below the left knee of $\mathcal{C}_1$. The fast equations force cell 2 to also jump up to the active phase. Note that this piece takes place on the fast time scale. Hence, on the slow time scale, both cells jump up at precisely the same time.

The second piece of the singular solution is when both oscillators lie on the right branch of $\mathcal{C}_1$. Note that the ordering in which the oscillators track along the left and right branches is now reversed. While in the silent phase cell 1 was ahead of cell 2. In the active phase, cell 2 leads the way. The oscillators remain on the right branch of $\mathcal{C}_1$ until cell 2 reaches the right knee.

The next piece of the singular solution is when the oscillators jump down to the silent phase. Cell 2 is the first to jump down. When $v_2(t)$ crosses θ_{syn}, s_2 switches from 1 to 0 on the fast time scale. This lowers the cubic corresponding to cell 1 from $\mathcal{C}_1$ to $\mathcal{C}_0$. If, at this time, cell 1 lies above the right knee of $\mathcal{C}_1$, then cell 1 must jump down to the silent phase.

The final piece of the singular solution is when both oscillators move down the left branch of $\mathcal{C}_0$ until cell 1 reaches the left knee. This completes one complete cycle.

In order to prove that the synchronous solution is stable, we must show that the cells are closer to each other after this cycle; that is, there is a compression in the 'distance' between the cells. There are actually two ways to show this compression; these correspond to two different ways to define what is meant by the 'distance' between the cells. One way is to consider a Euclidean metric between the cells when they lie on the same branch of a cubic. That is, if both cells lie on the same branch with the coordinates of cell i given by (v_i, w_i), then the distance between the cells is defined as simply $|w_1 - w_2|$. Note that this metric remains invariant as the cells jump up or jump down. This is because the jumps are horizontal and the values of w_i do not change. One can show, however, that the metric decreases as the cells evolve on the left and right branches if one imposes some mild assumptions on the nonlinearities in (2.1.1). We do not present this analysis here; see [34].

A second way to prove compression is to consider a 'time-metric'. This is defined as follows. Suppose that both cells are in the silent phase with cell 2 trailing cell 1. Then the distance between the cells is the time it takes for cell 2 to evolve on the left branch of $\mathcal{C}_0$ until it reaches the position of cell 1. The distance between the cells on the right branch is defined in a similar manner. We note that, unlike before, this metric is invariant as the cells evolve along the same branch. The compression takes place as the cells jump up and jump down. Again, some assumptions are required on the nonlinear functions in (2.1.1). This analysis is given in [26], where this synchronization mechanism is referred to as *fast threshold modulation.* Analysis for larger networks is given in [34].

3.2. Desynchrony with inhibitory synapses. We next consider two mutually coupled cells with inhibitory synapses. As before, we assume that the synapses are direct and fast. Here we assume, however, that each cell is excitable for each fixed level of input. We show that this network typically leads to antiphase behavior. Of course, this requires some assumptions on the nonlinearities in the network; these will follow naturally from the analysis.

As before, we begin with both cells in the silent phase with cell 1 at the left knee of $\mathcal{C}_0$ ready to jump up. Cell 2 is some small distance behind cell 1. The first piece of the singular solution begins when cell 1 jumps up. When v_1 crosses θ_{syn}, $s_1 \to 1$ and the cubic corresponding to cell 2 switches from $\mathcal{C}_0$ to $\mathcal{C}_1$. Note, however, that because the synapses are inhibitory, $\mathcal{C}_1$ lies *below* $\mathcal{C}_0$. Hence, cell 2 jumps from the left branch of $\mathcal{C}_0$ to the left branch of $\mathcal{C}_1$; it remains in the silent phase.

The second piece of the singular solution is when cell 1 is in its active phase and cell 2 is silent. Since the cells are assume to be excitable for all fixed levels of input, cell 2 approaches the stable fixed point that lies on the left branch of $\mathcal{C}_1$. During this time, cell 1 moves up the right branch of $\mathcal{C}_0$ until it reaches the right knee. Cell 1 is then ready to jump down to the silent phase.

When cell 1 jumps down and v_1 crosses θ_{syn}, s_1 switches from 1 to 0. This raises cell 2's cubic from $\mathcal{C}_1$ to $\mathcal{C}_0$. If cell 2 lies below the left knee of $\mathcal{C}_0$, then it jumps up to the active phase when cell 1 jumps down to $\mathcal{C}_1$. This mechanism, by which one cell jumps up upon being released from inhibition, is often referred to as *post inhibitory rebound.*

Cell 1 now moves down the left branch of $\mathcal{C}_1$ while cell 2 moves up the right branch of $\mathcal{C}_0$. This continues until cell 2 reaches the right branch and jumps down. If, at this time, cell 1 lies below the left knee of $\mathcal{C}_0$, then cell 1 jumps up due to post inhibitory rebound. One can then continue this solution to obtain a solution which behaves in an antiphase manner. The cells take turns lying in the active phase; one cell jumps up when it is released from inhibition by the other cell jumping down. Note that this antiphase behavior arises if the fixed point on the left branch of $\mathcal{C}_1$ lies below the left knee of $\mathcal{C}_0$ and the time the cells spend in the active phase is sufficiently long. If this active phase is too short, then the active cell will jump down before the silent cell is able to lie below the left knee of $\mathcal{C}_0$. In this case, the network will 'crash'; both cells will approach the stable quiescent state.

4. Synchrony with inhibitory synapses. Here we review the results in [31] concerning two mutually coupled cells with slow inhibitory synapses. The main result is that the synchronous oscillations may be stable if the synapses are indirect and sufficiently slow. We assume throughout this section that $\beta = \epsilon K$ where K is some fixed constant.

4.1. Singular solutions. We derive slow subsystems for when the cells lie in either the silent or active phase. There are several cases to consider and we only discuss two of these in detail. Here, we only consider direct synapses; the derivation of the slow equations for indirect synapses is very similar and will be commented on at the end of this subsection.

If both cells are silent, then $v_i < \theta_{syn}$ and the first term in (2.2.2) is zero. Hence, after letting $\tau = \epsilon t$ and setting $\epsilon = 0$, (2.2.1), (2.2.2) become

$$\begin{aligned} 0 &= f(v_i, w_i) - s_j g_{syn}(v_i - v_{syn}) \\ \dot{w}_i &= g(v_i, w_i) \\ \dot{s}_i &= -K s_i \end{aligned} \tag{4.1.1}$$

where $j \neq i$. This system can be simplified as follows. We write the left branch of $\mathcal{C}_s$ as $v = \Phi_L(w, s)$ and let $G_L(w, s) = g(\Phi_L(w, s), s)$. Each (w_i, s_j) must then satisfy the system

$$\begin{aligned} \dot{w} &= G_L(w, s) \\ \dot{s} &= -K s \end{aligned} \tag{4.1.2}$$

These equations determine the evolution of the slow variables while in the silent phase.

If both cells are active, then s_i is a fast variable. The only slow variables are the w_i. Instead of (4.1.1), the slow variables are now

$$\begin{aligned} 0 &= f(v_i, w_i) - g_{syn}(v_i - v_{syn}) \\ \dot{w}_i &= g(v_i, w_i) \\ 1 &= s_i \end{aligned} \tag{4.1.3}$$

This can be reduced to a system for just the slow variables as before. Denote the right branch of $\mathcal{C}_s$ by $v = \Phi_R(w, s)$ and let $G_R(w, s) = g(\Phi_R(w, s), s)$.

Then each w_i satisfies the scalar equation

$$\dot{w} = G_R(w, 1) \tag{4.1.4}$$

In a similar fashion, we can derive the slow subsystem for when one cell is active and the other is silent. For indirect synapses, there are further cases depending on if the x_i-variables have crossed their threshold θ_x or not. This is discussed in detail in [31].

4.2. Existence of synchronous oscillations. The singular synchronous solution is easily constructed; here we consider the case of direct synapses. We begin with both cells at the right knee of the right branch of $\mathcal{C}_1$. It is from this point that the cells jump down to the silent phase. While in the silent phase, the slow variables evolve according to (4.1.2). The cells can only leave the silent phase once they reach a left knee of one of the left branches. If the cells are able to reach such a point, then they will jump up to the active phase and return to the starting point. Hence, the existence of the synchronous solution depends on whether the cells can reach one of the jump-up points while in the silent phase.

If the cells are oscillatory, then the synchronous trajectory must reach one of the jump-up points. This is demonstrated in [31], where it is also shown that a synchronous solution can exist even though both cells, without any coupling, are excitable. This will be the case if the rate K of decay of inhibition is small enough and the cells are oscillatory for some fixed values of $s \in (0, 1)$. This is not possible if K is too large, since then the system decays fast and behaves in the slow regime like the system with $s = 0$.

The construction of the synchronous solution for the case of indirect synapses is very similar. There are some additional complications due to the additional slow variables x_i. The complete analysis is given in [31].

4.3. Instability of the synchronous solution for direct synapses. We show in this section that the synchronous solution is not stable when the synapse is direct. In the next section, we state results which show that the synchronous solution may be stable if the synapses are indirect. The stability depends on whether certain relationships between the parameters in (2.2.1), (2.2.3) are satisfied.

Suppose that the synapse is direct. We start with both oscillators in the silent phase, and assume that cell 1 reaches the jump-up curve $w_L(s)$ first. When cell 1 jumps, cell 2 begins to feel inhibition as v_1 crosses θ_{syn}. For the direct synapse, this inhibitory conductance, s_1, jumps instantly (in the slow time scale) to $s_1 = 1$. The effect of this on the v_2, w_2 equations is to instantly move the second cell away from its threshold. The amount that cell 2 is removed from the threshold stays bounded away from zero no matter how close the second cell was originally to the first one. Thus, infinitesimally small perturbations get magnified at this stage of the dynamics to finite size. Though the points can get somewhat closer as they transverse the rest of the cycle, they remain a finite distance apart over the cycle. This shows that small perturbations are expanded by the dynamics, and hence the synchronous solution is not stable to infinitesimal perturbations.

4.4. Statement of the main result. We now consider indirect synapses and show that the synchronous solution can be stable in some parameter ranges. We shall show that there are two combinations of parameters that govern the stability. Furthermore, only one of those two combinations matters to the stability.

For this result, it is necessary to make some further assumptions on the nonlinearities and parameters in (2.2.1), (2.2.3). It will be necessary to assume that

$$f_w < 0, \quad g_v > 0, \quad \text{and} \quad g_w < 0 \tag{4.4.1}$$

near the $v-$nullcline. For the main Theorem, we also assume that $f(v, w)$ is given by

$$f(v, w) = f_1(v) - g_c w(v - v_R) \tag{4.4.2}$$

where $g_C > 0$ and $v_R \leq v_{syn}$ represents a maximal conductance and reversal potential, respectively. This includes the well-known Morris-Lecar equations [20]. The analytical framework we develop, however, also applies to more general nonlinearities which satisfy (4.4.1). Some technical assumptions are also required on the nonlinear function $g(v, w)$. We need to assume that g_v is not too large near the right branches of the cubics $\mathcal{C}_s$.

We assume that the parameters α_x and β_x are sufficiently large, and $\dfrac{\alpha_x}{\alpha_x + \beta_x} > \theta_x$. This guarantees that each x_i can cross its threshold in order to turn on the inhibition. Precise conditions on how large α_x and β_x must be are given in [31].

We need to introduce some notation. Let a_- be defined as the minimum of $-\partial g/\partial w$ over the synchronous solution in the silent phase. Note from (4.4.1) that $a_- > 0$. Let $(w^*, s^*) = (w_L(s^*), s^*)$ be the point where the synchronous solution meets the jump-up curve, and let $\lambda = w_L'(s^*)$ be the reciprocal slope of the jump-up curve at this point. Finally, let a_+ denote the value of $g(v, w)$ evaluated on the right hand branch at the point where the synchronous solution jumps up. The main result is then the following. It holds for ϵ sufficiently small.

THEOREM 1. *Assume that the nonlinear functions and parameters in (2.2.1) and (2.2.3) satisfy the assumptions stated above. If $K < a_-$ and $Ks^* < a_+/|\lambda|$, then the synchronous solution is asymptotically stable.*

REMARK 1. The first condition in Theorem 1 is consistent with the numerical simulations of [42], who obtained synchronized solutions when the synapses recover at a rate slower than the rate at which the neurons recover in their refractory period.

REMARK 2. In order to interpret the second condition in Theorem 1, note that Ks^* is the rate of change of s at the point at which the synchronous solution jumps. a_+ is the rate of change of w on the right hand branch right after the jump. Since $\lambda = dw_L/ds$, multiplication by $1/|\lambda|$ transforms changes in w to changes to s. Thus, the second condition is analogous to the compression condition that produces synchrony between relaxation oscillators coupled by fast excitation as described in the previous section. $|\lambda|$ may be thought of as giving a relationship between the time constants of inhibitory decay and recovery of the individual cells; a larger $|\lambda|$ (corresponding to a flatter curve) means that a fixed increment of decay of inhibition (Δs) has a larger effect on the amount of recovery that a cell must undergo before reaching its (inhibition-dependent) threshold for activation.

REMARK 3. The two conditions needed in the statement of Theorem 1 correspond to two separate cases considered in the proof of Theorem 1. These two cases correspond to whether the two cells preserve their orientation (Case 1) or reverse their orientation (Case 2) on the right branch of the $s = 1$ cubic after one cycle. Theorem 1 says that, whatever case the synchronous trajectory falls into, if both conditions hold, then the synchronous solution is stable. Note, however, that the different cases require different conditions. Case 1 requires $K < a_-$ and Case 2 requires $Ks^* < a_+/|\lambda|$. Thus, by changing a parameter that switches the system between Case 1 and Case 2, one can change which combinations of time scales and other parameters control the stability of the synchronous solution. In particular, the system can be changed between stable and unstable without changing any time constants.

4.5. Outline of the proof of the main result. Our goal here is not to give a complete proof of Theorem 1; this is given in [31]. Instead we briefly illustrate where the compression of trajectories come from and explain the role of the second slow variable in producing this compression. We must first define what is meant by the 'distance' between the two cells, since compression always refers to a decrease of some notion of distance. Different notions of distance are needed depending on if the cells are silent or active.

Suppose that both cells are active. Then the cells lie on the right branch of either $\mathcal{C}_1$ or $\mathcal{C}_0$ depending on whether the variables x_i have crossed threshold or not. To simplify the discussion, we assume here that the rate at which a slow variable w_i evolves in the active phase does not depend on which cubic it lies on; that is, $G_R(w, s)$ is independent of s. We can then define the distance between the cells as the time it takes for the w-value of the lagging cell to reach the w-value of the leading cell. We denote this distance at time τ by $\rho_R(\tau)$. Note that this metric is invariant as long as the cells remain in the active phase.

Now suppose that both cells are silent and the cells lie near the synchronous trajectory. This will be the case if we consider perturbations of the synchronous solution for a stability analysis. Let (w^*, s^*) be the point on the curve $w = w_L(s)$ where the synchronous solution jumps up to the active phase. Fix τ and consider the physical translate of the jump-up curve where we translate the point (w^*, s^*) until it meets $(w_1(\tau), s_1(\tau))$. The distance is defined as the time it takes for cell 2 to evolve according to (4.1.2) until it crosses this translated curve. We denote this distance by $\rho_\lambda(\tau)$. Note that if τ is chosen so that cell 1 is on the curve of knees and cell 2 is still silent, then the $\rho_\lambda(\tau)$ is the time it takes for cell 2 to reach the curve of knees; that is, $\rho_\lambda(\tau)$ is the time from when one cell fires until the other cell does so.

This metric need not be invariant; however, the analysis in [31] shows that it must decrease if the rate K of decay of inhibition is sufficiently small. This follows from an explicit calculation for the derivative $\rho_\lambda{}'(\tau)$.

We now show that if K is sufficiently small, then there is compression as the cells evolve from the silent to the active phase. This is a key step in the proof of Theorem 1. We start with both cells in the silent phase and show that after they both jump up, the distance between them must decrease. There are actually several cases to consider; here we only consider one of them. For this case, we assume that when $\tau = 0$, both cells are in the silent phase with $s_1(0) > s_2(0)$. Let $T_i > 0$ be the time when cell i jumps up. We assume that $T_1 < T_2$ and $w_1(T_2) > w_2(T_2)$. That is, cell 1 is the first cell to jump up and after both cells jumps up, cell 1 is 'ahead' of cell 2. We claim that $\rho_R(T_2) < \rho_\lambda(0)$.

To prove this, note that both cells are silent for $0 < \tau < T_1$. Since $\rho_\lambda(\tau)$ decreases, for K sufficiently small, $\rho_\lambda(T_1) \equiv T_2 - T_1 < \rho_\lambda(0)$ and it suffices to prove that $\rho_R(T_2) < T_2 - T_1$. The first step in the proof of this last statement is to note that our assumption $s_1(0) > s_2(0)$ implies that $s_1(\tau) > s_2(\tau)$ as long as the cells are silent; in particular, $s_1(T_1) > s_2(T_1) > s_2(T_2)$. Since the curve of knees $w_L(s)$ has positive slope, we conclude that $w_1(T_1) < w_2(T_2)$. Together with our assumption that that $w_1(T_2) > w_2(T_2)$, this implies that the time on the right branch from $w_2(T_2)$ to $w_1(T_2)$ is less than the time on the right branch from $w_1(T_1)$ to $w_1(T_2)$. The first of these times is $\rho_R(T_2)$ and the second is $T_2 - T_1$. This is precisely what we wished to show. As mentioned above, there are several other cases to consider. A complete analysis is given in [31].

REMARK 4. The role of the slow synapse in our geometric analysis is to introduce a second slow variable while the cells are in the silent phase. This adds a dimension to both the phase plane of the slow silent subsystem and the manifold of jump-up points. The additional dimensions are important for both the existence and the stability of the synchronous solution. For existence, they permit the synchronous trajectory to 'escape' along the jump-up curve although the uncoupled cell is not able to do so. As for stability, the slow decay of inhibition leads to compression while the slow variables evolve in the slow silent phase plane. A second source of compression results from the interaction between the geometry of the jump-up curve $w_L(s)$ and the slow manifolds corresponding to the slow processes. In particular, the cells are able to jump up at different points along $w_L(s)$. Our analysis demonstrates that the slope of this curve leads to compression in the time-metric during the jumping up process.

5. Desynchrony with excitatory synapses. We now briefly discuss results in [2] which demonstrate that excitatory coupling can lead to *almost synchronous solutions* if the active phase of a single cell is much shorter than the silent phase; that is, the cells correspond to spiking neurons. This work is motivated by simulations in [21] who showed that there could be a stable state close to synchrony when the fully synchronized state was unstable.

Here we write the equations for a single oscillator as

$$\begin{aligned} v' &= f(v,w) \\ w' &= \frac{\epsilon g(v,w)}{\tau_\infty(v)} \end{aligned} \tag{5.1}$$

where the function $\tau_\infty(v)$ is given by

$$\tau_\infty(v) \;=\; 1 \quad \text{if} \quad v \;<\; v_\theta \qquad \text{and} \qquad \epsilon/\gamma \quad \text{if} \quad v \;\geq\; v_\theta$$

The parameter v_θ is the threshold for entering the active phase and γ governs the rate of passage through the active phase. We assume that v_θ lies between the two knees of the cubic $f = 0$, which we denote by $\mathcal{C}_0$. The functions f and g are as before.

The equations for two mutually coupled oscillators are

$$\begin{aligned} v_i' &= f(v_i,w_i) - g_{syn} H(v_j - v_{st})(v_i - v_{syn}) \\ w_i' &= \frac{\epsilon g(v_i,w_i)}{\tau_\infty(v_i)} \end{aligned} \tag{5.2}$$

and v_{st} is the synaptic threshold. v_{syn} is chosen to be high so that the synapse is excitatory; that is, $v_i - v_{syn} < 0$ along the singular solutions. The v-nullcline is the same as that of the uncoupled cell when $v_j < v_{st}$. For $v_j > v_{st}$, the effect of the coupling term is to raise the nullcline and change its shape; if g_{syn} is not too large, then the v-nullcline is still qualitatively cubic. We denote this cubic by C_1. Note that there are two thresholds v_θ and v_{st} in the equations. To simplify the discussion, we assume here that $v_{st} \;=\; v_\theta$.

As in previous sections, we analyze the solutions by treating ϵ as a small, singular perturbation parameter. We construct singular solutions after formally setting $\epsilon = 0$. The first step in constructing the singular solution is to use fast and slow time scales to derive equations for the fast and slow flows. An interesting feature of the model considered here is that the structure of these flows is quite different depending on if cells are silent or active.

We first demonstrate how to construct the singular trajectory for a single uncoupled cell. When ϵ is set equal to 0 in (5.1), we obtain the equations for the fast flow

$$\begin{aligned} v' &= f(v,w) \\ w' &= 0 \end{aligned} \tag{5.3}$$

if $v < v_\theta$ and

$$\begin{aligned} v' &= f(v,w) \\ w' &= \gamma g(v,w) \end{aligned} \tag{5.4}$$

if $v \geq v_\theta$. By introducing the rescaling $\tau = \epsilon t$ into (5.1) and then setting $\epsilon = 0$, we obtain equations for the slow flow in the silent phase,

$$\begin{aligned} 0 &= f(v,w) \\ \dot{w} &= g(v_L(w),w) \end{aligned} \tag{5.5}$$

if $v < v_\theta$, where $v_L(w)$ is obtained by solving $0 = f(v,w)$ along the left branch of the cubic $\mathcal{C}_0$.

Note that the equations (5.3) and (5.5) are simply scalar equations; in (5.3), the variable w serves as a parameter in the v' equation. The full two-dimensional system (5.1) has been reduced to two one-dimensional equations and solutions of these are easy to characterize. However, (5.4) is not reduced. In other fast-slow systems of the form (5.1), typically the entire v-nullcline consists of rest points for the fast flow. In this case, however, only the portion of $\mathcal{C}_0$ with $v < v_\theta$ consists of rest points.

The singular periodic orbit for a single uncoupled cell is constructed as follows. We begin the orbit begins at the left knee of $\mathcal{C}_0$, which we denote by (v_{LK}, w_{LK}). The first part of the singular orbit is a solution of (5.3) which connects (v_{LK}, w_{LK}) and (v_θ, w_{LK}). The second part is a solution of (5.4) which connects (v_θ, w_{LK}) to (v_θ, w_h), where $w_h > w_{LK}$; this corresponds to the active phase. The third part is a solution to (5.3) which connects (v_θ, w_h) back to $\mathcal{C}_0$ at some point $(v_L(w_h), w_h)$. The fourth and final part is a solution to (5.5) which connects $(v_L(w_h), w_h)$ to (v_{LK}, w_{LK}).

The construction of a synchronous solution for the coupled system is done in the same manner as that of the periodic solution for the uncoupled cell; the only difference is that the dynamics are changed as the voltage passes across $v_{st} = v_\theta$. Thus, if each $v_i > v_{st}$, then (5.4) is replaced by

$$\begin{aligned} v' &= f(v,w) - g_{syn}(v - v_{syn}) \\ w' &= \gamma g(v,w) \end{aligned} \tag{5.6}$$

Note that the synchronous solution is not the same as the uncoupled periodic solution. This is because they satisfy different equations while in the active phase.

There is no difficulty in proving that these singular periodic solutions perturb, for $\epsilon > 0$ small, to actual solutions of either (5.1) or (5.2). See, for example, [19]. The actual periodic orbits for ϵ small lie $O(\epsilon)$ close to the singular orbits, except near the left knee of $\mathcal{C}_0$.

To construct the almost-synchronous solutions, we again work with singular solutions. There are now more cases to consider, depending on which of the cells is silent or active. If both cells are silent (that is, each $v_i < v_\theta$), then the fast flow corresponding to each cell is given by (5.3), while the slow flow is given by (5.5). If both cells are active, then the fast flow for each cell is (5.6); there is no slow flow. Finally, suppose that one of the cells, say cell 1, is active and cell 2 is silent. Then the fast flow for cell 1 is (5.4) and the fast flow for cell 2 is

$$\begin{aligned} v' &= f(v,w) - g_{syn}(v - v_{syn}) \\ w' &= 0 \end{aligned} \tag{5.7}$$

There is no slow flow for this case.

We now give a heuristic argument to explain why there may exist a stable almost synchronous solution. We describe the construction of a singular solution of (5.2) in which one of the cells, say cell 1, begins at the left knee of $\mathcal{C}_0$ and cell 2 lies on the left branch of $\mathcal{C}_0$ just above cell 1. We follow the cells around in phase space until one of the cells returns to the left knee. If the other cell returns to the initial position of cell 2, then this will correspond to an almost synchronous solution. As in the previous constructions, the singular solution consists of several pieces.

The first piece is when cell 1 leaves the left knee; it satisfies (5.3) moving horizontally until it crosses v_θ. Suppose that this is when $t = t_1$. At this time, $H(v_1 - v_\theta)$ switches from 0 to 1 and cell 2 then satisfies (5.7). If cell 2 initially lies below the left knee of C_1, then it will jump up, continuing to satisfy (5.7) until v_2 crosses v_θ. Suppose that this is when $t = t_2$. For $t_1 < t < t_2$, cell 1 satisfies (5.4). Note that its trajectory is no longer horizontal. We assume in this heuristic argument that $v_1 > v_\theta$ for $t_1 < t < t_2$. When $t = t_2$, $H(v_2 - v_\theta)$ switches from 0 to 1 and cell 1 then satisfies (5.6). Assume that cell 1 is the first cell to cross v_θ again and this is when $t = T_1$. Then cell 2 satisfies (5.6) until it crosses v_θ, say when $t = T_2$. Once a cell crosses v_θ, it moves horizontally to a point on the left branch of $\mathcal{C}_0$. The cells then evolve along the left branch of $\mathcal{C}_0$ until one of the cells returns to the left knee and this completes the cycle. We note that it is possible for either of the cells to be the first to reach the left knee.

The singular flow naturally gives rise to a one-dimensional map, which we denote by Π. More precisely, if (v_2, w_2) is the initial position of cell 2 on the left branch of $\mathcal{C}_0$ and $(\hat{v}, \hat{w})$ is the position of the trailing cell after the other cell has returned to the left knee, then $\Pi(w_2) = \hat{w}$. This map is well defined if $w_2 - w_{LK}$ is sufficiently small; in particular, cell 2 should lie below the left knee of the excited cubic C_1. A fixed point of Π corresponds to a periodic singular solution. The orbit may be orientation preserving or orientation reversing, depending on if the first cell to reach the left knee of $\mathcal{C}_0$ is cell 1 or cell 2, respectively.

In [2], we give conditions which guarantee that Π defines a uniform contraction on some interval. We also prove that the resulting periodic singular solution perturbs, for ϵ small, to a stable periodic solution of (5.2). This periodic solution is referred to as an $O(\epsilon)$-synchronous solution for the following reason. The analysis demonstrates that if $\hat{w}$ is the fixed point of Π, then this must lie below the left knee of the excited cubic C_1. Hence, if cell 1 is the first to jump up, then cell 2 will jump up as soon as cell 1 crosses v_θ. The difference between the times the cells reach v_θ is therefore $O(1)$ in t-time. Since the total period in t–time is $O(1/\varepsilon)$, the normalized time difference ($\Delta time/period$), is $O(\varepsilon)$.

We conclude by stating two Theorems. The first is concerned with the existence and stability of almost-synchronous singular solutions; these correspond to fixed points of the map Π. The second Theorem demonstrates that the almost-synchronous singular solutions perturb for small, positive ϵ to $O(\epsilon)$–synchronous solutions of (5.2)

Before stating the Theorems, it is necessary to introduce some notation. Assume that the right knee of $\mathcal{C}_1$ is at (v_{RK}, w_{RK}) and choose β so that $g(\beta, w_{RK}) = 0$. We say that constants δ_1 and δ_2, which depend on γ, are $O(\gamma)$ apart if there exist K_1 and K_2 such that $K_1\gamma < |\delta_1 - \delta_2| < K_2\gamma$ for γ small.

THEOREM 2. *Suppose that the constants v_θ and β satisfy one of the following conditions:*

(R) Either $v_\theta > \beta$, , or $v_\theta < \beta$ and $\beta - v_\theta$ is sufficiently small.

(P) $v_\theta < \beta$ and g_{syn} is sufficiently small.

Then, for γ sufficiently small, there exist $\Delta_1 < \Delta_2$ such that Π defines a uniform contraction from the interval (Δ_1, Δ_2) into itself. If w_* is the resulting asymptotically stable fixed point of Π, then w_* and w_{LK} are $O(\gamma)$ apart. If (P) is satisfied, then Π is orientation preserving on (Δ_1, Δ_2), while if (R) is satisfied, then Π is orientation reversing on (Δ_1, Δ_2).

THEOREM 3. *The asymptotically stable singular periodic solution given by Theorem 2 perturbs, for $\epsilon > 0$, to an asymptotically stable $O(\epsilon)$-solution of (5.2).*

6. Applications. The techniques presented earlier generalize to larger networks which arise in important applications. Here we review some of these recent results.

In [33], a network is introduced as a model for oscillations in the CA3 region of the hippocampus. This network is motivated by models considered in Traub and Miles [38]. The network contains three types of cells, these being excitatory and both fast and slow inhibitory neurons. The network exhibits partially synchronized rhythms, by which we mean that only a small percentage of the cells participate in each event of neural activity. The number of cells which fire during each event is roughly constant; moreover, the cells which fire together in one event need not fire together in later events. The analysis in [33] clarifies the role that each type of cell and each parameter plays in obtaining partially synchronized rhythms. A symbolic scheme is derived which accurately determines the firing patterns for a large class of networks. The symbolic scheme predicts which cells fire during each event of activity and demonstrates why simple networks may give rise to multiple stable firing patterns.

In [34, 40], a model for image segmentation based on locally excitatory and globally inhibitory oscillator networks is proposed and analyzed. In this model, the phases of oscillators encodes the binding of pixels. Rigorous mathematical analysis and extensive computational studies demonstrate the network's ability to both synchronize the oscillators within a particular block corresponding to some object and desynchronize the oscillators within different blocks corresponding to different objects. The model has also been extended to treat the problem of noisy fragments within an image. The network is capable of effectively removing those oscillators which correspond to noisy fragments without affecting the other oscillators. An algorithm based on the network's dynamics is proposed and applied to some real images.

Biophysical models for the spindle and delta sleep rhythm are considered in [32]. This network consists of three types of cells; two of these lie in the thalamus while the other in the cortex. Each of these cell populations display different firing patterns during each stage of sleep and each of these patterns change from one stage of sleep to the other [7–9, 28, 30]. The focus of [32] is to understand how the intrinsic and synaptic properties of these different cells interact to produce the different rhythms. It is demonstrated that a change in the conductance of one kind of cell effectively rewires the thalamocortical circuit, leading to the transition from the spindling to the delta rhythm. This study illustrates the different roles both fast and slow inhibition can play in producing a particular rhythm.

The propagation and activity patterning in a network of locally connected model neurons that are coupled by fast inhibitory synapses and that have postinhibitory rebound capability is considered in [23]. A localized stimulus to the resting network can initiate activity that slowly propagates by recruiting successive neighboring cells. However, the resulting wave propagation depends very much on the footprint type in the network. With an on-center footprint, the activity spreads in discrete groups to give the wave a lurching appearance; an off-center footprint leads to a wave that propagates in a continuous manner. The recruitment activity also depends dramatically on the sign of the shunting synapses. If the synapses are excitatory instead of inhibitory then the speed of propagation is orders of magnitude faster.

Acknowledgments. Research for this paper was supported in part by the NSF grant DMS-9423796.

REFERENCES

[1] T. BAL AND D.A. MCCORMICK, *What stops synchronized thalamocortical oscillations ?*, Neuron **17** (1996), 297–308.

[2] A. BOSE, N. KOPELL AND D. TERMAN, *In Preparation.*

[3] T.R. CHAY AND J. RINZEL, *Bursting, beating, and chaos in an excitable membrane model*, Biophys. J. **47** (1985), 357–366.

[4] A.H. COHEN, S. ROSSIGNOL, AND S. GRILLNER, EDS., *Neural Control of Rhythmic Movements in Vertebrates*, John Wiley and Sons, Inc., New York (1988).

[5] A. DESTEXHE, A. BABLOYANTZ, AND T.J. SEJNOWSKI, *Ionic mechanisms for intrinsic slow oscillations in thalamic relay and reticularis neurons*, Biophys. J. **65** (1993), 1538–1552.

[6] A. DESTEXHE, T. BAL, D.A. MCCORMICK, AND T.J. SEJNOWSKI, *Ionic mechanisms underlying synchronized oscillations and propagating waves in a model of ferret thalamic slices*, J. Neurophysiol. **76** (1996), 2049–2070.

[7] A. DESTEXHE AND T.J. SEJNOWSKI, *Synchronized oscillations in thalamic networks: insights from modeling studies*, in Thalamus; Edited by M. Steriade, E.G. Jones, and D.A. McCormick (1996).

[8] D. GOLOMB, X.-J. WANG AND J. RINZEL, *Propagation of spindle waves in a thalamic slice model*, J. Neurophys **75** (1996), 750–769.

[9] D. GOLOMB, X.-J. WANG AND J. RINZEL, *Synchronization properties of spindle oscillations in a thalamic reticular nucleus model*, J. Neurophys. **72** (1994), 1109–1126.

[10] C.M. GRAY, *Synchronous oscillations in neuronal systems: Mechanisms and functions*, J. Comput. Neurosci. **1** (1994), 11–38.

[11] C.M. GRAY AND D.A. MCCORMICK, *Chattering Cells: Superficial pyramidal neurons contributing to the generation of synchronous oscillations in the visual cortex*, Science **274** (1996), 109–113.

[12] R.M. HARRIS-WARRICK AND E. MARDER, *Modulation of neural networks for behavior*, Ann. Rev. Neurosci. **14** (1991), 39–57.

[13] A.L. HODGKIN AND A.F. HUXLEY, *A quantitative description of membrane current and its application to conduction and excitation in nerve*, J. Physiol. (London) **117** (1952), 500–544.

[14] J.W. JACKLET, ED., *Neuronal and Cellular Oscillators*, Marcel Dekker Inc., New York (1989).

[15] J.G.R. JEFFREYS, R.D. TRAUB AND M.A. WHITTINGTON, *Neuronal networks for induced '40 Hz' rhythms*, Trends Neurosci. **19** (1996), 202–207.

[16] U. KIM, T. BAL, AND D.A. MCCORMICK, *Spindle waves are propagating synchronized oscillations in the ferret LGNd in vitro*, J. Neurophysiol. **74** (1995), 1301–1323.

[17] N. KOPELL AND G. LEMASSON, *Rhythmogenesis, amplitude modulation and multiplexing in a cortical architecture*, Proc. Nat. Acad. Sci (USA) **91** (1994), 10586–10590.

[18] R.R. LLINÁS, *The intrinsic electrophysiological properties of mammalian neurons: Insights into central nervous system function*, Science **242** (1988), 1654–1664.

[19] E.F. MISHCHENKO AND N.KH. ROZOV, *Differential Equations with Small Parameters and Relaxation Oscillations*, Plenum Press, New York and London (1980).

[20] C. MORRIS AND H. LECAR, *Voltage oscillations in the barnicle giant muscle fiber*, Biophysical J. **35** (1981), 193–213.

[21] P. Pinsky and J. Rinzel, *Intrinsic and network rhythmogenesis in a reduced Traub model for CA3 neurons*, J. of Comp. Neuroscience **1** (1994), 39–60.

[22] J. Rinzel, *A formal classification of bursting mechanisms in excitable systems*, Proceedings of the International Congress of Mathematics, (A.M. Gleason, ed.) AMS (1987), 1578–1593.

[23] J. Rinzel, D. Terman, X.-J. Wang, and B. Ermentrout, *Propagating Activity Patterns in Large-Scale Inhibitory Neuronal Networks*, Science **279** (1998), 1351–1355.

[24] J. Rinzel, *Bursting oscillations in an excitable membrane model*, in Ordinary and Partial Differential Equations (B.D. Sleeman and R.J. Jarvis, eds.), Springer-Verlag, New York (1985), 302–316.

[25] J. Rubin and D. Terman, *Geometric analysis of population rhythms in synaptically coupled neuronal networks*, Submitted.

[26] D. Somers and N. Kopell, *Rapid synchronization through fast threshold modulation*, Biol. Cyber. **68** (1993), 393–407.

[27] M. Steriade, M. Deschenes, *The thalamus as a neuronal oscillator*, Brain Res. Rev. **8** (1984), 1–63.

[28] M. Steriade, D.A. McCormick and T.J. Sejnowski, *Thalamocortical oscillations in the sleeping and aroused brain*, Science **262** (1993), 679–685.

[29] M. Steriade, D. Contreras, F. Amzica, *Synchronized sleep oscillations and their paroxysmal developments*, TINS **17** (1994), 199–208.

[30] M. Steriade, E.G. Jones and R.R. Llinás., *Thalamic Oscillations and Signaling*, Wiley, New York, 1990.

[31] D. Terman, N. Kopell and A. Bose, *Dynamics of two mutually coupled inhibitory neurons*, Physica D **117** (1998), 241–275.

[32] D. Terman, N. Kopell and A. Bose, *Functional reorganization in thalamocortical networks: Transition between spindling and delta sleep rhythms*, P.N.A.S. **93** (1996), 15417–15422.

[33] D. Terman and E. Lee, *Partial synchronization in a network of neural oscillators*, SIAM J. Appl. Math. **57** (1997), 252–293.

[34] D. Terman and D.L. Wang, *Global competition and local cooperation in a network of neural oscillators*, Physica D **81** (1995), 148–176.

[35] D. Terman, *Chaotic spikes arising from a model for bursting in excitable membranes*, SIAM J. Appl. Math. **51** (1991), 1418–1450.

[36] D. Terman, *The transition from bursting to continuous spiking in an excitable membrane model*, J. of Nonlinear Science **2** (1992), 1418–1450.

[37] R.D. Traub, R. Miles, and K.S. Wong, *Model of the origin of rhythmic population oscillations in the hippocampal slice*, Science **243** (1989), 1319–1325.

[38] R.D. Traub and R. Miles, *Neuronal Networks of the Hippocampus*, Cambridge University Press, New York (1991).

[39] M. von Krosigk, T. Bal and D.A. McCormick, *Cellular mechanisms of a synchronized oscillation in the thalamus*, Science **261** (1993), 361–364.

[40] D.L. Wang and D. Terman, *Locally excitatory globally inhibitory oscillator networks*, IIIE Trans. Neural Nets. **6** (1995), 283–286.

[41] X.-J. Wang and J. Rinzel, *Oscillatory and bursting properties of neurons*, in The Handbook of Brain Theory and Neural Networks; Edited by Michael A. Arbib (1995), 686–691.

[42] X.-J. Wang and J. Rinzel, *Alternating and synchronous rhythms in reciprocally inhibitory model neurons*, Neural Comp. **4** (1992), 84–97.

METASTABLE DYNAMICS AND EXPONENTIAL ASYMPTOTICS IN MULTI-DIMENSIONAL DOMAINS*

MICHAEL J. WARD†

Abstract. Certain singularly perturbed partial differential equations exhibit a phenomenon known as dynamic metastability, whereby the solution evolves on an asymptotically exponentially long time interval as the singular perturbation parameter ϵ tends to zero. This article illustrates a technique to analyze metastable behavior for a range of problems in multi-dimensional domains. The problems considered include the exit problem for diffusion in a potential well, models of interface propagation in materials science, an activator-inhibitor model in mathematical biology, and a flame-front problem. Many of these problems can be formulated in terms of non-local partial differential equations. This non-local feature is shown to be essential to the existence of metastable behavior.

1. Introduction. Certain time-dependent singularly perturbed partial differential equations exhibit a phenomenon known as dynamic metastability, whereby the solution evolves on an asymptotically exponentially long time interval as the singular perturbation parameter ϵ tends to zero. Metastable dynamics has been observed and analyzed over the past decade for certain classes of problems in a *one-spatial* dimensional setting (eg. [1, 8, 10, 12, 22, 31, 32, 40]). In this article we give examples of various problems that exhibit metastable dynamics in *multi-spatial* dimensional domains and we outline an asymptotic technique, known as the projection method, to analyze the metastable dynamics. The problems considered include the exit problem for diffusion in a potential well, models of interface propagation in materials science, an activator-inhibitor model in mathematical biology, and a flame-front problem.

There are several common features to many of these problems. The first common feature is that in the limit $\epsilon \to 0$, each problem admits an asymptotic quasi-equilibrium solution $u_q(\mathbf{x}, \epsilon\,; \alpha)$, in terms of some unknown parameter vector $\alpha \in S$, where S is some parameter set. This quasi-equilibrium solution satisfies the equilibrium problem up to exponentially small terms as $\epsilon \to 0$. The associated linearization of the nonlinear problem around u_q is exponentially ill-conditioned as $\epsilon \to 0$ for $\alpha \in S$. For some of the nonlinear problems considered below, this ill-conditioning is a consequence of an exponential localization of a radially symmetric canonical solution combined with a near translation invariance. In this situation and in an N-dimensional domain D, α is an N-vector giving the coordinates of the center $\mathbf{x_0}$ of the localized structure in D. As a result of the exponential ill-conditioning, exponential asymptotics is required to calculate the correct value of α corresponding to a true asymptotic equilibrium so-

*This work was supported by NSERC grant 5-81541.

†Department of Mathematics, University of British Columbia, Vancouver, B.C. V6T 1Z2, CANADA.

lution. In addition, metastable dynamics of the canonical solution for the time-dependent problem will occur when the exponentially small eigenvalues are the principal eigenvalues associated with the linearization. This metastable dynamics is then given by $u \sim u_q(\mathbf{x}, \epsilon; \alpha(t;\epsilon))$, where $\alpha(t;\epsilon)$ satisfies a finite dimensional dynamical system. This dynamical system evolves on an exponentially slow time scale as $\epsilon \to 0$.

A critical common feature in many of the problems considered here, which allows for the existence of metastable dynamics in a multi-spatial dimensional setting, is a non-local or global condition, such as mass conservation. This non-local condition, which arises in a natural way for the problems considered below, is essential for ensuring that the exponentially small eigenvalues are the *principal* eigenvalues associated with the linearization. This allows us to seek a quasi-steady solution to the nonlinear problem in the form $u \sim u_q(\mathbf{x}, \epsilon; \alpha(t;\epsilon))$. Since the principal eigenvalues associated with the linearized problem are exponentially small, the solution to this quasi-steady linearization must satisfy limiting solvability conditions as $\epsilon \to 0$ that ensures that the 'residual" is orthogonal to the eigenspace associated with the exponentially small eigenvalues. From this projection step and from certain key asymptotic exponential estimates for eigenfunctions on the boundary of the domain, an explicit asymptotic differential equation for $\alpha(t;\epsilon)$, characterizing the metastable dynamics, is derived. For all the problems considered below, the dynamics of $\alpha(t;\epsilon)$ is reduced to a surface integral over the boundary of the domain D. An asymptotic evaluation of this integral shows that only certain special points on the boundary of D influence the metastable dynamics.

The outline of this paper is as follows. In §2, we consider the linear exit time problem for diffusion in a potential well modeled by a well-known singularly perturbed convection-diffusion equation (i.e. [27]). When the potential is radially symmetric, it is shown that the metastable dynamics are influenced very strongly by the points on the boundary of the domain that are closest to the minimum of the potential. In §3, we analyze the motion of a straight-line interface for the Allen-Cahn equation in the neck, or channel, region of a two-dimensional dumbbell-shaped domain as shown in Fig. 1 below. The motion of such an interface is metastable and depends very strongly on the local behavior of the boundary of the domain near the connection points between the channel and the two lobes of the dumbbell (see [6, 19, 35]). In §4 we illustrate the metastable dynamics of bubble solutions for the constrained Allen-Cahn equation, which conserves mass and has applications to material science. In §5 we analyze metastable spike dynamics for an activator-inhibitor model in the limit of large inhibitor diffusivity, which yields the shadow problem (see [29]). The problems in §4 and §5 are both non-local and involve radially symmetric canonical solutions u_q whose center $\alpha = \mathbf{x_0}(t)$ drifts exponentially slowly towards the nearest point on the boundary of the domain. Finally, in §6 we briefly describe a flame-front model exhibiting metastable dynamics.

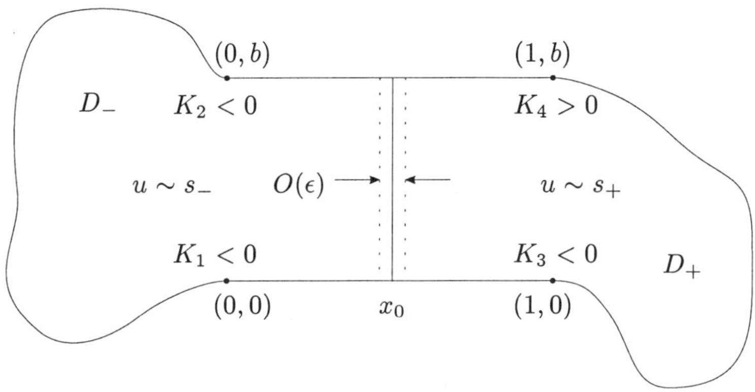

FIG. 1. *A dumbbell-shaped domain D and an interface centered at x_0.*

2. Metastable dynamics for the exit problem. The following linear convection-diffusion equation arises in determining the exit time distribution for a Brownian particle confined by a potential well $\Psi(\mathbf{x})$ ([24, 25, 27]):

$$u_t = \epsilon \triangle u - \nabla\Psi \cdot \nabla u\,, \qquad \mathbf{x} \in D\,, \quad t > 0\,, \tag{1a}$$

$$u = u_b(\mathbf{x})\,, \quad \mathbf{x} \in \partial D\,; \qquad u(\mathbf{x}, 0) = u_0(\mathbf{x})\,, \tag{1b}$$

where $\mathbf{x} = (x_1, x_2)$, $u = u(\mathbf{x}, t)$, $\epsilon \to 0$, D is a bounded two-dimensional domain, and $u_b(\mathbf{x})$ and $u_0(\mathbf{x})$ are smooth. Let $\bar{D} = D \cup \partial D$. It is assumed that the potential $\Psi(\mathbf{x})$ has a unique global minimum on $\bar{D}$ at some interior point $\mathbf{x_0} \in D$, where

$$\Psi(\mathbf{x_0}) = 0\,, \qquad \nabla\Psi(\mathbf{x_0}) = 0\,, \qquad H\left[\Psi(\mathbf{x_0})\right] > 0\,. \tag{2}$$

Here $H(\Psi) \equiv \Psi_{x_1x_1}\Psi_{x_2x_2} - \Psi^2_{x_1x_2}$ is the Hessian. We also assume that $\nabla\Psi(\mathbf{x}) \neq 0$ for $\mathbf{x} \neq \mathbf{x_0}$ and that $\nabla\Psi \cdot \hat{n} > 0$ on ∂D, where $\hat{n}$ is the unit outward normal to ∂D. As shown in [38], this problem provides the simplest example of metastable behavior in a multi-dimensional setting. Below we sketch the outline of the metastability analysis given in [38].

The following eigenvalue problem associated with (1a), (1b) is central to the analysis:

$$L_\epsilon \phi \equiv \epsilon \triangle \phi - \nabla\Psi \cdot \nabla\phi = -\lambda\phi\,, \quad \mathbf{x} \in D\,; \quad \phi = 0\,, \quad \mathbf{x} \in \partial D\,, \tag{3a}$$

$$(\phi, \phi)_w \equiv \int_D \phi^2 w \, dx = 1\,, \qquad w \equiv e^{-\Psi/\epsilon}\,. \tag{3b}$$

The eigenvalues λ_j for $j \geq 0$ are real with $\lambda_j > 0$ and $(\phi_j, \phi_k)_w = \delta_{jk}$ for $j, k = 0, 1, \ldots$ Suppose that the minimum value of Ψ on ∂D is taken at N distinct points $\mathbf{y_j} \in \partial D$ for $j = 1,..,N$, and that these minima are non-degenerate. Then, as is shown in [27], the principal eigenvalue λ_0 of (3a) is exponentially small as $\epsilon \to 0$ and has the asymptotic estimate

$$\lambda_0 \sim (2\pi\epsilon)^{-1/2} \left(H\left[\Psi(\mathbf{x}_0)\right]\right)^{1/2} e^{-\Psi^*/\epsilon} \sum_{j=1}^{N} |\nabla\Psi(\mathbf{y_j})| r_j^{-1/2} . \tag{4}$$

Here $\Psi^* \equiv \Psi(\mathbf{y_j})$ for $j = 1,..,N$, and

$$\begin{aligned} r_j \equiv |\nabla\Psi|^{-2} \Big[& \Psi_{x_1x_1}\Psi_{x_2}^2 - 2\Psi_{x_1x_2}\Psi_{x_1}\Psi_{x_2} \\ & + \Psi_{x_2x_2}\Psi_{x_1}^2 + \kappa_j |\nabla\Psi|^3 \Big] \Big|_{\mathbf{x}=\mathbf{y_j}} , \end{aligned} \tag{5}$$

where $\kappa_j < 0$ is the curvature of ∂D at $\mathbf{y_j}$.

The reciprocal of λ_0 determines the expected time for a particle initially located at $\mathbf{x} = \mathbf{x_0}$ to leave D. The most likely points of exit of the particle are those that minimize Ψ on D. Therefore, if Ψ is radially symmetric about $\mathbf{x_0}$, the asymptotic estimate for λ_0 depends critically on certain local information near the points on the boundary that are closest to ∂D. This feature will also occur for all of the problems considered in the sections below.

In terms of a normalization constant M_0, the corresponding eigenfunction ϕ_0 has the boundary layer form

$$\phi_0 \sim M_0 \left(1 - e^{\gamma\eta/\epsilon}\right) , \qquad \text{for} \qquad \gamma = \gamma(s) \equiv \nabla\Psi \cdot \hat{n} \Big|_{\partial D} > 0 . \tag{6}$$

Here s denotes arclength along ∂D and $-\eta$ is the distance from $\mathbf{x} \in D$ to ∂D. Since $\gamma > 0$, $\phi_0 \to M_0$ as $\eta/\epsilon \to -\infty$.

For $\epsilon \to 0$, the equilibrium solution $U(\mathbf{x};\epsilon)$ to (1) is a constant away from ∂D and has a boundary layer form near ∂D. It is given to leading order by

$$U(\mathbf{x};\epsilon) \sim \tilde{u}^\epsilon\left[\mathbf{x}; A_{0e}\right] \equiv A_{0e} + \left(u_b(s) - A_{0e}\right) e^{\gamma\eta/\epsilon} , \tag{7}$$

for some undetermined constant A_{0e}. Here u_b is written in terms of the arclength s. The determination of A_{0e} requires exponential precision as a result of the exponential ill-conditioning of the operator as is evident from (4). For the time-dependent problem when $t \gg 1$, we look for a solution of the form $u(\mathbf{x},t) \sim \tilde{u}^\epsilon\left[\mathbf{x}; A_0(t)\right]$, where $A_0(t)$ is a function to be determined. As a result of the exponentially small principal eigenvalue λ_0, the function $A_0(t)$ approaches its equilibrium value A_{0e} only over an asymptotically exponentially long time interval as $\epsilon \to 0$. This exponentially slow relaxation to the equilibrium value is known as metastable dynamics.

A projection method, which is based on a limiting solvability condition, can be used to calculate an explicit ODE for $A_0(t)$ characterizing the metastable behavior (see [38]). To do so, we introduce a correction $v(\mathbf{x},t)$ by $u(\mathbf{x},t) = \tilde{u}^{\epsilon}[\mathbf{x}; A_0(t)] + v(\mathbf{x},t)$. Substituting in (1), we obtain that $v(\mathbf{x},t)$ satisfies

$$v_t = L_\epsilon v - \tilde{u}^{\epsilon}_t + L_\epsilon \tilde{u}^{\epsilon}\,, \quad \mathbf{x} \in D\,, \quad t > 0\,, \tag{8a}$$

$$v = u_b - \tilde{u}^{\epsilon}\,, \;\; \mathbf{x} \in \partial D\,, \tag{8b}$$

$$v(\mathbf{x},0) = u_0(\mathbf{x}) - \tilde{u}^{\epsilon}[\mathbf{x}; A_0(0)]\,, \;\; \mathbf{x} \in D\,. \tag{8c}$$

We then expand v in terms of the eigenfunctions ϕ_j of (3) as

$$v(\mathbf{x},t) = \sum_{j=0}^{\infty} c_j(t)\phi_j(\mathbf{x})\,, \tag{9}$$

where the coefficient $c_j(t)$ satisfies

$$c_j' + \lambda_j c_j = \left(\phi_j, L_\epsilon \tilde{u}^{\epsilon}\right)_w - \int_{\partial D} \epsilon w v\, \partial_n \phi_j \, ds - \left(\phi_j, \tilde{u}^{\epsilon}_t\right)_w\,, \tag{10}$$

together with the initial value

$$c_j(0) = \int_D \left(u_0(\mathbf{x}) - \tilde{u}^{\epsilon}[\mathbf{x}; A_0(0)]\right) \phi_j w\, dx\,. \tag{11}$$

Here $w \equiv e^{-\Psi/\epsilon}$, and, in (10), ∂_n denotes the outward normal derivative to ∂D.

We now impose the limiting solvability condition. To ensure that $v \ll \tilde{u}^{\epsilon}$ over exponentially long time intervals it is necessary that $c_0(t) \equiv 0$. Hence, (10) gives an ODE for $A_0(t)$ and (11) yields the initial value $A_0(0)$. Since w is exponentially localized and ϕ_0 is known asymptotically, the various terms in (10) and (11) can be estimated asymptotically as in [38] to obtain the following explicit ordinary differential equation for $A_0(t)$:

$$A_0' \sim -\lambda_0 A_0 + \beta e^{-\Psi^*/\epsilon} \sum_{j=1}^{N} u_b(\mathbf{y_j}) |\nabla \Psi(\mathbf{y_j})| r_j^{-1/2}\,, \quad A_0(0) \sim u_0(\mathbf{x}_0)\,. \tag{12}$$

Here, λ_0 is given in (4), r_j is defined in (5), $\mathbf{x}_0$ is the global minimum of Ψ in $\bar{D}$, and $\beta \equiv (H[\Psi(\mathbf{x}_0)]/(2\pi\epsilon))^{1/2}$, where H is the Hessian. Moreover, $\mathbf{y_j} \in \partial D$, for $j = 1, .., N$, are those points where Ψ is minimized on ∂D (with a non-degenerate minimum) with minimum value $\Psi^* \equiv \Psi(\mathbf{y_j})$ for $j = 1, .., N$.

In summary, the metastable dynamics for (1) for $t \gg 1$, is given by $u(\mathbf{x},t) \sim \tilde{u}^{\epsilon}[\mathbf{x}; A_0(t)]$, where $\tilde{u}^{\epsilon}$ is defined in (7) and $A_0(t)$ satisfies (12). As $t \to \infty$, $A_0(t) \to A_{0e}$, where A_{0e} is the weighted average over the minimum points $\mathbf{y_j}$ of Ψ on ∂D

$$A_{0e} = \sum_{j=1}^{N} u_b(\mathbf{y_j}) \frac{|\nabla \Psi(\mathbf{y_j})|}{r_j^{1/2}} \left(\sum_{j=1}^{N} \frac{|\nabla \Psi(\mathbf{y_j})|}{r_j^{1/2}} \right)^{-1} \tag{13}$$

This result for A_{0e} is given in [27].

2.1. An explicit example. To illustrate this result, we consider a simple specific example. Let D denote the interior of the ellipse $x_2^2+x_1^2/4 \le 1$ and take $u_b = -x_2 + x_2^2/2$ as the boundary data on ∂D. Let $\Psi = x_1^2 + x_2^2$. Then, it is clear that $\mathbf{x_0} = (0,0)$, $N = 2$, and $\mathbf{y_1} = (0,1)$, $\mathbf{y_2} = (0,-1)$. A simple calculation shows that $H\left[\psi(\mathbf{x_0})\right] = 4$, $\Psi^* = 1$, and $r_j = 8/(14\pi)^{1/2}$ for $j = 1,2$. Thus, we obtain from (12) that

$$A_0' \sim \left(\frac{8}{14\pi\epsilon}\right)^{1/2} e^{-1/\epsilon}\left(1 - 2A_0\right). \tag{14}$$

Thus, away from the boundary layer near ∂D and for t sufficiently large, we have that the outer limit of $u(\mathbf{x},t)$ satisfies

$$u(\mathbf{x},t) \sim A_0(t) = \frac{1}{2} + \left(A_0(0) - \frac{1}{2}\right) e^{-\nu t}, \quad \nu \equiv 2\left(\frac{8}{14\pi\epsilon}\right)^{1/2} e^{-1/\epsilon}. \tag{15}$$

3. A bistable nonlinearity in a dumbbell-shaped domain. The simplest model for the phase separation of a binary mixture is the Allen-Cahn equation

$$u_t = \epsilon^2 \Delta u + Q(u), \qquad \mathbf{x} \in D \subset \mathbf{R}^2, \tag{16a}$$

$$\partial_n u = 0, \qquad \mathbf{x} \in \partial D. \tag{16b}$$

Here $\epsilon \ll 1$ and $Q(u)$ is a bistable nonlinearity having three zeroes located at $u = s_- < 0$, $u = 0$, and $u = s_+ > 0$, with

$$Q(s_\pm) < 0, \quad Q(0) > 0, \quad V(s_+) = 0, \quad V(u) = -\int_{s_-}^{u} Q(\eta)\, d\eta. \tag{17}$$

Thus, $V(u)$ is a double-well potential with wells of equal depth at s_+ and s_-. The two-dimensional dumbbell-shaped domain D is taken to be of the form $D = R \cup D_- \cup D_+$ where R is the rectangle $[0,1] \times [0,b]$, and D_- and D_+ are the two attachments on its sides (see Fig. 1)

Starting from initial data, the solution to (16) quickly develops internal layers of width $O(\epsilon)$ that separate the two minima s_+ and s_- of the potential $V(u)$. The normal velocity v of such an interface is well-known to be $v \sim \epsilon^2\kappa$, where κ is the curvature of the interface (see [34]). In addition, if the interface intersects the boundary ∂D, it must do so orthogonally.

In this section, we will investigate the dynamics of (16) when the interface is initially the straight line segment $x = x_0$, with $0 < x_0 < 1$, that connects the two sides of R as shown in Fig. 1. In this case, $\kappa = 0$ and the velocity curvature law $v \sim \epsilon^2\kappa$ gives no information about the motion of the interface.

In this situation, the dynamics of the interface depends critically on the nature of the boundary of the domain at the corner points of R given by $(0,0)$, $(0,b)$, $(1,0)$, and $(1,b)$. It is assumed that the domain boundary is smooth and that near these corners points, ∂D has the local behavior $\partial D = \{(x,y) \,|\, y = \psi_i(x)\}$ where

$$\text{(18a)} \quad \text{near } (0,0)\,; \quad y = \psi_1(x), \qquad \psi_1'(x) \sim -K_1(-x)^{\alpha_1}, \qquad \text{as } x \to 0^-,$$

$$\text{(18b)} \quad \text{near } (0,b)\,; \quad y = \psi_2(x) + b, \quad \psi_2'(x) \sim K_2(-x)^{\alpha_2}, \qquad \text{as } x \to 0^-,$$

$$\text{(18c)} \quad \text{near } (1,0)\,; \quad y = \psi_3(x), \qquad \psi_3'(x) \sim K_3(x-1)^{\alpha_3}, \qquad \text{as } x \to 1^+,$$

$$\text{(18d)} \quad \text{near } (1,b)\,; \quad y = \psi_4(x) + b, \quad \psi_4'(x) \sim -K_4(x-1)^{\alpha_4}, \quad \text{as } x \to 1^+.$$

Here $\alpha_i > 0$, for each i, and the constant K_i is proportional to the curvature of the ith corner when $\alpha_i = 1$.

Since the straight-line internal layer solution decays exponentially for $|x - x_0| \gg O(\epsilon)$, exponential precision is required to determine the motion of the interface. The projection method was used in [35] to incorporate such exponentially small effects and to calculate metastable dynamics. The key steps of the method are as follows: First an equilibrium solution to (16) in R is constructed. Next, the spectral properties of the linearization of (16) around the equilibrium solution are analyzed asymptotically to show the existence of an exponentially small principal eigenvalue, which is responsible for the metastable behavior. Finally, a limiting solvability condition of Fredholm type is used to derive an explicit ODE for the interface location in the straight channel R. Below we summarize the results of the analysis of [36].

We first construct an equilibrium solution. In the limit $\epsilon \to 0$, the straight-line equilibrium internal layer solution for (16) depends only on x and is given by $u(x;\epsilon) = u_0\left[\epsilon^{-1}(x - x_0)\right]$, where x_0 is the center of the layer with $0 < x_0 < 1$, and $u_0(z)$ satisfies

$$\text{(19a)} \qquad u_0'' + Q(u_0) = 0, \quad -\infty < z < \infty,$$

$$\text{(19b)} \qquad u_0(0) = 0; \quad u_0(z) \sim s_\pm, \quad \text{as } z \to \pm\infty.$$

The far-field behavior of $u_0(z)$ is given by

$$\text{(20)} \qquad u_0(z) \sim \begin{cases} s_+ - a_+ e^{-\nu_+ z}, & z \to +\infty\,, \\ s_- + a_- e^{\nu_- z}, & z \to -\infty\,. \end{cases}$$

Here the positive constants $\nu_\pm$ and $a_\pm$ are defined by

$$\text{(21)} \qquad \nu_\pm = \left[-Q'(s_\pm)\right]^{1/2},$$

$$\text{(22)} \qquad \log a_\pm = \log(\pm s_\pm) + \int_0^{s_\pm} \left(\frac{\pm\nu_\pm}{[2V(\eta)]^{1/2}} + \frac{1}{\eta - s_\pm} \right) d\eta\,.$$

The eigenvalue problem associated with linearizing (16) about $u_0[\epsilon^{-1}(x - x_0)]$ is

$$\text{(23a)} \qquad L_\epsilon \phi \equiv \epsilon^2 \Delta\phi + Q'(u_0)\phi = \lambda\phi\,, \qquad \mathbf{x} \in D\,,$$

$$\partial_n \phi = 0\,, \qquad \mathbf{x} \in \partial D\,, \tag{23b}$$

$$(\phi, \phi) = \int_D \phi^2 \, d\mathbf{x} = 1\,. \tag{23c}$$

Here $(u, v) \equiv \int_D uv \, d\mathbf{x}$. The eigenpairs are labeled by λ_j, ϕ_j respectively for $j = 0, 1, \ldots$, with $\lambda_j \to -\infty$ as $j \to \infty$.

We assume that the distance from the interface to the corners of R is $O(1)$. To estimate the principal eigenpair, we notice that $L_\epsilon u_0'[\epsilon^{-1}(x - x_0)] = 0$, that u_0' has no nodal lines, and that u_0' fails to satisfy the boundary condition (23b) by only exponentially small terms. Thus, as shown in [35], the principal eigenfunction ϕ_0 has the form $\phi_0 \sim N_0 \, [u_0' + \phi_{L_0}]$, where N_0 is a normalization constant and ϕ_{L_0} is a boundary layer function localized near the curved parts of ∂D that is used to satisfy the boundary condition (23b). Then, Green's identity can be applied to (23a) and u_0' to yield

$$\lambda_0 (u_0', \phi_0) = -\epsilon^2 \int_{\partial D} \phi_0 \partial_n u_0' \, ds\,. \tag{24}$$

After calculating ϕ_0 on ∂D, this equation is used to calculate the exponentially small eigenvalue λ_0. An important observation is that $\partial_n u_0' = 0$ except along the two attachments D_+ and D_-. Hence, the surface integral on the right side of (24) vanishes identically over R. In addition, since u_0' and ϕ_0 are both exponentially decaying away from the internal layer region it is clear that this surface integral is dominated asymptotically by the contribution arising from the corner regions of the attachments D_+ and D_-.

The boundary layer correction ϕ_{L_0} is calculated in [35] using a local coordinate system defined near ∂D. This allows us to obtain the following expression for ϕ_0 on ∂D as needed in (24):

$$\phi_0 \sim \begin{cases} N_0 a_- \nu_- e^{\nu_- \epsilon^{-1}(x - x_0)} \, (1 - \nu_- n_x) & x < x_0\,, \\ N_0 a_+ \nu_+ e^{-\nu_+ \epsilon^{-1}(x - x_0)} \, (1 + \nu_+ n_x) & x > x_0\,. \end{cases} \tag{25}$$

Here $\hat{\mathbf{n}} = (n_x, n_y)$ is the outward unit normal vector to ∂D. The dominant contribution to the inner product on the left side of (24) arises from the region near $x = x_0$ and is given asymptotically by

$$(u_0', \phi_0) \sim \epsilon N_0 b \beta\,, \qquad \text{where} \qquad \beta \equiv \int_{-\infty}^{\infty} \left[u_0'(z)\right]^2 \, dz\,. \tag{26}$$

Substituting (26), (25) and (20) into (24) we get the following expression for λ_0

$$\begin{aligned} \lambda_0 \sim -\frac{1}{b\beta} \Bigg\{ &\int_{\partial D_-} a_-^2 \nu_-^3 e^{2\nu_- \epsilon^{-1}(x - x_0)} (1 - \nu_- n_x) n_x \, ds \\ &- \int_{\partial D_+} a_+^2 \nu_+^3 e^{-2\nu_+ \epsilon^{-1}(x - x_0)} (1 + \nu_+ n_x) n_x \, ds \Bigg\}\,. \end{aligned} \tag{27}$$

The dominant contribution to the two surface integrals above arises from the corner regions. In these regions, we can use (18) to calculate n_x in terms of K_i and α_i and then we apply Laplace's method. In this way, we obtain in [35] the following explicit asymptotic formula for the principal eigenvalue λ_0 as $\epsilon \to 0$:

$$\lambda_0 \sim \frac{1}{b\beta}\left\{a_-^2\nu_-^3 A_- e^{-2\nu_-\epsilon^{-1}x_0} + a_+^2\nu_+^3 A_+ e^{-2\nu_+\epsilon^{-1}(1-x_0)}\right\}. \tag{28}$$

Here $A_\pm$ are defined by

$$A_- = K_1\left(\frac{\epsilon}{2\nu_-}\right)^{\alpha_1+1}\Gamma(\alpha_1+1) + K_2\left(\frac{\epsilon}{2\nu_-}\right)^{\alpha_2+1}\Gamma(\alpha_2+1), \tag{29a}$$

$$A_+ = K_3\left(\frac{\epsilon}{2\nu_+}\right)^{\alpha_3+1}\Gamma(\alpha_3+1) + K_4\left(\frac{\epsilon}{2\nu_+}\right)^{\alpha_4+1}\Gamma(\alpha_4+1), \tag{29b}$$

and $\Gamma(z)$ denotes the Gamma function.

Next the projection method is used to determine an ODE for the interface location $x_0 = x_0(t)$ for (16) in the rectangle R. It is assumed that the initial data has the special form $u_0[\epsilon^{-1}(x - x_0^0)]$, so that $x_0(0) = x_0^0$. Introduce the correction w by

$$u(\mathbf{x}, t) = u_0[\epsilon^{-1}(x - x_0(t))] + w(\mathbf{x}, t), \tag{30}$$

where $\mathbf{x} = (x, y)$ and where we assume that $w \ll u_0$ and $w_t \ll \partial_t u_0$ uniformly in time. This yields the quasi-steady linearization of (16) about u_0

$$L_\epsilon w \equiv \epsilon^2 \Delta w + Q'(u_0)w = \partial_t u_0, \qquad \mathbf{x} \in D, \tag{31a}$$

$$\partial_n w = -\partial_n u_0, \qquad \mathbf{x} \in \partial D. \tag{31b}$$

The solution to (31) is then expanded in terms of the eigenfunctions of (23) as $w = \sum_{j=0}^{\infty} c_j\phi_j/\lambda_j$, where the c_j are given by

$$c_j = (\phi_j, \partial_t u_0) + \epsilon^2 \int_{\partial D} \phi_j \partial_n u_0 \, ds. \tag{32}$$

Since λ_0, as estimated in (28), is exponentially small as $\epsilon \to 0$, we require that $c_0 \to 0$ as $\epsilon \to 0$ in order to ensure that $w \ll u_0$ over exponentially long time intervals. This limiting solvability yields the ODE for $x_0(t)$

$$(\phi_0, \partial_t u_0) = -\epsilon^2 \int_{\partial D} \phi_0 \partial_n u_0 \, ds. \tag{33}$$

Finally, the terms in (33) are evaluated asymptotically as $\epsilon \to 0$ as in [35] to give the following explicit ordinary differential equation for $x_0(t)$:

$$x_0'(t) \sim \frac{\epsilon}{b\beta}\left\{a_+^2\nu_+^2 A_+ e^{-2\nu_+\epsilon^{-1}(1-x_0)} - a_-^2\nu_-^2 A_- e^{-2\nu_-\epsilon^{-1}x_0}\right\}. \tag{34}$$

Here, b is the width of R, $\nu_\pm$ and $a_\pm$ are defined in (21) and (22), β is defined in (26), and $A_\pm$ is defined in (29).

Thus, the motion of the interface is determined by the local behavior at the corners of R and by the distance from the interface to these corners. The interface will move according to (34) until a steady state is attained or until the interface has moved to one of the sides of R. In the latter case, the subsequent evolution of the interface after it enters one of the attachments is given by the usual velocity-curvature law. The result (34) agrees with that obtained in [6] and [19] using a different method.

3.1. Some explicit examples. When $A_+A_- > 0$, we see from (34) that there is a unique equilibrium interface location given by

$$x_0^e \sim \frac{\nu_+}{\nu_+ + \nu_-} + \frac{\epsilon}{2(\nu_+ + \nu_-)} \log\left(\frac{a_-^2 \nu_-^2 A_-}{a_+^2 \nu_+^2 A_+}\right). \tag{35}$$

A sufficient condition for this equilibrium solution to be stable is that $K_i < 0$ for $i = 1,..,4$, which corresponds to a domain that is non-convex near each of the corners. This yields an example of a stable non-constant equilibrium solution in a non-convex domain. Such an equilibrium solution is impossible when the domain is convex (see [26]). A sufficient condition for the equilibrium solution to be unstable is that $K_i > 0$ for $i = 1, \ldots, 4$. This corresponds to a domain that is convex near each corner. When $A_+A_- < 0$ there is no equilibrium solution.

We conclude by giving an example of the slow dynamics. Let $Q(u) = 2(u - u^3)$, for which we get from (19) that $u_0(z) = \tanh z$. Then, we calculate $a_\pm = 2$, $\nu_\pm = 2$, $s_\pm = \pm 1$ and $\beta = 4/3$. Suppose that $\alpha_i = \alpha > 0$ for each i. Then, the ODE (34), becomes

$$x_0' \sim \frac{12\epsilon^{\alpha+2}\Gamma(\alpha+1)}{b\,4^{\alpha+1}} \left[(K_3+K_4)e^{-4\epsilon^{-1}(1-x_0)} - (K_1+K_2)e^{-4\epsilon^{-1}x_0}\right], \tag{36}$$

which has the steady-state location

$$x_0^e \sim \frac{1}{2} + \frac{\epsilon}{8}\log\left(\frac{K_1 + K_2}{K_3 + K_4}\right). \tag{37}$$

If $K_i < 0$ for $i = 1, \ldots, 4$, then this steady state is stable. Alternatively, if $K_i > 0$ for $i = 1, \ldots, 4$, then the steady state is unstable.

Notice that when K_1, $K_2 < 0$ and K_3, $K_4 > 0$, then from (36), $x_0' > 0$ for all time. In this case, the interface location, x_0 will move monotonically but exponentially slowly towards $x = 1$ (see Fig. 2). Once the interface reaches $x = 1$, it will move in the attachment D_+ by the velocity curvature law until it collapses against the boundary of D_+. Similarly, when K_1, $K_2 > 0$ and K_3, $K_4 < 0$, we have $x_0' < 0$ and the interface location will monotonically $x = 0$.

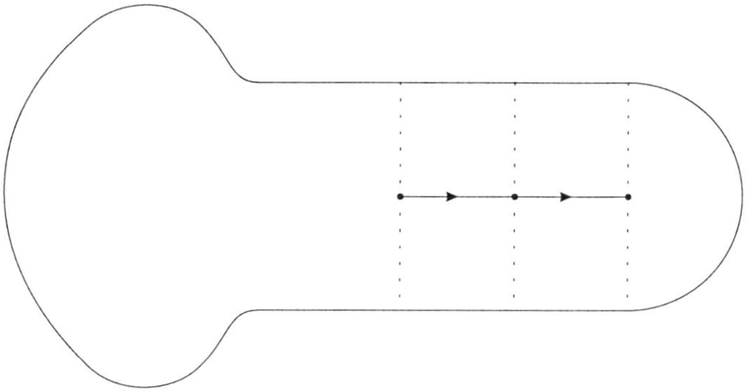

FIG. 2. *Since K_1, $K_2 < 0$ and K_3, $K_4 > 0$ for this domain, the interface moves to the right.*

4. The constrained Allen-Cahn equation. A simple model for the phase separation of a binary mixture is the Allen-Cahn equation with a mass constraint as introduced in [33],

$$u_t = \epsilon^2 \Delta u + Q(u) - \sigma\,, \qquad \mathbf{x} \in D \subset \mathbf{R}^2\,, \tag{38a}$$

$$\partial_n u = 0\,, \qquad \mathbf{x} \in \partial D\,, \tag{38b}$$

$$\int_D u(\mathbf{x}, t)\, d\mathbf{x} = M\,. \tag{38c}$$

Here $u = u(\mathbf{x}, t)$ is the concentration of one of the two species, $\mathbf{x} = (x, y)$, $\epsilon \ll 1$, D is a bounded two-dimensional domain, and the mass M is constant. We assume that $Q(u) = -V'(u)$, where $V(u)$ is a double-well potential with minima at $u = s_\pm$ where $V(s_\pm) = 0$. The function $Q(u)$ satisfies (17). The problem (38) is a non-local reaction-diffusion equation since, in order to satisfy the mass constraint (38c), we require

$$\sigma = \frac{1}{|D|} \int_D Q(u)\, d\mathbf{x}\,. \tag{39}$$

Here $|D|$ is the total area of D.

We now describe the different stages of the dynamics for (38) that have been studied in detail in [2–5, 33, 36] and [41]. The various stages of the dynamics of a single closed interface for (38) are shown qualitatively in Fig. 3.

The first stage is a transient phase. Starting from arbitrary initial data, in an $O(1)$ time interval the solution to (38) develops internal layers of width $O(\epsilon)$ separating the two minina of the potential well. Then, as shown by the asymptotic analysis of [33], for $\epsilon \to 0$ the normal velocity v of a single closed interface Γ satisfies the constrained velocity-curvature law

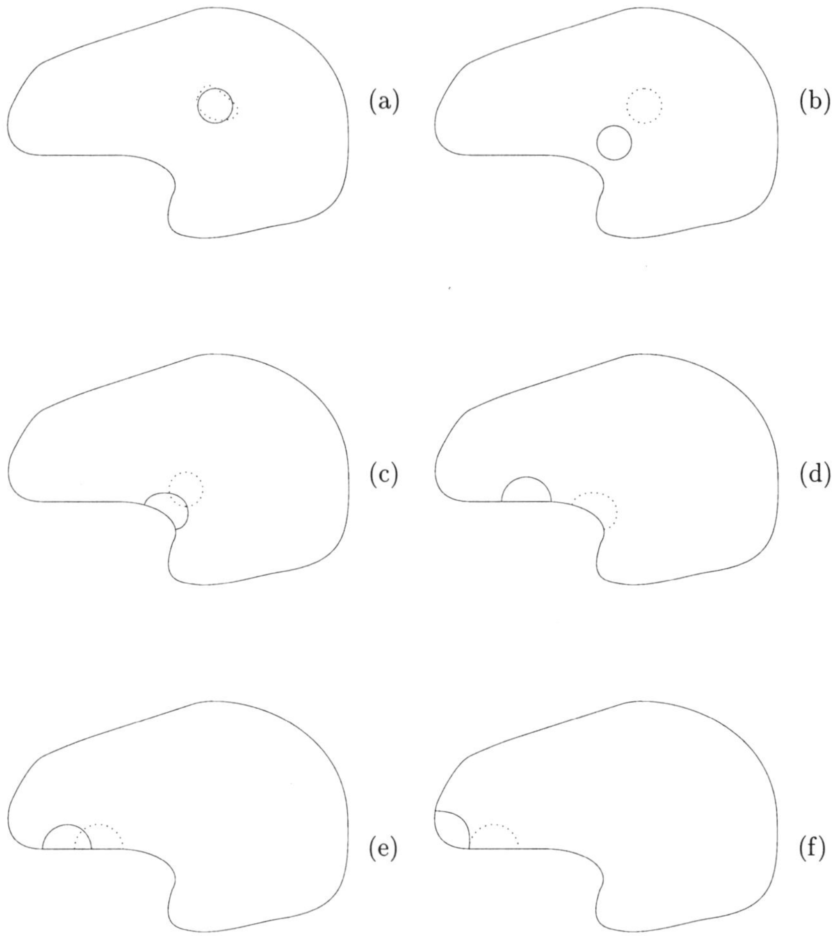

FIG. 3. *Evolution of a small convex interface inside a domain D. (a) The convex interface evolves by (40) into a circle. (b) The circular interface drifts, satisfying (41), towards the closest point on ∂D. (c) The interface attaches to ∂D, intersecting orthogonally. (d) The interface moves along ∂D satisfying (40). (e) If the interface encounters a flat portion of ∂D, it moves along this flat portion according to (43). (f) When a curved part of ∂D is reached, the interface again evolves by (40) until a steady state is attained, which is centered near a local maximum of the curvature of ∂D.*

$$v \sim \epsilon^2 \left(\kappa - \frac{1}{|\Gamma|} \int_\Gamma \kappa \, ds \right) . \tag{40}$$

Here κ is the curvature of Γ. This expression holds for interfaces in the interior of D and for interfaces that are connected to ∂D orthogonally. As shown in [13], a single closed convex interface evolving according to (40) will tend to a circle that encloses the same area.

The law of motion (40) gives no indication of the nature of the motion of a single circular interface contained in D. Such an interface is also referred to as a *bubble*. As shown in [2, 4, 5] and [41], the motion of such a bubble is metastable. The key feature that leads to this metastable behavior is the conservation of mass condition. As shown in [41], a bubble solution to (38) drifts exponentially slowly, without change of shape, towards the closest point on ∂D. This slow drift is due to an exponentially weak interaction between the tail of the bubble solution and the boundary ∂D. It was derived in [41] that the distance $r_m(t)$ between the center of the bubble and the closest point on the boundary of D, satisfies the asymptotic ODE

$$r_m'(t) \sim -\frac{\epsilon^{3/2}\zeta r_m^{-1/2}}{(1-\kappa_m r_m)^{1/2}} e^{-2\nu_+^\epsilon \epsilon^{-1}(r_m - r_b)} . \tag{41}$$

Here κ_m is the curvature of ∂D at the point closest to the bubble center. Here $\kappa_m > 0$ when D is convex at the closest point. In addition, r_b is the bubble radius, and ν_+^ϵ and ζ are defined below in (46b) and (52), respectively. This result, which is asymptotically valid only when $r_m(t) > r_b$, shows that the bubble will collapse against ∂D on an exponentially long time scale. In §4.1, we outline the key steps in the derivation of this result using the projection method.

Once the bubble hits ∂D, it quickly becomes attached to ∂D orthogonally and its boundary becomes a circular arc in order to minimize its perimeter. Its subsequent evolution is then given by (40). However, if the length scale of the interface is sufficiently small compared to the radius of curvature of ∂D, the interface will become approximately semi-circular in shape. The motion of such a semi-circular *drop* of radius $\delta \ll 1$ has been studied in [3]. It was shown in [3] that the center of a such a small drop of radius δ, with $\delta \ll 1$ but $0 < \epsilon < \delta^3$, satisfies the asymptotic ODE

$$\xi'(t) \sim \frac{4\epsilon^2\delta}{3\pi} K_D'(\xi(t)) . \tag{42}$$

Here ξ is an arclength parameter for ∂D, δ is the radius of the drop, and K_D is the curvature of ∂D (positive for a convex domain D). The drop will reach a stable equilibrium, where the interface can have minimum perimeter, at a local maximum of K_D (see [2, 3, 11]).

However, the relation (40) and the drop result (42) gives no indication of the motion of a semi-circular interface intersecting a flat portion of ∂D. This case, in which metastable behavior occurs, has been studied in [36]. In §4.2 we outline the metastability analysis for a semi-circular interface located on the straight-line boundary segment joining the points $(x_L, 0)$ and $(x_R, 0)$ as shown in Fig. 4. As derived in [36], the metastable motion of the center $x_0(t)$ of the semi-circular interface satisfies the asymptotic ODE

$$x_0'(t) \sim \frac{2\epsilon a_+^2 (\nu_+^\epsilon)^2}{\pi\beta} \left\{ \frac{K_R}{x_R - x_0} e^{-2\nu_+^\epsilon \epsilon^{-1}(x_R - x_0 - r_b)} \left(\frac{\epsilon}{2\nu_+^\epsilon}\right)^{\alpha_R+1} \Gamma(\alpha_R+1) \right. \tag{43}$$
$$\left. - \frac{K_L}{x_0 - x_L} e^{-2\nu_+^\epsilon \epsilon^{-1}(x_0 - x_L - r_b)} \left(\frac{\epsilon}{2\nu_+^\epsilon}\right)^{\alpha_L+1} \Gamma(\alpha_L+1) \right\} .$$

Here K_L, α_L and K_R, α_R are constants associated with the corner regions near $(x_L, 0)$ and $(x_R, 0)$, respectively. In addition, r_b is the radius of the semi-circular interface and a_+, ν_+^ϵ, and β are constants that depend on ϵ and can be calculated asymptotically for a given $Q(u)$.

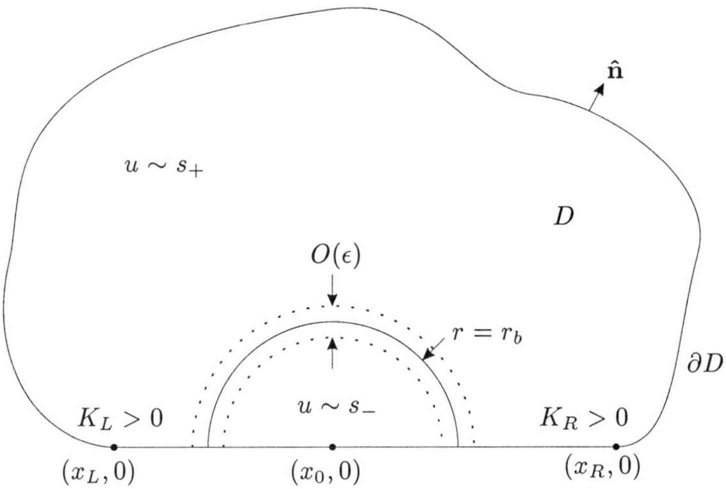

FIG. 4. *Plot of a two-dimensional domain D with a flat boundary segment and a semi-circular interface of radius $r = r_b$ centered at x_0.*

4.1. Metastable dynamics of interior bubbles. There are three basic steps in the projection method used in [41]. The first step is to construct a radially symmetric equilibrium bubble solution of radius r_b in all of $\mathbf{R}^2$. Then, we linearize (38) around this solution and analyze the spectrum associated with this linearization. As a result of the slight break in translation invariance, this spectrum contains exponentially small eigenvalues. Finally, we ensure that mass is conserved and that the solution to the quasi-steady linearized problem is orthogonal to the eigenspace associated with the exponentially small eigenvalues. This projection step yields an ODE for the center $\mathbf{x_0} = \mathbf{x_0}(t)$ of the bubble. We now briefly outline some of the details of this analysis.

The equilibrium bubble solution $u = U_b(r;\epsilon)$, $\sigma = \sigma_b(\epsilon)$ satisfies the radially symmetric version of (38). The solution to this problem can be constructed using the method of matched asymptotic expansions as in [41] with the result

$$U_b(r;\epsilon) \sim \begin{cases} S_+(\epsilon) - a_+(r_b/r)^{1/2} e^{-\nu_+^\epsilon \epsilon^{-1}(r-r_b)}, & r > r_b, \\ u_0(\rho) + O(\epsilon), \quad \rho = \epsilon^{-1}(r - r_b) = O(1), & \\ S_-(\epsilon) + a_-(r_b/r)^{1/2} e^{-\nu_-^\epsilon \epsilon^{-1}(r_b - r)}, & r < r_b, \end{cases} \tag{44}$$

and

$$\sigma_b(\epsilon) = \epsilon \frac{\beta}{(s_+ - s_-) r_b} + O(\epsilon^2). \tag{45}$$

Here $u_0(\rho)$ satisfies (19), and $a_\pm$, β are defined in (22), (26), respectively. The constants $S_\pm$ and $\nu_\pm^\epsilon$ are given by,

$$S_\pm(\epsilon) = s_\pm - \sigma_b \nu_\pm^{-2} + O(\epsilon^2), \tag{46a}$$

$$\nu_\pm^\epsilon = \nu_\pm \left[1 + \frac{\sigma_b}{2\nu_\pm^4} Q''(s_\pm) + O(\epsilon^2)\right], \tag{46b}$$

where $\nu_\pm$ is defined in (21).

The quasi-steady linearization of (38) is obtained by substituting

$$u(\mathbf{x}, t) = U_b\left[|\mathbf{x} - \mathbf{x_0}(t)|;\epsilon\right] + v(\mathbf{x}, t), \qquad \sigma(t) = \sigma_b(\epsilon) + \mu(t), \tag{47}$$

into (38), where $v \ll U_b$, $v_t \ll \partial_t U_b$ and $\mu \ll \sigma_b$. Here $\mathbf{x_0}(t)$ is the unknown trajectory of the center of the bubble. This leads to the following problem for v and μ:

$$L_\epsilon v \equiv \epsilon^2 \Delta v + Q'(U_b) v = \partial_t U_b + \mu, \qquad \mathbf{x} \in D, \tag{48a}$$

$$\partial_n v = -\partial_n U_b, \qquad \mathbf{x} \in \partial D; \qquad \int_D v \, d\mathbf{x} = 0. \tag{48b}$$

Let (λ_j, ϕ_j) for $j \geq 0$ be the eigenpairs of the associated eigenvalue problem

$$L_\epsilon \phi = -\lambda \phi, \ \mathbf{x} \in D; \quad \partial_n \phi = 0, \ \mathbf{x} \in \partial D; \quad (\phi, \phi) \equiv \int_D \phi^2 \, d\mathbf{x} = 1. \tag{49}$$

The principal eigenvalue satisfies $\lambda_0 < 0$ with $\lambda_0 = -\epsilon^2/r_b^2$ as $\epsilon \to 0$, and the corresponding eigenfunction ϕ_0 has the form $\phi_0 \sim N_0 \left(U_b' + \phi_{L0}\right)$. Here N_0 is a normalization constant and ϕ_{L0} is a boundary layer function, localized near ∂D, which allows $\partial_n \phi_0 = 0$ on ∂D to be satisfied. In addition, as a result of the near translation invariance and the exponential decay behavior (44), there are two exponentially small eigenvalues λ_1 and λ_2. The corresponding eigenfunctions are given asymptotically by

$\phi_j \sim N_j \left(\partial_{x_j} U_b + \phi_{Lj}\right)$ for some boundary layer functions ϕ_{Lj}, $j = 1,2$. Here we have set $x_1 = x$ and $x_2 = y$. A boundary layer analysis determines ϕ_{Lj} for $j = 0,1,2$ and hence we can obtain explicit asymptotic formulas for ϕ_j on ∂D for $j = 0,1,2$ (see [41]).

Next, we expand the solution $v(\mathbf{x},t)$ to (48) in terms of the eigenfunctions ϕ_j of (49) as $v(\mathbf{x},t) = \sum_{j=0}^{\infty} c_j\phi_j/\lambda_j$. The coefficients $c_j = c_j(t)$ for $j \geq 0$ are given by

$$c_j = -\left(\partial_t U_b, \phi_j\right) - \mu\left(1, \phi_j\right) - \epsilon^2 \int_{\partial D} \phi_j \partial_n U_b \, ds\,, \tag{50}$$

where $(f,g) \equiv \int_D fg\,d\mathbf{x}$. Since U_b and ϕ_j, for $j = 0,..,N$, are known when $\epsilon \ll 1$, we can calculate the inner products and the boundary integral in (50) asymptotically to determine c_j for $j = 0,1,2$.

The conditions to determine $\mu(t)$ and $\mathbf{x_0}(t)$ are as follows. First, we must ensure that $\int_D v\,d\mathbf{x} = 0$ in order to conserve mass. Since ϕ_0 is of one sign, this condition requires that v has no component in the direction of ϕ_0. Hence, $c_0(t) \equiv 0$, which then determines μ. Thus, the existence of the negative eigenvalue λ_0 for (49) does not lead to an instability of the bubble solution. We remark that if mass was not conserved, the bubble would shrink to a point under a mean curvature flow on a time scale $|\lambda_0^{-1}| = O(\epsilon^{-2})$ (see [14]). Next, since λ_j for $j = 1,2$ are exponentially small as $\epsilon \to 0$, we must also require that the limiting solvability conditions $c_j \to 0$ as $\epsilon \to 0$ for $j = 1,2$ be satisfied. These conditions yield a differential equation for $\mathbf{x_0}(t)$ that governs the metastable bubble motion.

In this way, it was obtained in [41] that for a bubble strictly contained in D, the center $\mathbf{x_0}(t)$ of the bubble satisfies the asymptotic ODE

$$\mathbf{x_0}' \sim \frac{\epsilon a_+^2 \nu_+^2}{\pi\beta} \int_{\partial D} r^{-1} e^{-2\nu_+^{\epsilon}(r-r_b)/\epsilon}\, \hat{\mathbf{r}}\, \left[1 + \hat{\mathbf{r}}\cdot\hat{\mathbf{n}}\right] \hat{\mathbf{r}}\cdot\hat{\mathbf{n}}\, ds\,. \tag{51}$$

Here $r = |\mathbf{x}(\xi) - \mathbf{x_0}(t)|$, $\hat{\mathbf{r}} = \left(\mathbf{x}(\xi) - \mathbf{x_0}(t)\right)/r$, and $\hat{\mathbf{n}} = \hat{\mathbf{n}}(\xi)$ is the unit outward normal to ∂D, where ξ parameterizes ∂D. Also, β and ν_+^{ϵ} are defined in (26) and (46b).

The (unstable) equilibrium location $\mathbf{x_{0e}}$ for the bubble center is obtained by setting $\mathbf{x_0}' = 0$ in (51). For a strictly convex domain, this yields that $\mathbf{x_{0e}}$ is located at an $O(\epsilon)$ distance from the center of the largest inscribed circle for D (see [41]). Finally, by using Laplace's method on (51) it is clear that the metastable bubble dynamics is determined largely in terms of local information from the point $\mathbf{x}(\xi_0)$ on ∂D nearest to $\mathbf{x_0}$. In this way, it is shown in [41] that the bubble drifts towards the closest point on ∂D and the distance $r_m(t) = |\mathbf{x}(\xi_0) - \mathbf{x_0}(t)|$ between the bubble center and the closest point satisfies (41), where ζ in (41) is given by

$$\zeta = \frac{2a_+^2\nu_+^2}{\pi\beta} \left(\frac{\pi}{\nu_+^{\epsilon}}\right)^{1/2}. \tag{52}$$

4.2. Metastable dynamics of attached bubbles. We now characterize the metastable dynamics of a semi-circular interface of radius r_b that intersects a flat portion of ∂D as shown in Fig. 4. The flat portion of ∂D is taken to be the straight-line segment between $(x_L, 0)$ and $(x_R, 0)$. The interface is centered around $\mathbf{x}_0 = (x_0, 0)$ where $x_L < x_0 < x_R$. We let $\partial D = \partial D_c \cup \partial D_s$ where ∂D_s refers to the straight-line segment of the boundary and ∂D_c denotes the remaining curved part of ∂D. The distance between the interface and ∂D_c is assumed to be a minima at either of the two corners $(x_L, 0)$ or $(x_R, 0)$.

Near the corner points, ∂D is assumed to have the following local behavior

$$\text{(53a)}\quad \text{near } (x_L, 0)\,;\ y = \psi_L(x),\ \ \psi_L'(x) \sim -K_L(x_L - x)^{\alpha_L}, \quad \text{as } x \to x_L^-,$$

$$\text{(53b)}\quad \text{near } (x_R, 0)\,;\ y = \psi_R(x),\ \ \psi_R'(x) \sim K_R(x - x_R)^{\alpha_R}, \quad \text{as } x \to x_R^+,$$

where $\alpha_L > 0$ and $\alpha_R > 0$. When $\alpha_L = \alpha_R = 1$, K_L and K_R are proportional to the curvatures of ∂D_c at the corners.

In [36] the projection method was used to determine the motion of the semi-circular interface. The outline of the analysis is as follows. First, we construct a radially symmetric equilibrium solution to (38) as in (44). This equilibrium solution is centered at $(x_0, 0)$ and it corresponds to the semi-circular interface. We then linearize the equation around U_b as in (47), where $\mathbf{x_0}(t) = (x_0(t), 0)$. This yields the linearized problem (48). The associated eigenvalue problem has a principal eigenvalue $\lambda_0 = -\epsilon^2/r_b^2$. In addition, due to the near translation invariance in the horizontal direction x, there is one exponentially small eigenvalue corresponding to the approximate eigenfunction $\phi_1 \sim \partial_x U_b$. The conservation of mass condition determines the correction μ to the Lagrange multiplier parameter σ. In addition, a limiting solvability condition, which requires that v be orthogonal to the eigenfunction associated with the exponentially small eigenvalue must be imposed. This yields the following asymptotic differential equation for $x_0(t)$:

$$\text{(54)}\qquad x_0'(t) \sim \frac{2\epsilon a_+^2 (\nu_+^\epsilon)^2}{\pi\beta} \int_{\partial D} \frac{(x - x_0)}{r^2} e^{-2\nu_+^\epsilon \epsilon^{-1}(r - r_b)} \hat{\mathbf{r}} \cdot \hat{\mathbf{n}}\, ds\,.$$

Here $\hat{\mathbf{r}} = (\mathbf{x} - \mathbf{x_0})/r$, $\mathbf{x_0} = (x_0, 0)$, $r = |\mathbf{x} - \mathbf{x_0}|$ and $\hat{\mathbf{n}}$ is the unit outward normal to ∂D. Also, β and ν_+^ϵ are given in (26) and (46b), respectively. Since, $\hat{\mathbf{r}}$ and $\hat{\mathbf{n}}$ are orthogonal on the straight-line segment ∂D_s of ∂D, the integral in (54) is an integral over the curved segment ∂D_c. For $\epsilon \to 0$, the dominant contribution to this integral arises from the corner regions $(x_L, 0)$ and $(x_R, 0)$. Laplace's method and the local information (53) can then be used to obtain the explicit differential equation (43) for $x_0(t)$.

Thus, the ODE (43) for $x_0(t)$ shows that the motion of the center of the semi-circular interface along the straight-line boundary segment between $(x_L, 0)$ and $(x_R, 0)$ is determined by the shape of the boundary at these corner points and by the distance from the interface to these points. The interface will move, without change of shape, until a stable steady state is reached or until the interface touches $(x_L, 0)$ or $(x_R, 0)$. If the interface reaches the curved part of the boundary, it will subsequently continue to evolve according to (40).

4.3. An explicit example of an attached bubble. When $K_L K_R > 0$, we obtain from (43) that there is a unique steady-state solution x_{0e}, satisfying $x_L < x_{0e} < x_R$. For $\epsilon \to 0$, a two-term approximation for x_{0e} is

$$(55) \quad x_{0e} \sim \frac{x_L + x_R}{2} + \frac{\epsilon}{4\nu_+^\epsilon} \log\left[\frac{K_L \Gamma(\alpha_L + 1)}{K_R \Gamma(\alpha_R + 1)} \left(\frac{\epsilon}{2\nu_+^\epsilon}\right)^{\alpha_L - \alpha_R}\right] + O(\epsilon^2).$$

This steady state is stable when $K_L < 0$, $K_R < 0$, and is unstable when $K_L > 0$, $K_R > 0$. Specifically, this shows that if D is convex near $(x_L, 0)$ and $(x_R, 0)$, then there is no stable equilibrium location on ∂D_s. There is no equilibrium solution if $K_L K_R < 0$.

To illustrate the slow dynamics, let $Q(u) = 2(u - u^3)$. Then, we calculate that

$$(56) \qquad \nu_\pm^\epsilon = 2[1 - \epsilon(4r_b)^{-1} + \cdots], \quad a_\pm = 2, \quad s_\pm = \pm 1, \quad \beta = 4/3.$$

Suppose that $\alpha_L = \alpha_R = \alpha$. Then, from (43) the center $x_0(t)$ of the semi-circular interface satisfies

$$(57) \quad x_0' \sim c\left[\frac{K_R}{x_R - x_0} e^{-2\nu_+^\epsilon \epsilon^{-1}(x_R - x_0 - r_b)} - \frac{K_L}{x_0 - x_L} e^{-2\nu_+^\epsilon \epsilon^{-1}(x_0 - x_L - r_b)}\right],$$

where

$$(58) \qquad c = \frac{6\epsilon^{\alpha+2}\Gamma(\alpha+1)}{4^\alpha \pi}.$$

The steady state location for x_0, when $K_L K_R > 0$, is

$$(59) \qquad x_{0e} \sim \frac{x_L + x_R}{2} + \frac{\epsilon}{8} \log\left(\frac{K_L}{K_R}\right).$$

For the initial value, $x(0) = x_0^0$, the qualitative properties of the dynamics associated with (57) are as follows. When $K_L > 0$ and $K_R > 0$, $x_0(t)$ moves exponentially slowly towards x_L if $x_0^0 < x_{0e}$, or towards x_R if $x_0^0 > x_{0e}$ (see Fig. 5). When $K_L < 0$ and $K_R < 0$, $x_0(t)$ will approach the stable steady state at x_{0e} for any initial condition (see Fig. 6). If $K_L < 0$ and $K_R > 0$, then $x_0(t)$ will move towards x_R as shown in Fig. 7. Finally, $x_0(t)$ will move towards x_L if $K_L > 0$ and $K_R < 0$. In each case, the interface moves in the direction that will allow its perimeter to decrease. When the interface reaches the corner points $(x_L, 0)$ or $(x_R, 0)$, the subsequent evolution of the interface is determined by (40).

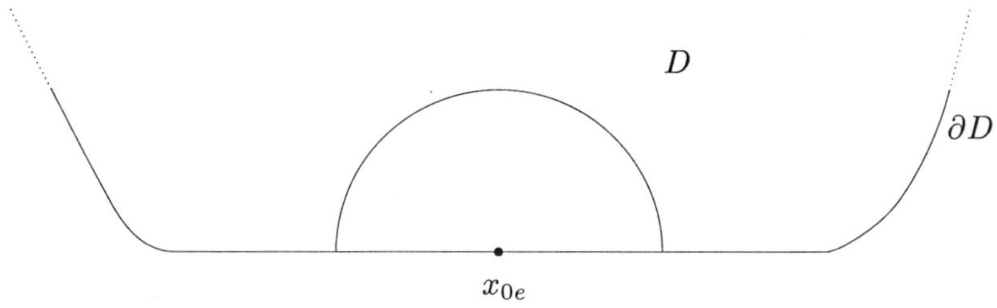

FIG. 5. *Plot of ∂D for which the center of the interface is at an unstable equilibrium. Here $K_L, K_R > 0$.*

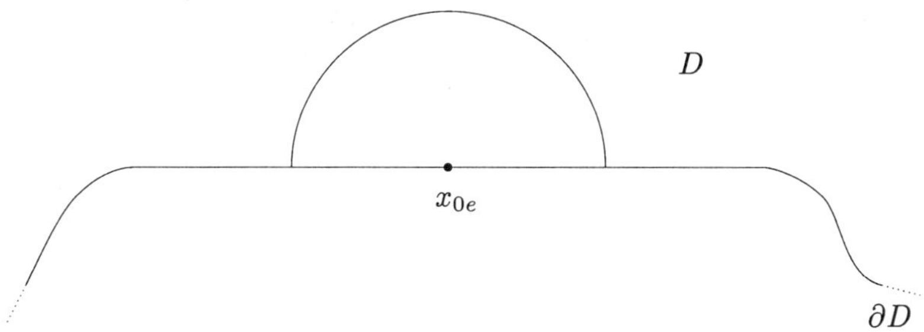

FIG. 6. *Plot of ∂D for which the center of the interface is at a stable steady state. Here $K_L, K_R < 0$.*

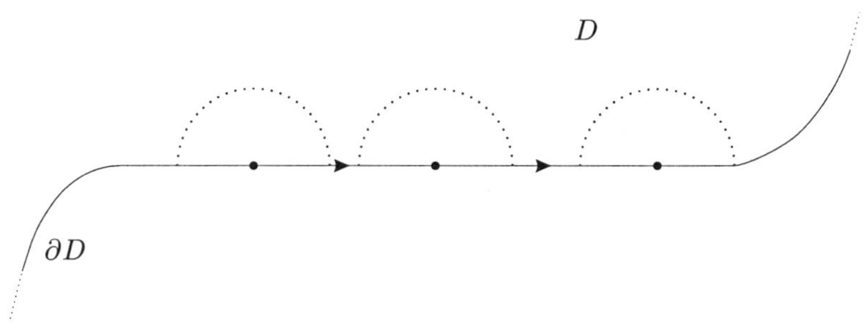

FIG. 7. *Plot of ∂D for which the center of the interface moves toward the right. Here $K_L < 0$ and $K_R > 0$.*

5. Slow spike motion for the shadow problem. Turing [39] proposed a reaction-diffusion system of activator-inhibitor type to mathematically model morphogenesis. From a linear stability analysis he suggested that such a system could have stable inhomogeneous solutions with isolated peaks in the inducer concentration. Subsequent studies (e.g. [15, 17]), which have involved large-scale numerical computations, have shown that robust spike-type patterns in the activator concentration are possible when the activator diffuses much more slowly than the inhibitor.

In [18], spike-type solutions are analyzed for an activator-inhibitor system known as the Gierer-Meinhardt model [15]. In scaled and dimensionless variables, this model is formulated as

$$a_t = \epsilon^2 \triangle a - a + \frac{a^p}{h^q}, \qquad \mathbf{x} \in D \subset \mathbf{R}^N, \tag{60a}$$

$$\tau h_t = k_h \triangle h - \mu h + \epsilon^{-N} a^m, \qquad \mathbf{x} \in D \subset \mathbf{R}^N, \tag{60b}$$

$$a_n = 0, \qquad h_n = 0, \qquad \mathbf{x} \in \partial D. \tag{60c}$$

Here $\epsilon \ll 1$, a and h are the activator and inhibitor concentrations, respectively, D is a bounded domain in $\mathbf{R}^N$, and the derivatives in (60c) are the normal derivatives to the boundary of D. Also, $\tau > 0$, $\mu > 0$ and the exponents satisfy

$$1 < p < 1 + mq, \qquad q > 0, \qquad m > 0.$$

There are only a few partial results for (60) in the fully nonlinear regime. A survey of some of these results are given in [29].

In [18], the weak coupling limit $k_h \gg 1$ for (60) is analyzed. This limit, together with the assumption that the dynamics of a are much slower than those of h, i.e. $\tau \ll 1$, leads to the following non-local problem for a, known as the shadow problem ([28, 29]):

$$a_t = \epsilon^2 \triangle a - a + \frac{a^p}{h^q} \qquad \mathbf{x} \in D \subset \mathbf{R}^N, \tag{61a}$$

$$h = \frac{\epsilon^{-N}}{\mu |D|} \int_D a^m \, d\mathbf{x}, \tag{61b}$$

$$a_n = 0, \qquad \mathbf{x} \in \partial D. \tag{61c}$$

Here $|D|$ is the volume of D.

For the equilibrium problem, we let $a = h^\gamma u$ in (61a), where $\gamma = q/(p-1)$, to obtain the following problem for u:

$$\epsilon^2 \triangle u - u + u^p = 0, \qquad x \in D; \qquad u_n = 0, \qquad \mathbf{x} \in \partial D. \tag{62}$$

Assume that $p < p_c(N)$, where $p_c(N)$ is the critical Sobolev exponent. Then (62) admits a solution with M spikes, where the spikes are located near some points $\mathbf{x_{0i}}$ for $i = 1, .., M$ in D that are to be determined. For a solution with a spike on the boundary of the domain, it was proved in [28] that the spike can be located at the point on ∂D where the mean curvature of ∂D is the largest. In [42], the projection method was used to construct a one-spike solution to (62) where the spike is contained strictly within D. It was shown that, for a convex domain, the spike is located at an $O(\epsilon)$ distance from the center of the largest inscribed sphere for D. This analysis required exponential precision since the spike-layer solution decays exponentially away from its core and hence interacts with the boundary only by exponentially small terms. More recently, the important papers of [7, 16] and [20] have proved that the problem for the determination of the locations of the spikes for a multi-spike solution is intimately connected with the geometric problem of the lattice packing of balls of equal radii inside D.

In [18] the projection method is used to show that (61) can support a metastable spike-layer solution. If the spike is initially localized near some point $\mathbf{x_0}(0)$ in D, then this spike will move exponentially slowly towards the closest point on the boundary ∂D. This result is critically dependent on the non-local nature of (61), which precludes the existence of an unstable mode with an $O(1)$ eigenvalue for the corresponding linearized problem. We now briefly outline the analysis in [18].

A one-spike equilibrium solution to (61) in $\mathbf{R}^N$ is represented by

$$a = a_E(\mathbf{x};\mathbf{x_0}) \equiv h^{\gamma} u_c(|\mathbf{x}-\mathbf{x_0}|) \qquad \gamma = q/(p-1)\,. \tag{63}$$

Here the radially symmetric function $u_c(\rho)$, called the canonical spike solution, satisfies the radially symmetric version of (62)

$$u_c'' + \frac{(N-1)}{\rho}u_c' - u_c + u_c^p = 0\,, \tag{64a}$$

$$u_c'(0) = 0 \quad \text{and} \quad u_c(\infty) = 0\,, \tag{64b}$$

$$u_c(\rho) \sim a\rho^{(1-N)/2}e^{-\rho}\,, \quad \text{as} \quad \rho \to \infty\,. \tag{64c}$$

In terms of this solution, h is given by

$$h = \left(\frac{\Omega_N}{\mu|D|}\int_0^\infty u_c^m \rho^{N-1}\,d\rho\right)^{\frac{p-1}{(p-1)-qm}}\,, \tag{65}$$

where Ω_N is the surface area of the unit N dimensional sphere. Since a_E satisfies the steady-state problem for (61a), but fails to satisfy the no flux boundary condition (61c) by only exponentially small terms for any value of $\mathbf{x_0} \in D$, we expect that the spectrum of the eigenvalue problem associated with the linearization about a_E contains exponentially small eigenvalues. The non-local nature of (61) will ensure that these exponentially small eigenvalues are the principal eigenvalues of the linearization.

The non-local eigenvalue problem for the linearization is obtained by introducing ϕ and η defined by

$$a(\mathbf{x},t) = a_E(\mathbf{x};\mathbf{x_0}) + e^{-\lambda t}\phi(\mathbf{x}), \tag{66a}$$

$$h(\mathbf{x},t) = h + e^{-\lambda t}\eta(\mathbf{x})\,. \tag{66b}$$

Here $\phi \ll a_E$ and $\eta \ll h$. From (61) this yields the non-local eigenvalue problem in D:

$$L_\epsilon \phi \equiv \epsilon^2 \Delta \phi + (-1 + p u_c^{p-1})\phi - \frac{mq\epsilon^{-N} u_c^p}{\beta_N \Omega_N} \int_D u_c^{m-1} \phi \, d\mathbf{x} = -\lambda \phi, \tag{67a}$$

$$\phi_n = 0 \quad \text{on} \quad \partial D \,. \tag{67b}$$

Here, β_N is defined by

$$\beta_N = \int_0^\infty u_c^{m-1} \rho^{N-1} \, d\rho \,. \tag{68}$$

Since u_c is localized near $\mathbf{x} = \mathbf{x_0}$, we only seek eigenfunctions of (67) that are localized near $x = x_0$. These eigenfunctions have the form

$$\tilde{\phi}(\mathbf{y}) = \phi(\mathbf{x_0} + \epsilon \mathbf{y}) \,, \qquad \mathbf{y} = \epsilon^{-\mathbf{1}}(\mathbf{x} - \mathbf{x_0}) \,. \tag{69}$$

They satisfy the corresponding infinite domain problem in $\mathbf{R^N}$,

$$\tilde{L}_\epsilon \tilde{\phi} \equiv \Delta_y \tilde{\phi} + (-1 + p u_c^{p-1})\tilde{\phi} - \frac{mq u_c^p}{\beta_N \Omega_N} \int_{\mathbf{R}^N} u_c^{m-1} \tilde{\phi} \, d\mathbf{y} = -\tilde{\lambda} \tilde{\phi}, \tag{70a}$$

$$\tilde{\phi} \to 0 \quad \text{as} \quad |\mathbf{y}| \to \infty \,, \tag{70b}$$

where $u_c = u_c(|\mathbf{y}|)$.

We now outline the key properties of the spectrum associated with (67) and (70). Define $\tilde{\phi}_i \equiv \partial_{y_i} u_c(|\mathbf{y}|)$. Then, from translation invariance and symmetry considerations, which shows that the integral in (70) vanishes when $\tilde{\phi} = \tilde{\phi}_i$, it follows that $\tilde{\phi}_i$ for $i = 1, .., N$ satisfies (70) with $\tilde{\lambda} = 0$. Here y_i is the ith coordinate of $\mathbf{y}$. Thus, (70) has a zero eigenvalue of multiplicity N with corresponding eigenfunctions $\tilde{\phi}_i = \partial_{y_i} u_c(|\mathbf{y}|)$ for $i = 1, .., N$. Since $\mathbf{x_0} \in D$, and u_c is exponentially localized, these eigenpairs will be perturbed by only exponentially small terms as a result of the finite domain in the eigenvalue problem (67). More precisely, it was shown in [18] that (67) has N exponentially small eigenvalues with corresponding eigenfunctions given by $\phi_i \sim \partial_{x_i} u_c + \phi_{Li}$, where ϕ_{Li} is a boundary layer function localized near ∂D that allows the boundary condition in (67b) to be satisfied. An analysis using Green's identity shows that these exponentially small eigenvalues for (67) are given for $\epsilon \to 0$ by

$$\lambda_i \sim -\frac{\epsilon a^2 N}{\hat{\beta}_N \Omega_N} \int_{\partial D} (x_i - x_{0i})^2 r^{-(1+N)} e^{-2r/\epsilon} (\hat{\mathbf{r}} \cdot \hat{\mathbf{n}})(1 + \hat{\mathbf{r}} \cdot \hat{\mathbf{n}}) \, dS \,, \tag{71}$$

for $i = 1, .., N$. Here $r = |\mathbf{x} - \mathbf{x_0}|$, $\hat{\mathbf{r}} = (\mathbf{x} - \mathbf{x_0})/r$, x_i is the ith coordinate of $\mathbf{x}$, and $\hat{\mathbf{n}}$ is the unit outward normal to ∂D. Also, a is given in (64c) and $\hat{\beta}_N$ is defined by

$$\hat{\beta}_N = \int_0^\infty \left[u_c'(\rho) \right]^2 \rho^{N-1} \, d\rho \,. \tag{72}$$

An asymptotic evaluation of (71) using Laplace's method then shows that λ_i is determined by local information from the point on ∂D that is closest to the center $\mathbf{x_0}$ of the spike.

The other key property of the spectrum concerns the principal eigenpair for (67). When the non-local term in (70) is absent, it is well-known that the local problem has an $O(1)$ negative eigenvalue $\tilde{\lambda}_0$ with a corresponding radially symmetric eigenfunction $\tilde{\phi}_0(\rho)$. In [18] we used a path-following method to numerically compute this eigenvalue in terms of a homotopy parameter δ, where δ with $0 \leq \delta \leq 1$, multiplies the non-local integral term in (70). In this way the effect of the non-local term is gradually introduced. It was shown in [18] that the eigenvalue branch as a function of δ, which emanates from $\tilde{\lambda}_0$, has crossed into the right-half plane when $\delta = 1$. Hence, the effect of the non-local term is to ensure that the principal eigenvalue of (70) is the zero eigenvalue of translation. In this way, it was shown that the principal eigenvalue of (67) is exponentially small as $\epsilon \to 0$.

The final step in the analysis of metastable spike motion is to linearize (61) around the quasi-equilibrium solution by writing

$$a(\mathbf{x}, t) = a_E(\mathbf{x}; \mathbf{x_0}(t)) + w(\mathbf{x}, t)\,, \tag{73}$$

where a_E is defined in (63), and $w \ll a_E$ and $w_t \ll \partial_t a_E$. The linearized problem is

$$L_\epsilon w = \partial_t a_E \quad \text{in} \quad D\,, \tag{74a}$$

$$\partial_n w = -\partial_n a_E\,, \quad \text{on} \quad \partial D\,, \tag{74b}$$

where L_ϵ is defined in (67a). By applying a limiting solvability condition that requires that w is orthogonal to the eigenfunctions associated with the exponentially small eigenvalues, the following ODE for the trajectory $\mathbf{x_0} = \mathbf{x_0}(t)$ was derived in [18]:

$$\mathbf{x_0}' \sim \frac{\epsilon N a^2}{\hat{\beta}_N \Omega_N} \int_{\partial D} \hat{\mathbf{r}}\, r^{1-N} e^{-2r/\epsilon} (1 + \hat{\mathbf{r}}\cdot\hat{\mathbf{n}})\, \hat{\mathbf{r}}\cdot\hat{\mathbf{n}}\, dS\,. \tag{75}$$

Finally, assume that there is a unique point on ∂D, labelled by $\mathbf{x_m}$ such that the distance from $\mathbf{x_0}$ to ∂D is minimized. Then, upon using Laplace's method on (75), it was shown in [18] that the spike-layer moves monotonically, but exponentially slowly, in a straight line towards the point $\mathbf{x_m}$ on ∂D and that the distance $r_m(t) \equiv |\mathbf{x_m} - \mathbf{x_0}(t)|$ satisfies the asymptotic ODE

$$r_m' \sim -\zeta r_m \left(\frac{\epsilon}{r_m}\right)^{(N+1)/2} H(r_m)\, e^{-2r_m/\epsilon}\,. \tag{76}$$

Here ζ is defined by

$$\zeta = \frac{2Na^2}{\Omega_N \hat{\beta}_N} \pi^{(N-1)/2}\,. \tag{77}$$

The function $H(r_m)$ is given in terms of the principal radii of curvature R_i, $i = 1, .., N-1$, of the boundary ∂D at the closest point $\mathbf{x_m} \in D$ by,

$$(78) \quad H(r_m) \equiv (1-r_m/R_1)^{-1/2} (1-r_m/R_2)^{-1/2} \dots (1-r_m/R_{N-1})^{-1/2} \,.$$

The main result (76), which is very similar to the result in §4 for the motion of bubble solutions for the constrained Allen-Cahn equation, shows that the spike will take an asymptotically exponentially long time to reach ∂D when $\epsilon \ll 1$.

6. A combustion problem. There are several other non-local models arising in different branches of science where localized structures evolve on exponentially long time scales. Two such problems are the evolution of hot spots in microwave heated ceramic materials (see [21] for the model) and the evolution of flame-fronts in a vertical channel.

In particular, in [30] an evolution equation describing the upward propagation of a flame-front in a vertical channel, subject to buoyancy effects, was derived. Suppose that the vertical channel has a constant cross-section D in the (x_1, x_2) plane and let the positive z axis be perpendicular to this cross-section and in the direction of propagation of the flame. Then, in terms of dimensionless variables, and in a particular asymptotic regime, the flame-front interface $z = \Phi(x_1, x_2, t)$ was found in [30] to satisfy the non-local problem

$$(79a) \qquad \Phi_t - \frac{1}{2}|\nabla\Phi|^2 = \epsilon\triangle\Phi + \Phi - \frac{1}{|D|}\int_D \Phi \, d\mathbf{x} \,, \qquad \mathbf{x} \in D \in \mathbf{R^2} \,,$$

$$(79b) \qquad \partial_n \Phi = 0 \,, \qquad \mathbf{x} \in \partial D \,.$$

Here $\epsilon \ll 1$, $\mathbf{x} = (x_1, x_2)$, $|D|$ is the area of the cross-section and ∂_n is the outward normal derivative. Some experimental pictures of the flame-fronts are shown in [23]. These pictures show that the flame-front assumes a roughly paraboloidal shape and that the tip of the front moves slowly.

The simpler one-dimensional problem in a slab geometry was studied in [9] and [37]. In this case, $\Phi = \Phi(x, t)$ and (79) reduces on the strip $0 < x < 1$, $t > 0$ to

$$(80) \qquad \Phi_t - \frac{1}{2}\Phi_x^2 = \epsilon\Phi_{xx} + \Phi - \int_0^1 \Phi \, dx \,, \qquad \Phi_x(0,t) = \Phi_x(1,t) = 0 \,.$$

Certain special classes of solutions of (80), corresponding to concave interface shapes were shown in [9] and [37] to be metastable. In particular, in [37] it was shown that the tip $x_0 = x_0(t)$ of a parabolic flame-front interface drifts exponentially slowly towards the closest point on the channel wall (i.e. either $x = 0$ or $x = 1$) according to the differential equation

$$(81) \quad x_0' \sim \sqrt{\frac{2}{\pi\epsilon}} \left[\left((1-x_0)^2 + \frac{\pi^2\epsilon}{3} \right) e^{-(1-x_0)^2/2\epsilon} - (x_0^2 + \frac{\pi^2\epsilon}{3}) e^{-x_0^2/2\epsilon} \right] .$$

The analysis in [37] leading to (81) was based on converting the non-local problem (80) to a local problem by applying the transformation $v = -\Phi_x$, which yields the Burgers-Sivashinsky equation

$$v_t + vv_x - v = \epsilon v_{xx}\,, \qquad v(0,t) = v(1,t) = 0\,. \tag{82}$$

The problem (79), and its one-dimensional counterpart (80), seem quite different in form from the quasi-linear problems considered in §3–5. However, this is really not the case. In fact, the transformation

$$\Phi = 2\epsilon \ln u \tag{83}$$

reduces (79) to the non-local problem

$$u_t = \epsilon \triangle u + u\left(\ln u - \frac{1}{|D|}\int_D \ln u \, d\mathbf{x}\right)\,, \qquad \mathbf{x} \in D\,, \tag{84a}$$

$$\partial_n u = 0\,, \qquad \mathbf{x} \in \partial D\,. \tag{84b}$$

This problem is similar to those considered earlier, but has yet to be analyzed.

Acknowledgements. It is my pleasure to acknowledge the contributions of my graduate students David Iron, Doug Stafford and Xiaodi Sun to some of the work described in this article. The support of NSERC grant 5-81541 is also gratefully acknowledged.

REFERENCES

[1] N. Alikakos, P.W. Bates, and G. Fusco, *Slow Motion for the Cahn-Hilliard Equation in One Space Dimension*, J. Diff. Equat. **90** (1991), 81–135.

[2] N. Alikakos, L. Bronsard, and G. Fusco, *Slow motion in the Gradient Theory of Phase Transitions via Energy and Spectrum*, Calc. Var. Partial Differential Equations **6**(1) (1998), 39–66.

[3] N. Alikakos, X. Chen, and G. Fusco, *Motion of a Drop by Surface Tension Along the Boundary*, Preprint.

[4] N. Alikakos and G. Fusco, *Some Aspects of the Dynamics of the Cahn-Hilliard Equation*, Resenhas **1**(4) (1994), 517–530.

[5] N. Alikakos and G. Fusco, *Slow Dynamics for the Cahn-Hilliard Equation in Higher Spatial Dimensions, Part 2: The Motion of Bubbles*, to appear, Arch. Rational Mech. Anal. (1998).

[6] N. Alikakos, G. Fusco, and M. Kowalczyk, *Finite Dimensional Dynamics and Interfaces Intersecting the Boundary: Equilibria and Quasi-Invariant Manifold*, Indiana Univ. Math. J. **45**(4) (1996), 1119–1155.

[7] P.W. Bates and G. Fusco, *Equilibria with Many Nuclei for the Cahn-Hilliard Equation*, Preprint (1998).

[8] P.W. Bates and J. Xun, *Metastable Patterns for the Cahn-Hilliard Equation: Parts I and II*, J. Diff. Equat. **111** (1994), 421–457; J. Diff. Equat. **117** (1995), 165–216.

[9] H. Berestycki, S. Kamin, and G. Sivashinsky, *Nonlinear Dynamics and Metastability in a Burgers Type Equation*, Compte Rendus Acad. Sci., Paris t. **321**, Serié 1 (1995), 185–190.

[10] J. Carr and R. Pego, *Metastable Patterns in Solutions of* $u_t = \epsilon^2 u_{xx} - f(u)$, Comm. Pure Appl. Math. **42** (1989), 523–576.

[11] X. Chen and M.M. Kowalczyk, *Existence of Equilibria for the Cahn-Hilliard Equation via Local Minimizers of the Perimeter*, Comm. Partial Differential Equat. **21**(7–8) (1996), 1207–1233.

[12] G. Fusco and J.K. Hale, *Slow Motion Manifolds, Dormant Instability and Singular Perturbations*, J. Dyn. Diff. Equat. **1** (1989), 75–94.

[13] M. Gage, *On an Area-Preserving Evolution For Plane Curves*, Contemp. Math. **51** (1986), 51–62.

[14] M. Gage and R.S. Hamilton, *The Heat Equation Shrinking Convex Plane Curves*, J. Diff. Geom. **23** (1986), 69–96.

[15] A. Gierer and H. Meinhardt, *A Theory of Biological Pattern Formation*, Kybernetik **12** (1972), 30–39.

[16] C. Gui, and J. Wei, *Multiple Interior Peak Solutions for some Singularly Perturbed Neumann Problems*, to appear, Journ. of Diff. Eq. (1999).

[17] D. Holloway, *Reaction-diffusion Theory of Localized Structures with Application to Vertebrate Organogenesis*, Ph.D Thesis (Chemistry), University of British Columbia, Vancouver (1995).

[18] D. Iron and M.J. Ward, *A Metastable Spike Solution for a Non-Local Reaction-Diffusion Model*, to appear, SIAM J. Appl. Math. (1998).

[19] M. Kowalczyk, *Exponentially Slow Dynamics and Interfaces Intersecting the Boundary*, J. Diff. Eq. **138**(1) (1997), 55–85.

[20] M. Kowalczyk, *Multiple Spike Layers in the Shadow Gierer-Meinhardt System: Existence of Equilibria and Approximate Invariant Manifold*, to appear, Duke Math. J. (1999).

[21] G. Kriegsmann, *Hot Spot Formation in Microwave Heated Ceramic Fibers*, IMA J. Appl. Math. **59** (1997), 123–148.

[22] J. Laforgue and R.E. O'Malley, *On the Motion of Viscous Shocks and the Super-Sensitivity of Their Steady-State Limits*, Methods and Appl. of Anal. **1** (1994), 465–487.

[23] E. von Lavante, and R.A. Strehlow, *The Mechanism of Lean Limit Flame Extinction*, Combustion and Flame **49** (1983), 123–140.

[24] D. Ludwig, *Persistence of Dynamical Systems under Random Perturbations*, SIAM Review **17** (1975), 605–640.

[25] R.S. Maier and D.L. Stein, *Limiting Exit Location Distributions in the Stochastic Exit Problem*, SIAM J. Appl. Math. **57**(3) (1997), 752–790.

[26] H. Matano, *Asymptotic Behavior and Stability of Semilinear Diffusion Equations*, Publ. Res. Inst. Math., Kyoto **15**, (1979), 401–451.

[27] B.J. Matkowsky and Z. Schuss, *The Exit Problem for Randomly Perturbed Dynamical Systems*, SIAM J. Appl. Math. **33** (1977), 365–382.

[28] W.M. Ni, *On the Shape of Least-Energy Solutions to a Semilinear Neumann Problem*, Comm. Pure Appl. Math. **XLIV** (1991), 819–851.

[29] W.M. Ni, *Diffusion, Cross-Diffusion, and their Spike-Layer Steady-States*, Notices of the AMS **45**(1) (1998), 9–18.

[30] Z. Rakib and G.I. Sivashinsky, *Instabilities in Upward Propagating Flames*, Combust. Sci. and Tech. **54** (1987), 69–84.

[31] L.G. Reyna and M.J. Ward, *On the Exponentially Slow Motion of a Viscous Shock*, Comm. Pure Appl. Math. **48** (1995), 79–120.

[32] L.G. Reyna and M.J. Ward, *Metastable Internal Layer Dynamics for the Viscous Cahn-Hilliard Equation*, Methods and Appl. of Anal. **2**(3) (1995), 285–306.

[33] J. Rubinstein and P. Sternberg, *Nonlocal Reaction-Diffusion Equations and Nucleation*, IMA J. Appl. Math. **48** (1992), 249–264.

[34] J. Rubinstein, P. Sternberg, and J. Keller, *Fast Reaction, Slow Diffusion, and Curve Shortening*, SIAM J. Appl. Math. **49**(1) (1989), 116–133.

[35] D. Stafford and M.J. Ward, *Metastable Dynamics and Spatially Inhomogeneous Equilibria in Dumbell-Shaped Domains*, to appear, Studies in Applied Math (1999).

[36] D. Stafford, M.J. Ward, and B. Wetton, *The Dynamics of Drops and Attached Interfaces for the Constrained Allen-Cahn Equation*, submitted, Europ. J. Appl. Math. (1999).

[37] X. Sun and M.J. Ward, *Metastability for a Generalized Burgers Equation with Applications to Propagating Flame-Fronts*, European J. Appl. Math. **10** (1999), 27–53.

[38] X. Sun and M.J. Ward, *Exponentially Ill-Conditioned Convection-Diffusion Equations in Multi-Dimensional Domains*, submitted, Methods and Appl. of Analy. (1999).

[39] A. Turing, *The Chemical Basis of Morphogenesis*, Phil. Trans. Roy. Soc. B **327** (1952), 37–72.

[40] M.J. Ward, *Metastable Patterns, Layer Collapses, and Coarsening for a One-Dimensional Ginzburg-Landau Equation*, Stud. Appl. Math. **91** (1994), 51–93.

[41] M.J. Ward, *Metastable Bubble Solutions for the Allen-Cahn Equation with Mass Conservation*, SIAM J. Appl. Math. **56**(5) (1996), 1247–1279.

[42] M.J. Ward, *An Asymptotic Analysis of Localized Solutions for some Reaction-Diffusion Models in Multi-Dimensional Domains*, Stud. Appl. Math **97**(2) (1996), 103–126.

LIST OF WORKSHOP PARTICIPANTS

- Shangbing Ai, Department of Mathematics, University of Pittsburgh
- Peter W. Bates, Department of Mathematics, Brigham Young University
- Daryl Bell, Department of Mathematics and Statistics, University of Nebraska
- Amitabha Bose, Department of Mathematics, NJIT
- Jared Bronski, Department of Mathematics, Stanford University
- Pavol Brunovsky, Department of Mathematics, Comenius University
- Won-Gyu Choe, Mathematics Department, Cornell University
- Michael J. Davis, Chemistry Division, Argonne National Laboratory
- Rafael de la Llave, Department of Mathematics, University of Texas
- Amadeu Delshams, Department Matematica Aplicada I, University Politecnica de Catalunya
- Francine Diener, Laboratoire CNRS de Mathematiques, Universite de Nice-Sophia-Antipolis (CNRS)
- Benoit Dionne, Department of Mathematics and Statistics, University of Ottawa
- Arjen Doelman, Mathematisch Inst., Utrecht University
- Freddy Dumortier, Limburgs Universitair Centrum
- Bernold Fiedler, Institute of Mathematics I, Free University of Berlin
- Paul Fife, Department of Mathematics, University of Utah
- Rob Gardner, Department of Mathematics, University of Massachusettes
- Nigel Goldenfeld, Department of Physics and Material Research Lab, University of Illinois- Urbana/Champaign
- John Guckenheimer, Department of Mathematics, Cornell University
- Richard Haberman, Department of Mathematics, Southern Methodist University
- Gyorgy Haller, Division of Applied Mathematics, Brown University
- Michael Hayes, Department of Mathematics, Boston University
- Daniel Henry, Lab. de Mecanique des Fluides et d'Acoustique Ecole Centrale de Lyon

- Yulij S. Ilyashenko, Department of Mathematics, Cornell University
- Miaohua Jiang Department of Mathematics and Computer Science Wake Forest University
- Mark Johnson
- Mike Jolly, Mathematics Department, Indiana University
- Christopher K.R.T. Jones, Division of Applied Mathematics, Brown University
- Tasso J. Kaper, Department of Mathematics, Boston University
- Yohannes Ketema, Department of Aerospace Engineering and Mechanics, University of Minnesota
- Alexander I. Khibnik, United Technologies Research Center
- Petar V. Kokotovis, Department ECE, University of California-Santa Barbara
- Gregor Kovacic, Department of Mathematical Science, Rensselaer Polytech Institute
- Bernd Krauskopf Department of Engineering Mathematics University of Bristol
- Martin (Maciej) Krupa, Inst. for Applied Mathematics, Technical Institute Vienna
- Norman Lebovitz, Department of Mathematics, University of Chicago
- Xiao-Biao Lin, Mathematics, North Carolina State University
- Kurt Lust, Center for Applied Mathematics, Cornell University
- Rolf-Martin Mantel, University of Minnesota Institute for Mathematics and its Applications
- David W. Mclaughlin, New York University-Courant
- Govind Menon, Division of Applied Mathematics, Brown University
- Konstantin Mischaikow, School of Mathematics, Georgia Institute of Technology
- Anatoly Neishtadt, Space Research Institute, Moscow
- Hinke Osinga, Control and Dynamical Systems, California Institute of Technology
- Marina V. Osipchuk, Institute for Mathematics and its Applications, University of Minnesota
- Linda R. Petzold, Department of Mechanical and Environmental Engineering, University of California-Santa Barbara
- Thomas A. Posbergh, Aerospace Engineering and Mechanics, University of Minnesota
- Keith Promislow, Department of Mathematics and Statistics, Simon Fraser University
- Arthur C. Ralfscs, Department of Mathematics, Ohio State University
- Sebastian Reich, Konrad-Zuse-Zentrum Berlin

- Kathleen Rogers, University of Minnesota Institute for Mathematics and its Applications
- Robert Roussarie, Department of Mathematics, University of Bourgogne
- M. Rudnev, Department of Mathematics/C1200, University of Texas-Austin
- Bjorn Sandstede, Department of Mathematics, Ohio State University
- Stephen Schecter, Department of Mathematics, North Carolina State University
- Teresa M. Seara, Departament de Matematica Aplicada I, Universitat Politecnica de Catalunya
- Jianzhong Su, Department of Mathematics, University of Texas at Arlington
- Peter Szmolyan, Angewandte Mathematik, Technische Universitat Wien
- David Terman, Department of Mathematics, Ohio State
- Michael Ward, Mathematics Department, University of British Columbia
- Shinya Watanabe, Department of Mathematical Sciences, Ibaraki University
- Martin Wechselberger, Technische Universitat Wien
- Warren Weckesser, Institute for Mathematics and its Applications, University of Minnesota
- Chonghun Zeng, Courant Institute, New York University
- Mohammed Ziane, Center for Turbulence Research, Stanford University

IMA SUMMER PROGRAMS

1987 Robotics
1988 Signal Processing
1989 Robust Statistics and Diagnostics
1990 Radar and Sonar (June 18–29)
New Directions in Time Series Analysis (July 2–27)
1991 Semiconductors
1992 Environmental Studies: Mathematical, Computational, and Statistical Analysis
1993 Modeling, Mesh Generation, and Adaptive Numerical Methods for Partial Differential Equations
1994 Molecular Biology
1995 Large Scale Optimizations with Applications to Inverse Problems, Optimal Control and Design, and Molecular and Structural Optimization
1996 Emerging Applications of Number Theory (July 15–26)
Theory of Random Sets (August 22–24)
1997 Statistics in the Health Sciences
1998 Coding and Cryptography (July 6–18)
Mathematical Modeling in Industry (July 22–31)
1999 Codes, Systems, and Graphical Models (August 2–13, 1999)
2000 Mathematical Modeling in Industry - A Workshop for Graduate Students (July 19–28)
2001 Geometric Methods in Inverse Problems and PDE Control (July 16–27)

IMA "HOT TOPICS" WORKSHOPS

- Challenges and Opportunities in Genomics: Production, Storage, Mining and Use, April 24–27, 1999
- Decision Making Under Uncertainty: Energy and Environmental Models, July 20–24, 1999,
- Analysis and Modeling of Optical Devices, September 9–10, 1999
- Decision Making under Uncertainty: Assessment of the Reliability of Mathematical Models, September 16–17, 1999
- Scaling Phenomena in Communication Networks, October 22–24, 1999
- Text Mining, April 17–18, 2000
- Mathematical Challenges in Global Positioning Systems (GPS), August 16-18, 2000
- Modeling and Analysis of Noise in Integrated Circuits and Systems, August 29–30, 2000
- Mathematics of the Internet: E-Auction and Markets, December 3–5, 2000
- Analysis and Modeling of Industrial Jetting Processes, January 10–13, 2001

SPRINGER LECTURE NOTES FROM THE IMA

The Mathematics and Physics of Disordered Media
Editors: Barry Hughes and Barry Ninham
(Lecture Notes in Math., Volume 1035, 1983)

Orienting Polymers
Editor: J.L. Ericksen
(Lecture Notes in Math., Volume 1063, 1984)

New Perspectives in Thermodynamics
Editor: James Serrin
(Springer-Verlag, 1986)

Models of Economic Dynamics
Editor: Hugo Sonnenschein
(Lecture Notes in Econ., Volume 264, 1986)

The IMA Volumes in Mathematics and its Applications

Current Volumes:

1 **Homogenization and Effective Moduli of Materials and Media** J. Ericksen, D. Kinderlehrer, R. Kohn, and J.-L. Lions (eds.)

2 **Oscillation Theory, Computation, and Methods of Compensated Compactness** C. Dafermos, J. Ericksen, D. Kinderlehrer, and M. Slemrod (eds.)

3 **Metastability and Incompletely Posed Problems** S. Antman, J. Ericksen, D. Kinderlehrer, and I. Muller (eds.)

4 **Dynamical Problems in Continuum Physics** J. Bona, C. Dafermos, J. Ericksen, and D. Kinderlehrer (eds.)

5 **Theory and Applications of Liquid Crystals** J. Ericksen and D. Kinderlehrer (eds.)

6 **Amorphous Polymers and Non-Newtonian Fluids** C. Dafermos, J. Ericksen, and D. Kinderlehrer (eds.)

7 **Random Media** G. Papanicolaou (ed.)

8 **Percolation Theory and Ergodic Theory of Infinite Particle Systems** H. Kesten (ed.)

9 **Hydrodynamic Behavior and Interacting Particle Systems** G. Papanicolaou (ed.)

10 **Stochastic Differential Systems, Stochastic Control Theory, and Applications** W. Fleming and P.-L. Lions (eds.)

11 **Numerical Simulation in Oil Recovery** M.F. Wheeler (ed.)

12 **Computational Fluid Dynamics and Reacting Gas Flows** B. Engquist, M. Luskin, and A. Majda (eds.)

13 **Numerical Algorithms for Parallel Computer Architectures** M.H. Schultz (ed.)

14 **Mathematical Aspects of Scientific Software** J.R. Rice (ed.)

15 **Mathematical Frontiers in Computational Chemical Physics** D. Truhlar (ed.)

16 **Mathematics in Industrial Problems** A. Friedman

17 **Applications of Combinatorics and Graph Theory to the Biological and Social Sciences** F. Roberts (ed.)

18 ***q*-Series and Partitions** D. Stanton (ed.)

19 **Invariant Theory and Tableaux** D. Stanton (ed.)

20 **Coding Theory and Design Theory Part I: Coding Theory** D. Ray-Chaudhuri (ed.)

21 **Coding Theory and Design Theory Part II: Design Theory** D. Ray-Chaudhuri (ed.)

22 **Signal Processing Part I: Signal Processing Theory** L. Auslander, F.A. Grünbaum, J.W. Helton, T. Kailath, P. Khargonekar, and S. Mitter (eds.)

23 **Signal Processing Part II: Control Theory and Applications of Signal Processing** L. Auslander, F.A. Grünbaum, J.W. Helton, T. Kailath, P. Khargonekar, and S. Mitter (eds.)

24 **Mathematics in Industrial Problems, Part 2** A. Friedman

25 **Solitons in Physics, Mathematics, and Nonlinear Optics** P.J. Olver and D.H. Sattinger (eds.)

26 **Two Phase Flows and Waves** D.D. Joseph and D.G. Schaeffer (eds.)

27 **Nonlinear Evolution Equations that Change Type** B.L. Keyfitz and M. Shearer (eds.)

28 **Computer Aided Proofs in Analysis** K. Meyer and D. Schmidt (eds.)

29 **Multidimensional Hyperbolic Problems and Computations** A. Majda and J. Glimm (eds.)

30 **Microlocal Analysis and Nonlinear Waves** M. Beals, R. Melrose, and J. Rauch (eds.)

31 **Mathematics in Industrial Problems, Part 3** A. Friedman

32 **Radar and Sonar, Part I** R. Blahut, W. Miller, Jr., and C. Wilcox

33 **Directions in Robust Statistics and Diagnostics: Part I** W.A. Stahel and S. Weisberg (eds.)

34 **Directions in Robust Statistics and Diagnostics: Part II** W.A. Stahel and S. Weisberg (eds.)

35 **Dynamical Issues in Combustion Theory** P. Fife, A. Liñán, and F.A. Williams (eds.)

36 **Computing and Graphics in Statistics** A. Buja and P. Tukey (eds.)

37 **Patterns and Dynamics in Reactive Media** H. Swinney, G. Aris, and D. Aronson (eds.)

38 **Mathematics in Industrial Problems, Part 4** A. Friedman

39 **Radar and Sonar, Part II** F.A. Grünbaum, M. Bernfeld, and R.E. Blahut (eds.)

40 **Nonlinear Phenomena in Atmospheric and Oceanic Sciences** G.F. Carnevale and R.T. Pierrehumbert (eds.)

41 **Chaotic Processes in the Geological Sciences** D.A. Yuen (ed.)

42 **Partial Differential Equations with Minimal Smoothness and Applications** B. Dahlberg, E. Fabes, R. Fefferman, D. Jerison, C. Kenig, and J. Pipher (eds.)

43 **On the Evolution of Phase Boundaries** M.E. Gurtin and G.B. McFadden

44 **Twist Mappings and Their Applications** R. McGehee and K.R. Meyer (eds.)

45 **New Directions in Time Series Analysis, Part I** D. Brillinger, P. Caines, J. Geweke, E. Parzen, M. Rosenblatt, and M.S. Taqqu (eds.)

46 **New Directions in Time Series Analysis, Part II**
D. Brillinger, P. Caines, J. Geweke, E. Parzen, M. Rosenblatt, and M.S. Taqqu (eds.)

47 **Degenerate Diffusions**
W.-M. Ni, L.A. Peletier, and J.-L. Vazquez (eds.)

48 **Linear Algebra, Markov Chains, and Queueing Models**
C.D. Meyer and R.J. Plemmons (eds.)

49 **Mathematics in Industrial Problems, Part 5** A. Friedman

50 **Combinatorial and Graph-Theoretic Problems in Linear Algebra**
R.A. Brualdi, S. Friedland, and V. Klee (eds.)

51 **Statistical Thermodynamics and Differential Geometry of Microstructured Materials**
H.T. Davis and J.C.C. Nitsche (eds.)

52 **Shock Induced Transitions and Phase Structures in General Media** J.E. Dunn, R. Fosdick, and M. Slemrod (eds.)

53 **Variational and Free Boundary Problems**
A. Friedman and J. Spruck (eds.)

54 **Microstructure and Phase Transitions**
D. Kinderlehrer, R. James, M. Luskin, and J.L. Ericksen (eds.)

55 **Turbulence in Fluid Flows: A Dynamical Systems Approach**
G.R. Sell, C. Foias, and R. Temam (eds.)

56 **Graph Theory and Sparse Matrix Computation**
A. George, J.R. Gilbert, and J.W.H. Liu (eds.)

57 **Mathematics in Industrial Problems, Part 6** A. Friedman

58 **Semiconductors, Part I**
W.M. Coughran, Jr., J. Cole, P. Lloyd, and J. White (eds.)

59 **Semiconductors, Part II**
W.M. Coughran, Jr., J. Cole, P. Lloyd, and J. White (eds.)

60 **Recent Advances in Iterative Methods**
G. Golub, A. Greenbaum, and M. Luskin (eds.)

61 **Free Boundaries in Viscous Flows**
R.A. Brown and S.H. Davis (eds.)

62 **Linear Algebra for Control Theory**
P. Van Dooren and B. Wyman (eds.)

63 **Hamiltonian Dynamical Systems: History, Theory, and Applications**
H.S. Dumas, K.R. Meyer, and D.S. Schmidt (eds.)

64 **Systems and Control Theory for Power Systems**
J.H. Chow, P.V. Kokotovic, R.J. Thomas (eds.)

65 **Mathematical Finance**
M.H.A. Davis, D. Duffie, W.H. Fleming, and S.E. Shreve (eds.)

66 **Robust Control Theory** B.A. Francis and P.P. Khargonekar (eds.)

67 **Mathematics in Industrial Problems, Part 7** A. Friedman

68 **Flow Control** M.D. Gunzburger (ed.)

69 **Linear Algebra for Signal Processing**
A. Bojanczyk and G. Cybenko (eds.)

70 **Control and Optimal Design of Distributed Parameter Systems**
J.E. Lagnese, D.L. Russell, and L.W. White (eds.)

71 **Stochastic Networks** F.P. Kelly and R.J. Williams (eds.)

72 **Discrete Probability and Algorithms**
D. Aldous, P. Diaconis, J. Spencer, and J.M. Steele (eds.)

73 **Discrete Event Systems, Manufacturing Systems, and Communication Networks**
P.R. Kumar and P.P. Varaiya (eds.)

74 **Adaptive Control, Filtering, and Signal Processing**
K.J. Åström, G.C. Goodwin, and P.R. Kumar (eds.)

75 **Modeling, Mesh Generation, and Adaptive Numerical Methods for Partial Differential Equations** I. Babuska, J.E. Flaherty, W.D. Henshaw, J.E. Hopcroft, J.E. Oliger, and T. Tezduyar (eds.)

76 **Random Discrete Structures** D. Aldous and R. Pemantle (eds.)

77 **Nonlinear Stochastic PDEs: Hydrodynamic Limit and Burgers' Turbulence** T. Funaki and W.A. Woyczynski (eds.)

78 **Nonsmooth Analysis and Geometric Methods in Deterministic Optimal Control** B.S. Mordukhovich and H.J. Sussmann (eds.)

79 **Environmental Studies: Mathematical, Computational, and Statistical Analysis** M.F. Wheeler (ed.)

80 **Image Models (and their Speech Model Cousins)**
S.E. Levinson and L. Shepp (eds.)

81 **Genetic Mapping and DNA Sequencing**
T. Speed and M.S. Waterman (eds.)

82 **Mathematical Approaches to Biomolecular Structure and Dynamics**
J.P. Mesirov, K. Schulten, and D. Sumners (eds.)

83 **Mathematics in Industrial Problems, Part 8** A. Friedman

84 **Classical and Modern Branching Processes**
K.B. Athreya and P. Jagers (eds.)

85 **Stochastic Models in Geosystems**
S.A. Molchanov and W.A. Woyczynski (eds.)

86 **Computational Wave Propagation**
B. Engquist and G.A. Kriegsmann (eds.)

87 **Progress in Population Genetics and Human Evolution**
P. Donnelly and S. Tavaré (eds.)

88 **Mathematics in Industrial Problems, Part 9** A. Friedman

89 **Multiparticle Quantum Scattering With Applications to Nuclear, Atomic and Molecular Physics** D.G. Truhlar and B. Simon (eds.)

90 **Inverse Problems in Wave Propagation** G. Chavent, G. Papanicolau, P. Sacks, and W.W. Symes (eds.)

91 **Singularities and Oscillations** J. Rauch and M. Taylor (eds.)

92 **Large-Scale Optimization with Applications, Part I: Optimization in Inverse Problems and Design**
L.T. Biegler, T.F. Coleman, A.R. Conn, and F. Santosa (eds.)

93 **Large-Scale Optimization with Applications, Part II: Optimal Design and Control**
L.T. Biegler, T.F. Coleman, A.R. Conn, and F. Santosa (eds.)

94 **Large-Scale Optimization with Applications, Part III: Molecular Structure and Optimization**
L.T. Biegler, T.F. Coleman, A.R. Conn, and F. Santosa (eds.)

95 **Quasiclassical Methods**
J. Rauch and B. Simon (eds.)

96 **Wave Propagation in Complex Media**
G. Papanicolaou (ed.)

97 **Random Sets: Theory and Applications**
J. Goutsias, R.P.S. Mahler, and H.T. Nguyen (eds.)

98 **Particulate Flows: Processing and Rheology**
D.A. Drew, D.D. Joseph, and S.L. Passman (eds.)

99 **Mathematics of Multiscale Materials** K.M. Golden, G.R. Grimmett, R.D. James, G.W. Milton, and P.N. Sen (eds.)

100 **Mathematics in Industrial Problems, Part 10** A. Friedman

101 **Nonlinear Optical Materials** J.V. Moloney (ed.)

102 **Numerical Methods for Polymeric Systems** S.G. Whittington (ed.)

103 **Topology and Geometry in Polymer Science** S.G. Whittington, D. Sumners, and T. Lodge (eds.)

104 **Essays on Mathematical Robotics** J. Baillieul, S.S. Sastry, and H.J. Sussmann (eds.)

105 **Algorithms For Parallel Processing** M.T. Heath, A. Ranade, and R.S. Schreiber (eds.)

106 **Parallel Processing of Discrete Problems** P.M. Pardalos (ed.)

107 **The Mathematics of Information Coding, Extraction, and Distribution** G. Cybenko, D.P. O'Leary, and J. Rissanen (eds.)

108 **Rational Drug Design** D.G. Truhlar, W. Howe, A.J. Hopfinger, J. Blaney, and R.A. Dammkoehler (eds.)

109 **Emerging Applications of Number Theory** D.A. Hejhal, J. Friedman, M.C. Gutzwiller, and A.M. Odlyzko (eds.)

110 **Computational Radiology and Imaging: Therapy and Diagnostics**
C. Börgers and F. Natterer (eds.)

111 **Evolutionary Algorithms** L.D. Davis, K. De Jong, M.D. Vose, and L.D. Whitley (eds.)

112 **Statistics in Genetics** M.E. Halloran and S. Geisser (eds.)

113 **Grid Generation and Adaptive Algorithms** M.W. Bern, J.E. Flaherty, and M. Luskin (eds.)

114 **Diagnosis and Prediction** S. Geisser (ed.)

115 **Pattern Formation in Continuous and Coupled Systems: A Survey Volume**
M. Golubitsky, D. Luss, and S.H. Strogatz (eds.)

116 **Statistical Models in Epidemiology, the Environment, and Clinical Trials**
M.E. Halloran and D. Berry (eds.)

117 **Structured Adaptive Mesh Refinement (SAMR) Grid Methods**
S.B. Baden, N.P. Chrisochoides, D.B. Gannon, and M.L. Norman (eds.)

118 **Dynamics of Algorithms**
R. de la Llave, L.R. Petzold, and J. Lorenz (eds.)

119 **Numerical Methods for Bifurcation Problems and Large-Scale Dynamical Systems**
E. Doedel and L.S. Tuckerman (eds.)

120 **Parallel Solution of Partial Differential Equations**
P. Bjørstad and M. Luskin (eds.)

121 **Mathematical Models for Biological Pattern Formation**
P.K. Maini and H.G. Othmer (eds.)

122 **Multiple-Time-Scale Dynamical Systems**
C.K.R.T. Jones and A. Khibnik (eds.)

Forthcoming Volumes:

1998–1999: Mathematics in Biology

Computational Modeling in Biological Fluid Dynamics

Membrane Transport and Renal Physiology

Mathematical Approaches for Emerging and Reemerging Infectious Diseases

Decision Making under Uncertainty: Energy and Environmental Models

1999 Summer Program: Codes, Systems, and Graphical Models

1999–2000: Reactive Flow and Transport Phenomena

Fire

Confinement and Remediation of Environmental Hazards and Resource Recovery

Dispersive Corrections to Transport Equations, Simulation of Transport in Transition Regimes, Multiscale Models for Surface Evolution and Reacting Flows